Computers and Medicine

Editor
Bruce I. Blum

Information Systems for Patient Care

Edited by
Bruce I. Blum

Applied Physics Laboratory
The Johns Hopkins University
Laurel, Maryland, U.S.A.

With 107 Illustrations

Springer-Verlag
New York Berlin Heidelberg Tokyo

Bruce I. Blum
Applied Physics Laboratory
The Johns Hopkins University
Hopkins Road
Laurel, MD 20707
U.S.A.

Library of Congress Cataloging in Publication Data
Main entry under title:
Information systems for patient care.
(Computers and medicine)
Bibliography: p.
Includes index.
1. Medicine—Data processing. 2. Medical care—Data processing. 3. Information storage and retrieval systems—Medicine. 4. Information storage and retrieval systems—Medical care. I. Blum, Bruce I. II. Series: Computers and medicine (New York, N.Y.) [DNLM: 1. Computers—Congresses. 2. Information systems—Congresses. 3. Medicine—Congresses. W26.5 I429 1977–82]
R858.I545 1984 362.1′068 83-19655

Softcover reprint of the hardcover 1st edition 1984

Typeset by University Graphics, Inc., Atlantic Highlands, New Jersey.

9 8 7 6 5 4 3 2 1

ISBN-13: 978-1-4612-9742-0 e-ISBN-13: 978-1-4612-5218-4
DOI: 10.1007/978-1-4612-5218-4

Series Preface

Computer technology has impacted the practice of medicine in dramatic ways. Imaging techniques provide non-evasive tools which alter the diagnostic process. Sophisticated monitoring equipment presents new levels of detail for both patient management and research. In most of these high technology applications, the computer is embedded in the device; its presence is transparent to the user.

There is also a growing number of applications in which the health care provider directly interacts with a computer. In many cases, these applications are limited to administrative functions, e.g., office practice management, location of hospital patients, appointments, and scheduling. Nevertheless, there also are instances of patient care functions such as results reporting, decision support, surveillance, and reminders.

This series, Computers and Medicine, will focus upon the direct use of information systems as it relates to the medical community. After twenty-five years of experimentation and experience, there are many tested applications which can be implemented economically using the current generation of computers. Moreover, the falling cost of computers suggests that there will be even more extensive use in the near future. Yet there is a gap between current practice and the state-of-the-art.

This lag in the diffusion of technology results from a combination of two factors. First, there are few sources designed to assist practitioners in learning what the new technology can do. Secondly, because the potential is not widely understood, there is a limited marketplace for some of the more advanced applications; this in turn, limits commercial interest in the development of new products.

In the next decade, one can expect the field of medical information science to establish a better understanding of the role of computers in medicine. Furthermore, those entering the health care professions already will have had some formal training in computer science. For the near term, however, there is a clear need for books designed to illustrate how computers can assist in the practice of medicine. For without these collections, it will be very difficult for the practitioner to learn about a technology which certainly will alter his or her approach to medicine.

And that is the purpose of this series: the presentation of readings about the interaction of computers and medicine. The primary objectives are to describe the current state-of-the-art and to orient medical and health professionals and students with little or no experience with computer applications. We hope that this series will help in the rational transfer of computer technology to medical care.

Laurel, Maryland
1984

BRUCE BLUM

Preface

In 1952 there were fewer than 20 electronic computers and the first satellite was still six years off. Today, one generation later, home computing is a hobby and home satellite antennae are available. One would expect this technological revolution to have an impact on medicine. The nature of this impact, however, only now is becoming apparent.

The first attempts to apply computers to medical care were sound in concept but, unfortunately, were not supported by reliable or inexpensive systems. The result was overpromise, underdelivery, and rejection. In many cases the life of a system was shorter than the publication cycle of the paper which described it. Yet, by the mid-1970's, it was common to read quantified reports on five years' experience with an operational system. Clearly, it was being demonstrated, the computer was able to play a role in patient care.

Information systems now have been accepted as an integral part of the modern health care facility. They provide three overlapping classes of service:

- Cost reduction—This implies that the cost of operation with a computer is less than that without one. Typically, cost reductions target scarce resources and do not necessarily result in a net saving. For example, the HIS can reduce the clerical burden on nursing, thereby allowing fewer nurses to provide the same level of patient care.
- Operational scope—Many computer applications simply replicate a manual operation on a scale which is beyond the capability of human organizations. In this way they enable a facility to perform functions it could not otherwise do. The modern clinical laboratory system is an illustration.
- Patient care—It is now being recognized that the information system provides an excellent tool for the management of the patient record and the routine surveillance of patient data by predefined decision criteria. This book provides many examples.

Decisions to install information systems for cost reduction or operation scope are based upon well-established criteria. Consequently, there are many commercially-supported systems to choose from. Unfortunately, the role of information systems in patient care is less well understood. The physician demand is limited; the available products are few. Nevertheless, we now know how these systems can help. Therefore, a primary goal for the near term must be the development of realistic health provider expectations which will create a demand so that the next generation of commercial medical information systems will include more comprehensive patient care services. For it is in this area that the major contribution of computer technology will be made.

The objective of this book is to bring together a collection of papers designed to educate physicians and other health care providers about the use of information systems in patient care. Most of the papers in this collection were previously published in the *Proceedings of the Symposium on Computer Applications in Medical Care* (SCAMC). Several chapters were written for this collection; others are revisions of earlier papers.

The first symposium was held in 1977 and attracted an audience of under 200. By 1982, the symposium audience had grown to almost 2,400. This interest in SCAMC is a reflection of both the curiosity of potential users and a professional maturity among those actively participating in the field. This book was conceived as a bridge between the voluminous SCAMC proceedings and the previous generation of out-of-print texts.

The organization of a book like this requires the active support of many people. In particular I should like to acknowledge the assistance of the contributors. Many took the time to update and revise their earlier papers. Dr. Morris Collen and Ms. Elizabeth McColligan each contributed new chapters. I also would like to thank the SCAMC Board of the Directors and Secretariat for their assistance. Without them, this project would never have been started.

At a personal level I would like to mention Dr. Richard Johns who helped me learn enough to undertake this task, my secretary, Edna Maciejczyk, who managed to get this all organized, and the people at Springer-Verlag who converted the manuscript into a fine book. But most of all, I have to thank Harriet. She moderates my irrational compulsiveness and makes it all fun.

Contents

Contributors

The following is a list of contributors to this volume. The contributed chapter number is indicated in parentheses.

J. Barclay Adams (19)
John M. Ashton, M.D. (27)
Jack Atkinson (6)
Marion J. Ball, Ed. D. (4)
Jeff Balsam (27)
G. Octo Barnett, M.D. (19)
Peter D. Beaman (19)
Donald D. Beck (14)
Sanford Benett, M.S. (14)
E. S. Bergan (9)
Lonnie Blevins (16)
Bruce I. Blum (1) (31)
Evelyn Bradley, Dr.P.H. (35)
B. A. Brantley, M.D., Ph.D. (15)
Earl W. Brian, M.D. (5)
David M. Brickman, M.D. (27)
Helen C. Brieden, R.N. (6)
Cesar A. Caceres, M.D. (2)
James G. Campbell (35)
M. G. Chadwick (9)
James K. Childerston (5)
P. D. Clayton, Ph.D. (8)
Morris F. Collen, B.E.E., M.D. (28)
Barbara P. Deason, M.S. (24)
Erica L. Drazen (34)
Richard M. DuBois, Ph.D. (33)
Harold P. Erdman, Ph.D. (29)
S. J. Feagin (15)
Richard H. Fiddleman (21)
Daniel J. Forward, M.D. (25)
G. P. Gafke (9)
R. M. Gardner, Ph.D. (8)
Tull Glazener (16)
Larry Goldstein (32)
Jaqueline K. Greenlie (19)
John H. Greist, M.D. (29)
Jeff Haas (16)
Dorcas R. Hardy (5)
W. E. Hammond, Ph.D. (15)
Betty G. Hudson (6)
Stanley E. Jacobs, Ph.D. (4), (13)
James W. Jefferson, M.D. (29)
Frederick R. Jelovsek, M.D. (24)
Norma S. Justice (19)
S. A. Kahn (9)
Alan Kaplan (5)
Barbara D. Kerlin, Ph.D. (20)
David R. Kerr, D.D.S., M.B.A. (36)
Chester King (32)
Diane Kittredge, M.D. (23)
Marjorie H. Klein, Ph.D. (29)
George Kolenaty (10)
Ingeborg M. Kuhn, Ph.D. (14)
R. Kunstaetter M.D. (22)
Larry Lemmon (16)
Raymond E. Lenhard, M.D. (31)
Larry Manire (32)
Vincent F. Maturi, Ph.D. (33)
Elizabeth E. McColligan, M.S., M.P.H. (18), (31)

G. E. McCue (9)
Clement McDonald, M.D. (16)
Jim Meeks-Johnson (16)
David J. Mishelevich, M.D., Ph.D. (6)
Elaine I. Mize (6)
Norman M. Neches, M.D. (23)
A. Eugene Osburn, D.O. (23)
Monica S. Parent (19)
T. A. Pryor, Ph.D. (8)
Diane M. Ramsey-Klee, Ph.D. (14)
Hope Richard (24)
James Robertson, Jr. (6)
Anna L. Robinson, R.N. (6)
Jonathan E. Rodnick, M.D. (14)
James E. Rusnak (12)
David D. Schlager (26)
Alan R. Shapiro, M.D. (17)
G. Edward Shissler, M.D. (23)
D. W. Simborg, M.D. (9)
J. S. Sobolewski, Ph.D. (11)
Michael E. Somand (19)
W. W. Stead, M.D. (15)
R. L. Stewart (9)
M. J. Straube (15)
Robert M. Strong, Ph.D. (32)
D. R. Studney, M.D. (22)
S. G. Tolchin, Ph.D. (9)
E. Van Brunt, M.D., Sc.D. (3)
Fredric R. Van Deusen (19)
Donald Van Slyke (6)
H. R. Warner, M.D., Ph.D. (8)
Bruce D. Waxman (19)
Lawrence L. Weed, M.D. (7)
Q. E. Whiting-O'Keefe, M.D. (9)
Gio Wiederhold, Ph.D. (14)
Edward M. Young (5)
Rita D. Zielstorff (30)

I. INTRODUCTION

This book is about the use of information systems for patient care. While a great deal is known about this subject, there are many physicians and other health care professionals who have little knowledge about this field. It is for them that this collection was compiled.

In choosing the introduction, I had two alternatives. The first was to write my own. This would provide the perspective of someone with experience in information system development but no medical background. Alternatively, I might have chosen among the writings of the pioneers in the field. This would combine a mature judgement with a historical perspective, and thereby present insights that I could not be expected to have.

As you will see, I chose both options and have included three introductory chapters. In a field as new as this, it is useful to have multiple viewpoints in order to formulate individual judgements.

1

Information Systems for Patient Care*

Bruce I. Blum

Introduction

The term "computer literacy" is now quite common. It implies the existence of some esoteric technology combined with a need to understand how it works. Nonsense. Computers are tools, and we are all literate in their use. We use them to deposit and withdraw money at our banks. Handheld calculators have replaced the slide rule. It is impossible to make travel arrangements without them. We welcome and expect these applications of computers. We need know no more about computers to use them than we now know about internal combustion engines to drive our cars.

Whether or not we recognize it, computers also have become an integral part of the delivery of health care. Federal and state funding for services could not be managed without them. Every provider expects timely laboratory reports that are printed by a computer. While there are justifiable complaints when "the computer is down," who would accept the restrictions of manually produced reports in a nonautomated clinical laboratory?

Clearly, the computer has become an effective tool for patient care. Yet its potential only now is being recognized. The purpose of this introductory review is twofold. First, it presents a historical overview of the development of medical systems. This provides a perspective for seeing where we are and what trends can be anticipated.

The second part of this chapter surveys the benefits of clinical information systems. To a very large extent, the application of computers to medical care has been a cottage industry with many individual products independently produced. Within the last ten years, however, standardized, exportable systems have been developed. Careful study has identified measurable benefits. With the proliferation of very inexpensive equipment we run the danger of ignoring existing knowledge and returning to the cottage industry stage. The reinvention of the wheel is always gratifying—even if the result is elliptical.

*This chapter was prepared especially for this collection.

So, in what follows I offer a view of how we got to where we are and what we have learned in getting there. An underlying motivation of this chapter (and indeed the entire book) is to present the clinical information system (CIS) as a useful tool for the delivery of care. This should result in greater demand for care-oriented facilities that, in turn, should produce a greater variety of available products. Perhaps this is really what we mean by "computer literacy": The existence of a user community that realistically understands how computer technology can aid them in their work and that actively seeks the support of appropriate products.

Historical Overview

The application of computers to medicine began as the first computers became commercially available. Naturally, the history of medical computing has been closely tied to the growing capability of computers. In general, there has been a lag of five years between a technical advance in the computer field and its use in medical computing. It is convenient, therefore, to divide computer development into phases by decade and medical applications into ten year phases which span the decades. By following this convention, it is possible to describe the state-of-the-art in medical computing in the context of its current phase and also to project—based upon trends in computer technology—what the next phase may bring.

Phases in Computer Development

The computer technology of the 1950's saw the transition from vacuum tubes to transistors. Computers were large, required considerable support in both maintenance and environmental control, and were generally limited to running one job at a time. Operating systems were developed to facilitate the use of the computer by programmers and operators. Tools to write programs grew from assembly languages to high-order languages such as COBOL and FORTRAN. In the larger facilities, computers were linked to isolate the slower operations (such as card reading and printing of output) by relegating them to smaller, less expensive computers; transfer between machines was normally by magnetic tape.

This first phase of computer system use was characterized by a high cost for equipment, restricted availability of computer time, and a heuristic approach to problem solving. The applications were limited to those of computational complexity and the routine processing of high volume data. (Of course, the term Electronic Data Processing during this phase also included punch card sorters and plug board printers.) There was a very limited community of developers, and most training came on the job.

The 1960's saw two major, parallel developments: the use of integrated circuits and the growth of computer science. The result was the manufacture of large, reliable computers which were too powerful to justify dedication to a single task. Consequently, multiprogramming operating systems were developed which allowed a single computer to execute several tasks concurrently. This same technology provided an environment for interactive applications.

In addition to these larger computers, smaller computers (minicomputers) were also manufactured. Generally considered special purpose or process control devices, few had much support by way of operating systems and languages. In many cases, smaller devices were now electronically linked to larger mainframe computers. The cost of equipment still was high, but the range of applications had grown dramatically, and the cost of program preparation was becoming a significant factor in the total cost of a system. Computer science established itself as a formal discipline proving an intellectual foundation for the development and evaluation of computer tools and systems. Academic programs were established to train both programmers and scientists.

The 1970's witnessed an unanticipated revolution in very large scale integration (VLSI) which produced computers of great reliability, small size and low cost. The principal mainframe computers remained functionally the same; but their increased availability and reduced prices resulted in proliferation. Applications which were previously impractical became cost-effective.

During this phase it was clear that the greatest cost component in a system had become the cost of software development. While the upgrading of hardware could be easily accomplished by the direct replacement of components with identical interfaces, modifications to software development methodology was far more complex. During this phase the field of software engineering addressed these problems and established new methodologies for implementation. Although development tools still were deficient, a framework for improvement had been established. Yet social and economic factors inhibited the rate of change. Maintenance absorbed about half the life cycle cost of an application, and massive investments in obsolescent training and products limited the rate of change.

While the 1980's have only just started, some of the characteristics in this phase are obvious. First, computers are no longer confined to the computer room; they are integral components of watches, cameras, and kitchen appliances. Packaged as microcomputers (personal computers), it is now possible for the hobbyist to own the equivalent computer power of the largest systems of the 1950's. Because of the lowered cost and smaller packaging, computer architectures also are undergoing change. Networks allow the integration of multiple systems, and the Japanese are involved in a major effort to produce a fifth generation computer.

As the cost of hardware falls, the relative cost of software development continues to grow. Yet, on the other hand, computer literacy also is growing. In just two decades, we have gone from the slogan "do not bend, fold or multilate" to a population which uses computer terminals at work, in the bank, and even in the supermarket. Programming now is a skill taught in third grade; virtually every college student is routinely exposed to more computer science than was available in the early 1960's. Every new engineer and scientist is also a programmer.

Table 1. Historical Overview of Computing

Phases in Computer Development:		Phases in Medical Computing:	
1950's	A New Technology	1955–1965	Experimentation and Orientation
1960's	Foundations and Export	1965–1975	Data Processing Success
1970's	Revolution in Technology	1975–1985	Information Processing Success
1980's	A Ubiquitous Tool	1985–	Knowledge Processing Success

Phases in Medical Computing

In providing a historical overview of medical computing, it is convenient to identify three processing elements:

Data. These are the individual elements which can be quantified. Most mathematical computation operates on data.

Information. This consists of elements derived from data either through analysis or conglomeration.

Knowledge. This is the set of rules that define the relationships among information (and by extension, data).

While this is an oversimplification, Table 1 uses this hierarchy to label the phases of medical computing.

The first phase of medical computing spanned the period from 1955 to 1965. The emphasis was on experimentation and an understanding of the technology. Few practical applications were implemented. The technology could not reliably and economically support the concepts of the visionaries. While the tendency to overpromise and underdeliver led to a period of general rejection and skepticism, this phase did produce a rethinking of needs and an orientation to the new technology.

The next phase, from the mid-1960's to the mid-1970's was primarily characterized by accomplishments in the processing of medical data. The ability of the computer to provide reliable, rapid computation produced a new environment for the collection and analysis of data. The results were a growth in the biomedical engineering disciplines and experimentation with embedded computers in devices such as the CAT scanner. This emphasis was (and continues to be) reflected in the table of contents in the leading computer-oriented technical journals.

Successes in information processing during this phase were limited. The Lockheed experience of the late 1960's resulted in the Technicon MIS and a more sobering appraisal of the complexity of clinical information systems. While simple business systems and special process control applications (such as a laboratory system) were within the state-of-the-art, most attempts to implement information systems met with limited success. The half life of a system during this phase was roughly equal to the publication cycle, and many documented systems were no longer operational when reported.

By the time of the third phase—from the mid-1975 to date—computer technology was able to support the information processing requirements of the medical community. Consequently, 1975 presented a watershed in the scientific literature. Reports were now available which evaluated five years' operational experience with systems. While most applications required some outside funding, systems were produced which are used and paid for out of operational revenues. Families of transportable systems also are being developed, and medical information systems has become a billion dollar industry.

The next section will discuss the established benefits of the present phase of medical computing; nevertheless, it is useful to suggest briefly the nature of the next phase. One can anticipate that many of the problems associated with the management of clinical information will be identified and resolved. The availability of powerful, inexpensive computers (with network access to other data bases) will bring comprehensive information systems to all who need them. Emphasis will now be placed upon the use of this information in a consistent, reliable manner, ie., the application of knowledge. And so we will have progressed from thinking about how we might better manage clinical laboratory data, to the implementation of systems which printed test results, to the use of systems which integrated the data for each patient in a timely and accurate manner, to systems which will apply rules to process this information and recommend actions.

Computer Applications in Patient Care

At the plenary session of the Sixth Annual Symposium on Computer Applications in Medical Care (SCAMC-82), Don Lindberg used an agricultural analogy to suggest three stages of development in a technology. The first stage is the introduction of measures whose response can be recognized quickly. For example, the use of a pesticide can be correlated with dead locusts. The second category involves changes which require more time to achieve a benefit, for example, the use of fertilizer in a single growing season. The third stage entails more complex activities which span many seasons, for example, genetic improvement of crops. (Of course, there is a fourth stage in which the advances of the other levels are evaluated in the context of a larger universe. This may be an issue in agriculture; it is not yet one in medical computing.)

When we consider the first stage of improvement, the dead locusts are as common as CRT's in our health care facilities. No modern clinical laboratory could satisfy the volume demands without computers; no moderate-sized hospital could manage its administrative and financial responsibilities without computers. In these instances, computers go well beyond cost-saving; the processes themselves have become so complex that it would not be feasible to operate at the current scale without automated assistance.

At the second level of innovation, successes are also easy to find. We have clearly documented how information systems reduce errors and improve the timeliness of reporting. The Hospital Information Systems illustrate how a generalized technology can be adopted to an individual institution; they enhance communications and improve the environment (and professional satisfaction) for skilled practitioners.

Finally, at the third level, there are examples of systems which have become an integral component in the process of patient care—a factor which indicates both widespread computer literacy and skill in the implementation of sophisticated applications. Systems are used to provide surveillance, print reminders, manage and display data, and apply knowledge bases to support decision-making.

The remainder of this chapter considers how clinical information systems (CIS) impact patient care. We define a CIS as an application which manages clinical data, retains a long-term data base, and is used in the health care process. The classification of individual systems is somewhat arbitrary. For this book we shall exclude financial systems (no clinical data), embedded computer applications such as the CAT scanner (no data base), and physiological monitoring systems (no long-term data base). The benefits of a CIS are organized into three basic categories.

Administrative support. The process of patient care involves many administrative functions which range from those involving patient interaction (e.g., appointments and schedules) to those of a strictly clerical nature (e.g., charting). The CIS provides benefits to the practitioners and patients by performing the same functions either more efficiently or more effectively.

Medical decision-making assistance. A computer has the ability to apply limited rules routinely to vast bodies of structured data without fatigue. The practitioner, on the other hand, has the ability to integrate a large number of analogue sources. By joining these two capabilities, the CIS can improve patient care.

Cost impact. A key factor in the evaluation of a system is its cost for the benefits provided. There are many difficulties in establishing the cost of a system; some results and analysis methodologies are considered.

Administrative Support

The process of administering health care is information intensive. Some of the information consists of well-defined elements such as patient identification, demographic, and financial data; other information is so subtly organized that it defies simple classification. Clearly, the former is suitable for automated support and the latter must remain in the domain of the provider.

In a classic study performed in the mid-1960's, Jydstrup and Gross [1] analyzed the work flow in three hospitals. They found that about one-fourth of all activity was devoted to information processing which they defined as any action involving the recording, reading, or transmitting of information. While much of this information processing cannot be performed by automation, the recognition of this information dependence strongly suggests that a CIS can be effective in

Modifying Work Flow;
Improving Timeliness;
Reducing Errors;
Providing Access to New Functions.

We examine each in turn.

Modifying Work Flow. The ability of the computer to organize, centralize, and extract information provides a potential for the reorganization of work flow in order to

(1) reduce redundancy of information processing; and
(2) reassign clerical tasks which are normally performed by health care professionals.

That this is accomplished has been demonstrated in several studies. Schmitz [2] examined the impact of an automated system on hospital employee work patterns and found that there was reduction in time on the telephone and in paperwork. Similar effects were reported by Flagle [3] in a series of studies using operations research methods. With respect to laboratory systems, he concluded that the users' subjective preference for the system outweighed the quantified improvement in speed and accuracy. Finally, a comprehensive analysis of the impact of a hospital information system on the cost of operation found that the CIS

> improved the efficiency of nursing care and, through staff attrition and care for a greater number of patients, reduced nursing costs per patient at El Camino Hospital by 5 percent [4].

Improving Timeliness. The ability of a CIS to instantly transmit information once it has been recorded provides a means to improve access to information. Where automated radiology reporting systems with physician query stations are used, reports are available immediately after the radiologist's examination [5, 6]. In the case of laboratory systems, the same phenomenon of instant reporting (and consequently reduction in telephone queries) has been documented. For example, the turnaround for a routine complete blood count was reduced from 5.6 to 1.6 hours after a computer was installed [7].

The ability to improve timeliness extends beyond the realm of order entry/results reporting. By better organizing the process and eliminating redundancy, functions such as patient admitting have been found to be less time-consuming thereby resulting in improved patient and worker satisfaction [8].

Reducing Errors. Any information intensive function will be prone to error. A major advantage of a CIS is its ability to examine and reject data which are not consistent with preestablished criteria. Since this section deals with administrative support, we shall consider only administrative errors which impact patient care.

Simborg installed and evaluated two different automated systems in a 30-bed medical unit in the early 1970's. The first of these, a ward information management system, computerized the physicians' orders. A study of the automated and a similar, but unautomated, wards was conducted [9]. A summary of the magnitude and types of errors follows:

	Automated	Unautomated
Number orders examined	856	857
Transcription errors (rate)	1.7%	7.3%
Communication errors (total)	9	70
Uninterrupted communication errors	6	38
Failure to exactly carry out order (rate)	5.8%	14.7%

The second application, a unit dose pharmacy system, resulted in a similar reduction in the failure to carryout orders [10]. The errors of commission in the normal system were 4.6 times more frequent. A summary of the error distribution follows:

	Automated	Unautomated
Observations	1,234	1,428
Wrong route	1	2
Wrong dose	7	76
Extra dose	3	8
Unordered drug	5	18
Wrong form	4	1

Another class of error relates to the absence or incorrectness of information in the medical record. In one study, the charts of 355 outpatients were examined for the completeness and accuracy of drug orders [11]. Twenty-one percent omitted the name of one or more drugs and 62 percent contained inaccuracies regarding dosage or directions. These errors are eliminated when this information is recorded in an integrated CIS because the main record and the pharmacy record are the same. Further, because of the extensive input error-checking, the quality of the CIS data is high. Studies of two independent systems each found that the error rates were 1% or less and were significantly higher when manual systems were used [12, 13].

Providing Access to New Functions. In many cases, manual operations cannot support functions as size increases. For example, a manual clinical laboratory has a ceiling on the number of procedures it can perform. Given the power of the multichannel analyzers and the increasing number of procedures ordered, no large scale laboratory can operate effectively without some automated support. A more clinically oriented example involves the organization of data to assist medical decision-making. When the Johns Hopkins Oncology Center had a limited number of beds, certain key clinical parameters were normally plotted each day. After the Center was moved to a 56-bed facility, an automated system was required; the service would be impossible to provide with a manual operation [14].

New functions can also be developed by use of the data base created by the CIS. The information is especially effective for evaluation of the quality of care and feedback to the providers [15,16,17,18].

Medical Decision-Making Assistance

The previous section illustrated how a CIS can assist in the management of patient related administrative functions. This section considers the benefits of the CIS when it is used by the provider.

These CIS tools are grouped as follows:

Displaying Clinical Information;
Automatic Surveillance;
Automatic Reminders;
Decision-Making Support.

Displaying Clinical Information. The orderly display and organization of clinical data assures reliable access to a complete and accurate data base. It has been shown, for example, that the presence of a problem list in ambulatory records leads to a better provider recognition of the problem [19]. A few systems (e.g., COSTAR [20] and PROMIS [21]) maintain a completely automated medical record. Other systems provide computer access to limited parts of the patient record. Because of the difficulty in establishing a control group, there are few studies to show how the availability of a computer stored record effects patient care.

The Computerized Medical Record Summary System was used to evaluate the impact of an automated survey [22]. Four hundred and eighty-four clinic patients were randomly selected and divided into 241 summarized and 238 nonsummarized patients. While the study is still underway, preliminary results show that there is no difference between the two groups with respect to hospitalization. However, where patients were hospitalized during the first two years following the summarization, the average length of stay in the second year was less for the summarized patients. The study authors have not postulated a reason for this variation in hospitalization patterns. To date, no difference in outcome has been identified, but differences in the cost of care have been found.

Automatic Surveillance. The CIS can have immediate access to all of a patient's clinical data. Consequently, there is a potential for the CIS to review all new inputs and examine them with respect to a set of predetermined criteria. Johns provides an excellent model of the process [23]. The limiting factors, of course, are the cost and complexity of the matching (or surveillance) process.

Pharmacy systems provide an example of how this service can be implemented in a cost-effective manner.

The need to know the synergistic and antagonistic effects of drug combinations places a burden upon the prescribing physician. Studies have shown that errors in ordering contraindicated drugs lead to complications in patient therapy [24]. One system, using after the fact batch processing, found that 9% of the drug orders had interactions which they classified as 29.8% major, 27.9% moderate and 42.6% minor [25].

The HELP system installed at the LDS Hospital in Salt Lake City compares pharmacy orders against the patient's diagnosis and test results as well as the drug profile. An evaluation of over 13,000 patients monitored found that 5% had drug alerts and that 77% of these alerts resulted in the physician changing the patient's therapy [26]. Further, only 28.9% of these alerts were based upon drug–drug interactions. The remaining alerts were divided as follows: 44.9% drug–laboratory, 16.2% drug–allergy, 7.3% drug dose, 1.3% drug–diet, and 1.4% other causes. The evaluation of another pharmacy system produced similar results [27].

The CIS can also be used in other surveillance situations. In one application using COSTAR, a test was made to see if—after a positive throat culture was reported—there was a documented provider interaction with the patient [28]. When no such report was found, a reminder was sent to the provider. The result of using this system assured a virtually complete documentation record of provider response to each such event.

Automatic Reminders. Automatic reminders represent a specialized class of surveillance reports in which the provider routinely receives reminders for each patient encounter based upon a preestablished set of rules. (In most cases, the rules extend beyond the single dimension of a highly targeted objective such as a documented response to a positive throat culture.) For example, a computer based information system for patients on long term oral anticoagulants is able to properly manage 85% of the patients using only data from the previous three clinic visits [29].

The HELP system has shown that automated protocols for physician assistants provide better results than paper protocols [13]. The most complete analysis of the impact of patient reminders on physicians has been conducted by McDonald. Using a CIS for an adult diabetes clinic, encounter forms were provided and key clinical events (e.g., test results, drug prescriptions) were recorded. Protocols then were developed (1) to indicate if the last measurement of a test to monitor medication effect is outdated and (2) to suggest that current therapy should be changed. Half the clinicians were given reminders based upon these protocols; all clinicians were given the encounter form. The clinicians responded to 36% of the events of the first category with reminders and only 11% without. For reminders of the second category, they responded to 28% of the events with a reminder and 13% without [30].

A second study [31] used a crossover design in which the clinicians were divided into groups as follows:

First control and then study;
First study and then control.

If the response to events was learned, one would expect that the group participation in the study first would have better performance in the control period, i.e., there would be a residual carryover. One would also expect third year postgraduates to perform better than first year postgraduates. Yet there was no statistically significant difference among the clinicians either based upon order of participation or level of training. This led McDonald to conclude, "It appears that the prospective reminders do reduce errors, and that many of these errors are probably due to man's limitations as a data processor rather than to correctable human deficiencies." A third study examined the use of reminders with and without sampling relevant medical literature citations [32]. The presence of the citation did not significantly change the clinicians' response rate as compared to the reminders alone.

The use of prospective reminders also has been demonstrated effective as a consultant extender for cancer chemotherapy. In this case, a set of rules was prepared and distributed to 75 outreach physicians. One hundred and ninety-five patients were treated for 2,612 visits. By use of computer programmed rules, 97% of the patients received appropriate chemotherapy [33]. In another test, protocol compliance was 94% for the algorithm group and 64% in the nonuser group [34]. The ability to manage complex protocols has lead to the development of several automated protocol management systems for cancer therapy [35,36,37,38]. These systems have not been in use for a sufficient period of time to have been evaluated.

Decision-Making Support. Clearly, each of the above applications supports decision-making. Automation is also used in more direct ways. Clinical algorithms may be encoded as branching logic. Bleich has demonstrated the effectiveness of this technique in the evaluation of acid-base disorders [39]. Systems have been developed to integrate literature citations with the process of recording the medical record. Weed's PROMIS system is an example of this approach to providing a knowledge base [40]. The use of artificial intelligence has also demonstrated the ability to diagnose accurately an illness [41,42,43]. While few such systems are in routine clinical use, there are several commercially available minicomputer based systems which provide diagnostic facilities [44].

With the growth of comprehensive clinical data bases, several systems have been developed to use the data base to define a cohort for an individual patient. This is then analyzed to assist in the diagnosis or to predict outcome for alternate therapies. DeDombal has been using such an approach for the diagnosis of abdominal pain [45]. The HELP sectors of Warner's system allow the user to define the criteria in a programming-like language [46]. ARAMIS is a nationwide data base system designed for arthritis and rheumatic diseases [47]. It, too, allows the matching of an individual against a group with a similar profile. Finally, the Duke Prognostogram [48] can be used to predict the outcome of either surgical or medical treatment for patients with angina. Despite the relatively high cost of collecting the data for the computer analysis, this system has been sufficiently successful in reducing unnecessary surgery that third party payors have accepted charges for the computer studies as a procedure [49].

Cost Impact

The inability of the first generation of medical systems to achieve its goals resulted from technology limitations. As equipment becomes less expensive and more reliable, the range of cost effective applications will grow. However, for those interested in applying CIS technology in the near term, evaluations must be based upon the current state-of-the-art. This section reviews what has been established with respect to the cost impact of the CIS.

The market for hospital information systems in 1977 was $419 million and had grown to close to a billion dollars in 1982 [50]. While most of this investment is for administrative and financial applications, it should be noted that these procurement decisions are being made with little regard to demonstrated cost savings. Lindberg has stated that Medical Information Systems are being evaluated by the default method of simply observing the free market force at work [51]. A report to Congress in 1977 summarized the cost of the CIS as follows [52]:

> Medical information systems are an expensive technology. Operating costs for a hospital-based system range from $4 to $9 per patient per day. For systems based on ambulatory care sites, costs range from $0.50 to $14 per patient visit. Costs of implementation are high. Costs are, however, likely to decrease in the future, because of lower prices for computer hardware and higher volume. Moreover, a majority of both hospitals and ambulatory care facilities now using medical information systems report overall savings in institutional costs due to their computer systems. At least one study has documented cost savings. In particular, savings are experienced in labor expenses.

Because information systems are expensive, the institution usually has limited flexibility once a system has been selected and installed. That is, after the expenditure of startup costs, the institution can adapt itself to the system (or vice versa), or discontinue the use of the system; few can afford the cost of rapidly trying an alternate system. For this reason, much of the analysis for system selection is prospective in nature [53,54].

Drazen has examined and performed an analysis of the methods for evaluating the costs of automated hospital systems [55]. She concluded that

(1) few rigorous studies have been conducted;
(2) most of this work is prospective in nature; and
(3) little work has been done in validating the predictive methodologies.

She also suggested that improvements in information flow constituted the major motivation to implement an Automated Hospital Information System.

The CIS which has received the most detailed and comprehensive attention (and is the basis of the OTA summary cited above) is the Technicon HIS installed at the El Camino Hospital. The first cost study [56] developed three models for analysis:

(1) actual savings based upon before/after studies;
(2) potential savings based upon the assumption that all labor savings were realized; and
(3) an interhospital labor trend model in which cost growth over time was compared to the costs of other, similar local hospitals.

Only the third model demonstrated a savings, i.e., the use of a HIS was shown to reduce the rate of cost growth over time. A more recent study of the same system found that the system "was estimated to be approximately 60% self-supporting in 1975 [4]."

Despite this lack of firm cost-saving evidence, we must recognize the difficulty in evaluating the cost of this type of application. It is impossible to separate the effects of organizational change from the effects of the CIS. For example, a CIS in a cancer center was used to aid in the operation of a hemopheresis unit by

(1) matching blood product donors with patients; and
(2) managing the processes of donation and transfusion.

An analyses of one year's operation showed that the incidence of alloimmunization for acute leukemia patients was reduced to 26% compared to a national average of 65%. Moreover, the total cost to patients for blood products was reduced by $280,000 [57,58]. However, no attempt was made to allocate the benefits among a computer system, physicians, operational staff, and capital equipment. It is clear, nevertheless, that this success could never have been achieved without a CIS.

Conclusion

This chapter briefly indentified those aspects of CIS use which can impact patient care. The following conclusions can be drawn:

(1) Information systems have been demonstrated to have a beneficial impact with respect to patient care. That is, the state of knowledge is such that

we understand how computers can be used efficiently in a clinical setting. This statement could not have been made a decade ago.

(2) These demonstrated benefits are supported by only a small number of systems. Further, most applications are limited to teaching or research settings. Few commercial systems are designed to have an impact on the patient care process. In fact, most have been developed to support financial (e.g., billing), administrative (e.g., registration), process-oriented (e.g., laboratory work management), and communication (e.g., order entry) functions.

(3) Information systems for clinical settings have been widely accepted with only limited cost justification. New systems are more robust and flexible. The disappointments of the first generation of medical systems will not be repeated. However, the goals of the current users still may not be realized.

If one accepts these conclusions, then one should expect only limited applications of computer technology to patient care in the near term. Yet, several trends are obvious. Equipment costs are declining; the number of providers accustomed to computer usage is increasing; and experience is demonstrating which clinical applications can be effective. Consequently, one can hope that clinicians will place new demands on the commercial vendors. This, in turn, will lead to greater functionality in the support of the health care process. Given current implementation rates and a system half-life of five years, it is reasonable to expect a major change in the use of the clinical information system by the end of this decade.

References

[1] Jydstrup, R. A. and M. J. Gross. Cost of Information Handling in Hospitals. *Health Services Research,* Winter 1966, pp. 235–271.

[2] Schmitz, H. H. Productivity Effectiveness: It can be done in the health care field. *Proceedings of the Ninth Annual Society of Management Information Systems.* Chicago, Ill.: Society for Management Information Systems, September 1977, p. 1.

[3] Flagle, C. C. Operations Research with Hospital Computer Systems, in M. F. Collen Ed., *Hospital Computer Systems,* New York: Wiley, 1974, p. 427.

[4] Coffey, R. M. How a Medical Information System Affects Hospital Costs: The El Camino Hospital Experience. *Research Summary Series,* National Center for Health Services Research, DHEW Publication No. (PHS) 80-3265, March 1980.

[5] Wheeler, P. S. and D. W. Simborg. The Johns Hopkins Radiology Computer Reporting System. *Electromedica,* 2-3/75, pp. 104–108.

[6] Markivee, C. R. 1978 Update of the Missouri Automated Radiology System. *Proceedings of the 1978 MUMPS Users' Group Meeting,* Patricia J. Zimmerman, (Ed.), December 1978, pp. 102–105.

[7] Lincoln, T. L. and R. A. Korpman. Computers, Health Care, and Medical Information Science. *Science* **210,** No. 4467, October 1980, pp. 257–263.

[8] Heard, M. R. and J. C. Thomas. A Hospital Information System, Its Impact on Costs, Personnel and Patients in One Department, in *Cost Containment, Caps and Consumerism with the Health Care Delivery System* **2** Chicago: Center for Hospital Management Engineering of the American Hospital Association, 1978.
[9] Simborg, D. W. The Development of a Ward Information-Management System. *Meth. Inform. Med.* **12** No. 1, pp. 17–26, 1973.
[10] Simborg, D. W. and H. J. Derewicz. A Highly Automated Hospital Medication System Five Years' Experience and Evaluation. *Annuals of Internal Medicine* **83,** pp. 342–345, 1975.
[11] Monson, R. A. and C. A. Bond. The Accuracy of the Medical Record as an Index of Outpatient Drug Therapy. *JAMA* **240,** No. 10, November 1978, pp. 2182–2184.
[12] Studney, D. R., J. B. Adams, A. Gorbach, S. Guenthner, M. M. Morgan and G. O. Barnett. A Computerized Prenatal Record, *Obstetrics and Gynecology* **50,** No. 1, July 1977, pp. 82–82.
[13] Cannon, S. R. and R. M. Gardner. Experience with a Computerized Interactive Protocol System Using HELP. *Computers and Biomedical Research* **13,** 1980, pp. 399–409.
[14] Blum, B. I. and R. E. Lenhard, Jr. An Oncology Clinical Information System, *Computer Magazine,* November 1979, pp. 42–50.
[15] Martin, J. B. Towards Effective Computer-Aided Medical Care Evaluation Studies, *Inquiry* **XV,** June 1978, pp. 129–141.
[16] Given, C. W., L. Simoni, R. S. Gatlin, and R. J. Sprafka. The Use of Computer Generated Patient Profiles to Evaluate Resident Performance in Patient Care. *The Journal of Family Practice,* **5,** No. 5, 1977, pp. 831–840.
[17] Langill, G. and E. Finch. A Computer Assisted Nursing Adult Program. *Nursing* **55,** No. 3, March 1978, pp. 36–37.
[18] Jacinto, M. S., K. Kleinmann and J. Margolin. Pharmacist-Monitored, Computerized Drug Usage Review. *Am. J. Hosp. Phar.* **31,** May 1974, pp. 508–512.
[19] Starfield, B. H., D. W. Simborg, S. D. Horn, and S. A. Yourtee. Continuity and Coordination in Primary Care: Their Achievement and Utility. *Medical Care* **XIV,** No. 7, July 1976, pp. 625–636.
[20] Barnett, G. O. Computer-Stored Ambulatory Record (COSTAR). *Research Digest Series,* National Center for Health Services Research, DHEW Publication No. (HRA) 76-3145.
[21] Stratmann, W. C. A Demonstration of PROMIS, *Research Summary Series,* National Center for Health Services Research, DHEW Publication No. (PHS) 79-3247, September 1979.
[22] Rogers, J. L., O. M. Haring, R. A. Watson. Automating the Medical Record: Emerging Issues, *Proceedings of the Fourth Annual Symposium on Computer Applications in Medical Care,* November 1980, pp. 255–263.
[23] Johns, R. J. and B. I. Blum. The Use of Clinical Information Systems to Control Cost as Well as to Improve Care. *Transactions of the American Clinical and Climatological Association,* **90,** 1978, pp. 140–152.
[24] Simborg, D. W. Medication Prescribing on a University Medical Service—The Incidence of Drug Combinations with Potential Adverse Interactions. *The Johns Hopkins Medical Journal* **139,** 1976, pp. 23–26.
[25] Greenlaw, C. W. and D. D. Zellers, Computerized Drug-Drug Interaction Screening System. *Am. J. Hosp. Pharm.* **35,** May 1978, pp. 567–570.

[26] Halse, R. K., S. J. Clark, J. C. Jackson, H. R. Warner, and R. M. Gardner. Computerized Medication Monitoring System. *Am. J. Hosp. Pharm.* **13,** 1976, pp. 1061–1064.

[27] Bouchard, V. E., J. E. Bell, H. R. Freedy, and M. G. Duffy. A Computerized System for Screening Drug Interactions and Interferences. *Am. J. Hosp. Pharm.* **29,** July 1972, pp. 564–569.

[28] Barnett, G. O., R. Winickoff, J. L. Dorsey, M. M. Morgan, and R. S. Lurie. Quality Assurance Through Automated Monitoring and Concurrent Feedback Using a Computer-Based Medical Information System. *Med. Care,* **16,** 1978, pp. 962–969.

[29] Hoffer, E. P., K. D. Marble, P. M. Yurchak, and G. O. Barnett. A Computer-Based Information System for Managing Patients on Long-Term Oral Anticoagulants. *Computers and Biomedical Research* **8,** 1975, pp. 573–579.

[30] McDonald, C. J. Use of a Computer to Detect and Respond to Clinical Events: Its Effect on Clinician Behavior. *Annals of Internal Medicine* **84,** 1976, pp. 162–167.

[31] McDonald, C. J. Protocol-Based Computer Reminders, the Quality of Care and the Non-Perfectability of Man. *New England J. of Med.* **295,** No. 24, December 1976, pp. 1351–1355.

[32] McDonald, C. J., G. A. Wilson, and G. P. McCabe, Jr. Physician Response to Computer Reminders, *JAMA* **244,** No. 14, October 1980, pp. 1579–1581.

[33] Wirtschafter, D., J. T. Carpenter, and E. Mesel. A Consultant-Extender System for Breast Cancer Adjuvant Chemotherapy. *Annals of Internal Medicine* **90,** 1979, pp. 396–401.

[34] Wirtschafter, D. D., M. Scalise, C. Henke, and R. A. Gams. Do Information Systems Improve the Quality of Clinical Research? Results of a Randomized Trial in a Cooperative Multi-Institutional Cancer Group. *Computers and Biomedical Research,* No. 14. 1981, pp. 78–90.

[35] Friedman, R. B., S. M. Entine, G. M. Murray, D. Holladay, and C. Steinhart. An Integrated Protocol Management System. *Proceedings of the Third Annual Symposium on Computer Applications in Medical Care,* October 1979, pp. 81–84.

[36] Blum,. I., R. E. Lenhard, Jr., and E. E. McColligan. Protocol Directed Patient Care Using A Computer. *Proceedings of the Fourth Symposium on Computer Applications in Medical Care,* November, 1980, pp. 753–761.

[37] Horwitz, J., H. Thompson, R. H. Friedman, T. Concannon, J. G. Krikorian, and P. M. Gertman. Computer-Assisted Patient Care Management in Medical Oncology. *Proceedings of the Fourth Annual Symposium on Computer Applications in Medical Care,* November 1980, pp. 771–778.

[38] Esterhay, R. J., Jr., and E. S. Walton. Cancer Treatment Research Protocols and PROMIS. *Proceedings of the Fourth Annual Symposium on Computer Applications in Medical Care,* November 1980, pp. 762–770.

[39] Bleich, H. L. Computer Evaluation of Acid-Base Disorders. *J. of Clin. Investigation* **48,** 1969, pp. 1689–1695.

[40] PROMIS Laboratory, "Representation of Medical Knowledge" and PROMIS. *Proceedings of the Second Annual Symposium on Computer Applications in Medical Care,* November 1978, pp. 368–400.

[41] Shortliffe, E. H., B. G. Buchanan, and E. A. Feigenbaum. Knowledge Engineering for Medical Decision Making: A Review of Computer-Based Clinical Decision Aids. *Proc. IEEE* **67,** No. 9, September 1979, pp. 1207–1234.

[42] Pople, H. E. Jr., J. D. Myers and R. A. Miller. DIALOG: A Model of Diagnostic Logic for Internal Medicine. *Proc. of the 4th IJCAI* **2**,September 1975, pp. 849–855.
[43] Weiss, S., C. C. Kulikowski, and A. Safir. Glaucoma Consultation by Computer. *Computers in Biology and Medicine,* No. 8, 1978, pp. 25–40.
[44] Patrick, E. A., J. Fattu, and R. Uthurusamy, CONSULT I: Consulting and Diagnosis for Doctors, CONSULT II: Video Player Added. *Proc. of the Fifth Illinois Conf. on Medical Information Systems,* May 1979, pp. 1–13.
[45] DeDombal, F. T., D. J. Leaper, J. C. Horrocks, J. R. Staniland, A. P. McCann. Human and Computer Aided Diagnosis of Abdominal Pain: Further Report with Emphasis on Performance of Clinicians. *Brit. Med. J.* **1,** 1974, pp. 376–380.
[46] Warner, H. R. *Computer Assisted Medical Decision-Making.* New York: Academic Press, 1979.
[47] Fries, J. F. and D. McShane. ARAMIS: A National Chronic Disease Data Bank System. *Proceedings of Third Annual Symposium on Computer Applications in Medical Care,* October 1979, pp. 798–801.
[48] Rosati, R. A., F. McNeer, F. Starmer, *et. al.* A New Information System for Medical Practice. *Archives of Internal Medicine* **135,** August 1975.
[49] Lindberg, D. A. B. *The Growth of Medical Information Systems in the United States, Lexington, Massachusetts: Lexington Books, D. C. Heath and Company, 1975, p. 98.*
[50] Report by the U. S. General Accounting Office. Computerized Hospital Medical Information Systems Need Further Evaluation to Ensure Benefits from Huge Investments. AFMD-81-3, November 1980.
[51] *Lindberg, ibid,* p. 72.
[52] *Policy Implications of Medical Information Systems.* Office of Technology Assessment, November 1977.
[53] Collen, M. F. A Guideline Matrix for Technological System Evaluation. *J. of Medical Systems* **2,** No. 3, 1978, pp. 249–254.
[54] Klarman, H. E. Application of Cost-Benefit Analysis to Health Systems Technology. *J. Occupational Med* **15,** No. 3, March 1974, pp. 172–186.
[55] Drazen, E. and J. Metzger. Methods for Evaluating Costs of Automated Hospital Information Systems. *Research Summary Series,* National Center for Health Services Research, DHHS Publication No. (PHS) 81-3283, July 1981.
[56] Gall, J. E. *et al.* Demonstration and Evaluation of a Total Hospital Information System. NCHSR Research Digest Series 77-3188, March, 1977.
[57] Waymack, P. An Evaluation on Platelet Transfusion Practices in the Johns Hopkins Oncology Center. Internal Report, 1980.
[58] Lenhard, R. E., B. I. Blum, J. M. Sunderland, H. G. Braine, and R. Saral. The Johns Hopkins Oncology Clinical Information System. *Proceedings of the Sixth Annual Symposium on Computer Applications in Medical Care,* October 1982, pp. 28–43.

2

The Need for Automation in Health Care*

Cesar A. Caceres, M.D.

Many have applied computers to the problems of health and medicine. Solutions to many problems have been found. But the use of computers in health and medicine has not yet reached the level at which it has produced the answer to our nationally recognized problems of cost, accessibility and quality in health and medical care. Can a "computer" do that? That is a basic question.

A second question to consider refers to our continuing use of the word "computer" in medical care. Does the physician in practice have any reason to be concerned with a piece of hardware, such as a computer? The answer is no. Should anyone be surprised at that answer? Again, the answer is no.

What the physician in practice must be concerned with is the utilization of readily available technology for all aspects of routine health and medical care. That implies, as an immediate followup, that the physician be concerned with the utilization of "automation" in health and medical care. I would like to suggest that symposia and professional organizations perhaps change their names to more functional and less hardware-oriented ones. Physicians, like patients, are interested in the end result—so we must be interested in the useful output, and only as an item of interest, the tool and the application of that tool.

Automation is a process. Automation is not a process which involves only a computer. Automation involves the people that work with the computer: who give it input and who use its output. Because the automation process requires and involves people (and because it has useful health and medical output), the physician must be concerned with it even if he or she is not interested in computers.

Automation is a system which is concerned with practicality and reality, as well as personalities. In the field of health and medicine, those three items are essential considerations. So the concept presented here is that we not emphasize the hardware and its software, but the total process and how it is going to change health or medicine in relationship to quality, accessibility or cost.

There are problems in medical quality we know, but in this country we have a high quality health care comparable to the best in the world. There are problems with accessibility. But on the whole, I know of no circumstances in which we can say that health care is routinely unavailable. But we do know that we have problems of cost. Our Congress emphasizes these. Our patients are aware of the high cost, and the press, quite properly, does not let us forget that problem. So I would emphasize with priority the area of cost and ask the following question.

Does automation help in reducing costs? The answer should be obvious. Automation is used to achieve a defined level of quality and accessibility but, as every business knows, it is used to reduce costs or to allow a process or a product to become cost-effective. So another question can be asked. Why have we not answered the challenge of costs in health care through the utilization of automation? Is it that the applications have not been good? Is it that our utilization of computers in the automation process has been inappropriate? Or, is it simply that we have, perhaps as an act of omission, concerned ourselves with quality foremost and forgotten the basic necessities of cost and cost-effectiveness.

In this last year (1977), under a study grant form the Fannie E. Rippel Foundation, in a project for the Association for the Advancement of Medical Instrumentation, we have done searches through the National Library of Medicine in reference to various applications of computers in health and medical care. We have done some of these searches, limiting the code words to "costs" and "cost effectiveness." We've been surprised that there are so few papers in this area in literally every application. Ten or twelve in the last 2 years in any specific area is good.

What that means effectively is that when press, Congress, and patients look at us we appear not to be concerned with costs, even though we may be using technology, automation and computers which inherently will reduce costs. I believe it is time for us to emphasize the basic reason for the use of technology and automation. It is, I believe, to meet the challenge of the goals that we have set for ourselves—quality, accessibility, and emphatically, cost.

However, if we cannot find many references to studies on costs in the National Library of Medicine, I would point to foci of government like HEW, which has indicated that technology adds to the cost of health care and say that, if there are few references pointing to the cost reductions, there are fewer that would point to technology adding to cost. I am concerned that some of our health planners, and particularly some in administration, seem to ignore data in this regard and appear to be more interested in attempts to regulate for regulations sake and from their fear, not substantiated in the literature, that technology will increase costs. Technology, and particularly automation, have routinely decreased costs in other areas, hence, should certainly not be expected to increase them in health and medical care.

My first theme then is

(1) that we should emphasize technology as an answer to the challenge of costs in health and medical care: and
(2) to propose as a primary area for research, development, and analysis, review of the multiplicity of technology already available.

I agree that consideration of mode of payment for health care is important, and there is a need to consider regulation to assure quality and consistency with our national ethic, but just as computers are only hardware and not the only consideration when we think of automation or technology, the mode of payment is only a tool. Making the best computer or the best system to pay a bill will not improve health or medical care, only gradually correcting or improving the process that is used with that piece of hardware, or with the payment system will. This tells us that the mode of payment for health care will be more cost effective, if it pays for optimal health care delivery which, I believe, can be achieved in today's era, only through the utilization of technology, that is the utilization of the art and science of industry.

To best know what to improve or correct in the process, one must know the limitations. That is the next stage of significance and the second theme that I would like to emphasize. To utilize hardware, one must know the conceptual limitations of that hardware.

Types of Computer Limitations

Computer analysis of many medical parameters is now commonplace. However, the practicing physician should be aware of certain limitations of computer performance. There are four general orders of limitations.

First-Order Limitations

First-order limitations are those based on hardware (computer memory, etc.). The computer may not allow us to store, retrieve, or analyze as we would wish. This is generally more an economic than a real technical limitation.

Second-Order Limitations

Second-order limitations are those resulting from errors in programming. Sometimes errors are due to lack of comprehension on the part of the programmer; or, they may be errors similar to those made by a typesetter. Second-order limitations (if they are the result of a programmer's poor comprehension) are easily noticed in illogical output displays.

But sometimes output errors are subtle and can change the meaning of the output without our being aware of the error. This is particularly true when minor errors are made, so that one number or decimal can be shifted in a rarely occurring circumstance. Often months or years pass before it is possible to catch these errors. They may be found more often by opportunistic than deterministic events. The physician can do little more than know that such exist and (as for every laboratory test) exert his or her own quality control to avoid becoming subject to them.

Third-Order Limitations

A third-order limitation is dependent on the ability of the computer to abstract the data to be used in measuring; data such as electrophysiological signals. The method for determination of the waveforms in a medical signal might pose the limiting factor in such a case. This potential limitation is tied to the background knowhow of the programmer designer, suggesting that a professional programmer/engineer with an interdisciplinary mind should do the programming.

Avoidance of third-order limitations is the difficult area in the mechanical aspects of computers in medicine. Avoidance depends on an integrated knowledge of what is desired at the output criteria stage; the hardware characteristics of the computer; and methods of determination, which might involve statistics, mathematics, or pattern recognition. Clearly, until we know everything there is to know about the details of these aspects and other developing fields, there will be errors in output which may be deceptive.

What is desired is clinical accuracy and not research precision. Hence, to avoid third-order limitations the physician must ascertain

(1) that the programs are intended for routine clinical use, developed under guidance by experts who are aware of clinical practice;
(2) that they have been evaluated for clinical use; and
(3) that the programs have the qualities found in industrial products and not in research laboratories.

Fourth-Order Limitations

Third-order limitations could be markedly affected in a benefical way if fourth-order limitations were minimal. It is this last area where the physician/researcher should play the greatest role in the work of computer use in medicine. The fourth-order limitation is dependent on the clinical criteria used to handle the raw data. It is generally as a result of review of this area alone that the practicing physician will indicate that a program is right or wrong, good or bad.

This fourth area is the simplest area to redefine and to correct in the computer system if

(1) medicine has defined it; and
(2) the physician has realized his human variability when he operates on the same data, i.e., attempting to use the same criteria as the computer.

A number of fourth-order limitations will be rapidly obliterated as criteria become more standardized and agreed upon. The interface actions of the researcher with the practicing physician can markedly help in this and need reemphasis. Of course, some clinical problems are too complex to expect today's experience, judgment, or intuition to suffice for precise codification into any computer program.

Analysis Systems

Most automated analysis systems are examples of one of three classes of technology. The first is a manual system (such as the ECG) that supports the acquisition, recognition, interpretation, use, storage, transfer, and updating of patient-based data. This man–machine system, used for decades, now can be considered inadequate for the provison of medical measurements.

Semiautomated Systems

Introduction of a digital computer into the measurement system converts it into a semiautomated system. With a semiautomated system, the experience of specialists can be brought immediately to every patient's bedside. Medical signals can be transmitted directly to the processing system by telemetry devices or by hard wire and indirectly by tape or another transfer medium. This second class of system can analyze signals instantly, and results can be displayed on a variety of output devices.

Fully Automated Systems

The fully automated system is the third class of medical measurement system. Such a system, which can in part operate without direct human intervention, exists only in carefully controlled, developmental situations such as in the monitoring of postsurgical patients, where certain programmed decisions are initiated and (automatically triggered by patient emitted signals) properly recognized, and interpreted. For example, a fall in blood pressure, an increase in pulse rate, and a decline in urinary output are all signals that can be detected by automated instrumentation and can be interpreted as indicative of shock. This can lead automatically to an increase in the electrically controlled infusion rate of blood transfusion.

Computerization of ECG analysis led the way in automating data received directly from the patient. Now, there are perhaps 50 regional groups in the United States providing partially or completely automated electrocardiographic processing on a supplier basis to providers of care. There are numerous studies of diagnostic capability, accuracy, and reliability of automated ECG.

Of utmost significance is the value of computer ECGs in predicting disease. For example, the computer initially interpreted tracings in one population we reviewed as showing a 26% incidence of ischemic heart disease, an incidence far higher than that obtained by human readers' interpretation. Five-year followups of cases classified as infarct by computer and not by cardiologists showed very high predictive value. Thus, the computer was not false-positive, but prepositive in many cases. This suggests that one should take great care before dismissing a false-positive computer diagnosis.

The analysis of other physiologic signals can be automated in the same way. The ECG was chosen as the first signal for automated recognition and interpretation of patterns because physicians were familiar with it and because they had 50 years of experience relating its waveforms to specific clinical means; but almost any other physiological signal (e.g., electroencephalogram, spirogram, phonocardiogram) could have been chosen as an initial effort. Many have since demonstrated that each of these signals lends itself to automated measurement and interpretation.

Continued refinement of the computer software, and miniaturization techniques and electronic advances have led to the marketing of a variety of systems that incorporate many cumbersome steps into small, portable instruments that acquire, measure, and interpret the signal and produce a graphic or hard copy result. In additon, entirely new diagnostic and monitoring devices, such as computerized tomography, have within five years been developed and widely distributed.

Automated Signal Analysis Systems That Support Managerial Decisions

Most systems that support clinical managerial decisions are designed on the assumption that the critical deficiency under which clinicians operate is the lack of relevant information. However, the most important deficiency from which clincians suffer is rather an overabundance of irrelevant information. Failure to recognize this distinction in system design is critical. If one sees the clinician's need as the choice of optimal actions in various clinical situations, then the system design ought to change the emphasis from providing all the relevant data to discarding the irrelevant data. The most important function of the latter is the condensation of data by data reduction and filtration.

The application of computer-based technology to clinical medicine has been hampered in the past by inattention to the definition of discrete, project-oriented tasks, by the lack of a suitable basic science to support the managerial decision making processes required, and by a scarcity of medical personnel trained and educated in the science of automation.

Among potential benefits of automation are:

(a) improved legibility, accuracy, and precison of observations and results;
(b) standardization of terminology and format;

(c) increased accessibility and availability of patient data for the decisionmaker;
(d) reduction in observer variation; and
(e) provison of individual and collective patient records with multiple observations in manageable form so that patient derived data can be used to satisfy a wide range of administrative and clerical needs including billing, status reporting, and the provison of basic data for peer review.

Properly structured, an automated system can support internal efficiencies in staffing, skill transfer, and facility utilization that, together with the provision of an improved patient data base and continuity of clinical observations, can lead to a more effective product, improved patient care and cost effectiveness.

3

Some Problems of the 1970's: A Clinical Perspective*

E. Van Brunt, M.D., Sc.D.

The title of this review might give one pause; the *Oxford English Dictionary* cites the term clinical as referring to the sick or death bed. Let it be established quickly, then, that I do not view the application of computer technology in medical care as being in a diseased or terminal state. An allied definition is: based on observation rather than theory; it is in this spirit that this comment on medical computing during this past decade is presented.

The environment that has prevailed in the 1970's, and is likely to persist well into this decade, was outlined in 1972, at an international colloquium in San Francisco, on the subject of Technology and Health Care in the 1980's [1]. It was observed then that health services had gained a higher profile over the past few decades, and that health care would remain a prime political issue as capital intensive technology continues to develop in a traditionally labor intensive service environment. Some authorities expected a continuation in the United States of the cultural trend toward the growth and toleration of divergent lifestyles and values; the evolution of "mosaic cultures" in contrast to the mass society theories of the 1950's. Coupled with this was the observation that health care does not readily lend itself to the mass production approach of industry, i.e., higher costs cannot always be balanced by increasing productivity. It can be concluded, then, that the patterns of health care delivery will continue to vary and standard methods will be slow to develop in the practice of medicine. At the same time, more public attention is being given to structured systems of health care, such as Health Maintenance Organizations (HMO), and to methods of health care evaluation and public accountability, e.g., Professional Standards Review Organization (PSRO). There has been increasing use of technology, and it can be expected to continue, yet one of the problems associated with the application of some modern technologies in medical care is the fact that the more advanced or complex the technology, the greater the overall costs of health care. This is not a paradoxical effect, since it can be explained in part by the way medicine is practiced, the essence of which is the one-to-one provider–patient correspondence. Technology contributes by providing information and/or enhancing the provider's treatment capabilities, but it usually does not reduce the individual labor intensive character of the medical care process.

In 1972 it was predicted that expansion of information handling capabilities would tend to progress in a fragmented manner, and that advances in hardware would outpace those in software, wherein lie the major problems of design and implementation. Finally, it has been emphasized repeatedly over the years that successful implementation of data processing technology in the medical care environment requires a long-term commitment of large amounts of both capital and interdisciplinary personnel. Now, eight years later, all of these observations and conclusions appear still to be valid; nothing has occurred in this past decade to suggest any important departure from them.

The scene for the 1970's, then, and probably for a good part of the 1980's, is one of increasing public and government involvement in the philosophy, methodology and financing of medical care, of increasing use of technology despite costs and controversy and, while the basic physician–patient relationship changes but little, of probable increasing effects on physicians' modes of practice.

These comments, therefore, are directed to those who would employ computer technology as part of a system of medical care activities that depends upon the management of medical information for more than one function in more than one clinical setting. The concern here is not so much for those whose interests lie in highly specialized, so-called 'rifle-shot' applications of computers in medicine (thus not to the specialist in nuclear medicine, radiation therapy or wherever the remarkable CAT scanner resides, nor to the small medical group or hospital whose business is managed very efficiently with the aid of computers) as it is for those who would manage and use information derived from many such applications.

With this bias, a look across the decade of the seventies reveals that, while computer technology is being applied in a wide variety of areas in medicine and biology, there is a remarkable paucity of applications that directly support the delivery of medical care. It is possible to group the applications into three general "levels" of visibility.

1. Medical Care Financing and Hospital Administration

The visibility of applications in these two areas far and away dominates that of all of the others. I refer to the health insurance area, including government and privately financed third-party indemnity and prepayment programs, and to hospital administration, where the bulk of data processing services have been developed to support diverse accounting procedures for hospital administrative management and health care financing. These functions are not medical in the strict sense of the word, but they have become an inseparable part of the medical care process of today; in so doing, they highlight an organizational problem of first importance, and one that has plagued for years those involved in acquiring or developing computer support for their medically-based activities: should the approach be, on the one hand, to acquire a primary administrative support computer capability, and attempt to build-in or add-on various medical care support functions or, on the other hand, should the approach be to develop a primary medical care support computer capability which, by virtue of the information that is involved, should also satisfy virtually any administrative requirements? There is no doubt that the first has been the mode; it is less complex, it rests upon a base of established business experience and its costs are easier to justify. The 'administrative approach', however, typically has led to the development of systems that provide incomplete and/or inadequate support in the diverse clinical settings where medicine is practiced.

2. Medical Research

At a second level of visibility lies a large cluster of data processing applications that, directly or indirectly, support basic and clinical medical research. Complex statistical analyses are performed now with relative ease; computer supported biomathematical tools such as SPSS and BMD have become available; network facilities such as ARPANET and CYCLADE are available for appropriately scaled users, and a large and growing number of small, economical, specialized computers are appearing with the result that the very attitudes of the medical investigators are changing, with an accompanying increase in the range of problems which may be explored. Allied to the research laboratory is the process of information dissemination. The development of services such as MEDLARS, MEDLINE, TOXLINE and regional blood bank inventory systems surely herald the kinds of services that, one day, will become integrated progressively more into the process of medical care. In any case, from the points of view of economics and penetration into the health care field, the applications cited above dominate all others.

3. Clinical Medicine

A look at the domain of clinical medicine reveals the existence of some computer applications which can be characterized, despite their heterogeneity, by the fact that each has become fully integrated into the process of medical care; some are now considered indispensable. The archetype of this class of applications is computerized axial tomography; less dramatic but equally integrated are applications in nuclear medicine and radiation dosimetry. The remainder in this group are in varying stages of development and in use in many locations in this country; while not yet indispensable to clinical medicine, they show promise of becoming so in the future:

(1) Computer interpretation of electrocardiograms (ECG) has been in development since the mid-1960's, and is becoming an increasingly utilized service. In 1976 in the United States, an estimated 5 ± 1 million ECG's were processed by computer programs as a clinical service. This represented approximately 6% of the total number (90,000,000) of ECG's done that year. Parenthetically, it should be noted that this does not mean that 6% of physician time was conserved since, in many centers, an important fraction of the computer program interpreted ECG's also are interpreted by physicians.

(2) Closely allied to ECG analysis are specialized computer hardware and software applications which have demonstrated a potential for improving the quality of medical services in the areas of intestive care and anesthesia. Arrhythmia analysis, physiological determinations such as cardiac output or peripheral vascular resistance, and process control functions that monitor, for example, the integrity of complex ventilation therapy equipment, are representative of applications which most certainly will become integrated into institutional medical care procedures as soon as the ratio of costs to utility and reliability is sufficiently reduced.

(3) Most clinical laboratories and pharmacies today conduct their operations without the support of computer technology. However, much effort is being expended in attempts to develop and market automated *systems* of data handling in these two areas, in order to improve such functions as quality control, process control, message handling and report generation. As with ECG analysis, some developed systems have shown clearly that the use of computer technology in these areas can improve the quality and reliability of operations, but that the coexisting problems of costs and adaptability (both to different environment and changing technology) have not yet been resolved to the point where these data processing services are standard practice, let alone indispensable.

(4) Multiphasic Health Testing Services (MHTS) perhaps deserves special comment. MHTS, defined in its modern form as a *system* of health evaluation techniques, and employing instrumentation, computer and communications technology, originated in the early 1960's. The concept has been adopted in one or another form in both the western world and the orient. Multiphasic Health Evaluation methods have been shown to operate more economically than more traditional models [2]. The controversies that exist concerning this subject have more to do with the appropriateness of the medical information to be collected in projects with specific but different objectives than with the demonstrated efficiency of the methods employed. While the number of computer supported medical services of this type has increased, from a few dozen MHTS centers in the United

States in 1968 to over 350 in 1978, and while many of these centers process upwards of 100 patients per day, at which loads economy of scale and improvement in data handling have been demonstrated, the relative utilization of this modality is very low.

(5) There is that class of applications, still in early stages of research and development, which has been termed clinical decision support, ie., those processes resulting in decisions which we all recognize as diagnosis, prognosis and therapy. Whether the approach consists of the structuring of decision trees, using Bayesian estimates of likelihood for different nodes or elements, or whether more complex methods are employed, e.g., using techniques derived from artificial intelligence studies, they have to date focused on only a few model pathophysiological states. One of the current limitations to development is the difficulty in dealing with coexisting clinical conditions (the rule rather than the exception in the real world), and an important limitation to implementation in the real world will be, perhaps, a years-long process of orienting medical students and postgraduate physicians in training to these new capabilities as they emerge.

In summary, there has been an enormous amount of activity over the decade of the 1970's with a highly variable visibility of the different medical computing applications. The great majority are not involved with direct medical care. At the upper most level are extensive computer supported applications in medical care financing and hospital administration, and a small number of operating medical support service applications which are fully integrated into the medical care process. At a lower level are a lesser number of medical service applications which show great promise, but still are not sufficiently advanced and/or cost-effective for general acceptance. Finally, there are a small number of scattered special applications in the area of direct patient care that wait not only for more effective methods, but for a series of adjustments in medical education and medical care practices (Figure 1).

From a clinical point of view the problems that are revealed by this brief survey can be brought to focus in the interrelationships between three areas of computer applications that have been notable for their very low visibility. They are the so-called hospital or medical information systems (MIS), the medical record and medical education.

Definitions of MIS are many, some differing merely by nuance, others by underlying philosophy. Lindberg's reasonable definition, "a set of formal arrangements by which facts concerning the health or health care of individual patients are stored and processed in computers", would be viewed by most workers in the field as implying that an MIS, by virtue of managing information, is meant to facilitate medical management of the individuals whose information it contains [3].

In the past 20 years, efforts to develop medical information systems have resulted in a long list of unfulfilled or partially completed projects consisting in the aggregate of many dozens, possibly several hundreds, of millions of dol-

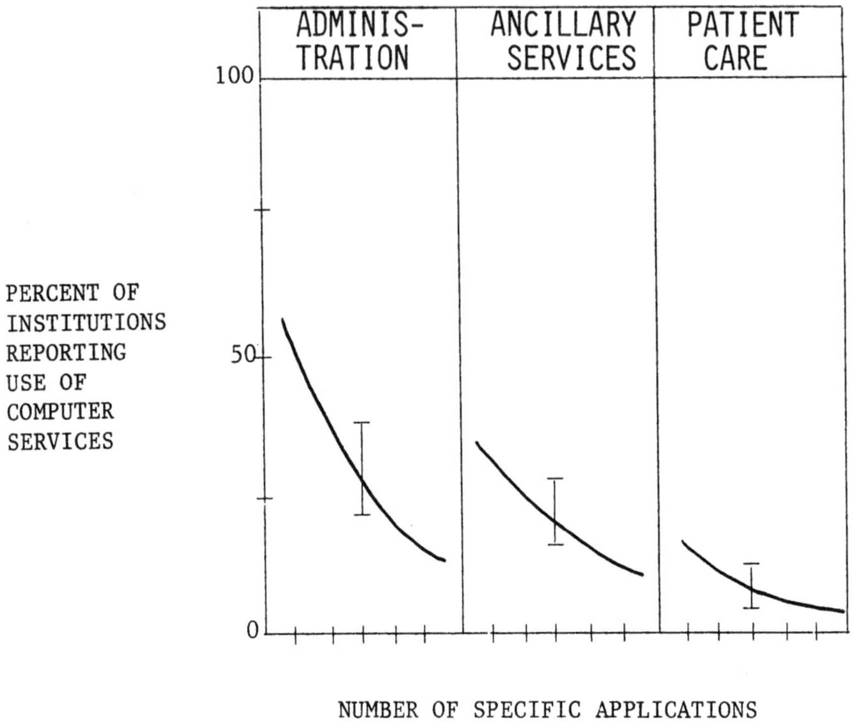

Figure 1. (Adapted from Spencer, W. A., 1976 [10]).

lars and thousands of man years of effort. The objectives of MIS projects usually have been cited as

(1) to reduce or contain that fraction (reportedly about one-third) of total medical center operating costs that are attributable to information handling;
(2) to improve the quality of hospital operations by improving specific process and quality control functions—both medical and administrative;
(3) to improve the quality of medical care by reducing those risks which are attributable to mismanagement of medical information; and
(4) to build an information base which would permit better analysis of hospital operations and better planning for future institutional and community health care requirements.

The question of how well these objectives have been met is controversial in the extreme.

At the present time in the United States there are no operational systems that meet the numerous requirements of an integrated institutional medical care data support system. One of the more successful commercial systems that is currently available has satisfied some MIS objectives (initially developed as the Lockheed Medical Information System; further developed and marketed by Technicon Corporation). It appears to perform very well for many hospital functions, and has been reported to have undergone favorable cost analysis in one of the institutions where it is in operation. It is also expensive, costing of the order of one million U.S. dollars per year for a medium-size general hospital; it lacks the capability of maintaining continuous medical records and does not adequately support ambulatory care. Although data vary with the source, the development of this system reportedly required of the order of 15 years and $20 million dollars to bring to an acceptable level of operation. Money aside, is this an inordinate time to develop an information service that, however well it meets its intended goals, still has important limitations? I do not think so.

In the past two decades it has been demonstrated clearly that even limited MIS developmental projects require much more time than expectations would have it, 'long range' planning of five, or even eight, years notwithstanding. The COSTAR project required an estimated eight years from its inception to the point where it was funded by nongovernment services, and an additional five to produce its first marketable product [4]. MEDITECH reportedly has evolved over a period of 10 to 14 years, and the very impressive but still limited HELP system continues developing and expanding after nearly 15 [5]. The Kaiser-Permanente Computer-stored Integrated Medical Record system was in development for nearly 10 years before the project was terminated [6].

Common to developmental efforts is the fact that the ability of the planners to conceptualize their objectives exceeds their ability to develop and implement specific functions that will satisfy the objectives. The concepts change but little, while system specifications and functions typically are developed stage by stage and changed frequently. The development of clinical medical applications of computers, therefore, has been an *evolutionary* process. The experiences of the seventies have served to underscore the admonitions of those workers in the field in the sixties who insisted that the basic requirement for success is a long term committment of capital and personnel [7]; in addition, it has afforded us a better appreciation for *how* long.

There are many reasons for this problem of the evolutionary nature of MIS development. One obvious one is the nature of the medical record and the use of medical information.

The computer-stored medical record is one of the controversial subjects in the field of medical computing, due partly to problems of definition. Many reports have been published proclaiming the use of computer-stored medical records. Virtually all such records are but partial or incomplete, and rarely are integrated for use for both inpatient and outpatient care. By today's standards, an individual's medical record reflects not only single encounters or classes of data, but contains a large variety of information concerning that individual's medical experience and health status over a *long period of time*. This is more important than ever today, with progressive trends toward combined hospital and ambulatory care, interfacility sharing of special services, and regionalization of various aspects of medical services and administration. The ramifications of these statements are considerable. Implied is the use of information not only by physicians and other persons involved directly in the process of providing individual patient care, but the ulterior use of the information by a variety of other persons, for purposes of quality control of ancillary services, teaching, and in clinical and health services research, not to mention many administrative uses.

In the United States, we have experienced progressive difficulties in medical records management in recent years. With increasing specialization, increasing variety of kinds of medically important data, increasing mobility of urban patients, and the demands of various government and private agencies, the availability and communicability of needed information seem to become increasingly slow and unreliable. Redundancies are created in the form of special records; file management is error prone; storage space is diminishing; the problem of legibility remains ever present, and the physical transportability of documents is less efficient. Faced with these difficulties, the obvious great potential of computer and communications technology for medical data storage, retrieval and communications has been and continues to be a strong attraction. What, then, has prevented a higher level of achievement in this most basic of medical application areas? There are many factors involved, but two seem to stand out: First, the implications of a continuous, or "lifetime", medical record, and the usual ways in which medical records are used, are not readily appreciated by most persons outside the medical field, and sometimes not even by some within the field. In an institutional setting the volume of data, system requirements for repeated updating, reliable error control procedures, hierarchical data structures, above all communicating appropriate and useful information to the provider of care *at the appropriate time,* constitute a formidable set of data processing problems. The eventual solution(s) will very likely have to do with technical advances such as distributed processing and distributed data base management, but in any case are predominantly economic. The feasibility of continuous medical record systems has been demonstrated; the costs to date simply have been unacceptably high.

The second and closely related fundamental factor is the problem of medical data collection. The sources of medical data are large in number, and validity of the data is uncomfortably variable. Computer technology is used in many medical school basic science laboratories, and medical students are therefore exposed to a limited number of specialized applications, but other clinical applications in medicine have been slow to be assimilated into medical school teaching programs. Very little is taught in such areas as medical decision processes, that which Dr. Homer Warner refers to as 'medicalware'. Accordingly, many methods for recording the observations, conclusions and decisions of doctors and other providers have been devised. None matches the elegant simplicity of pen and paper. While methodological improvements are forthcoming, the problem is not only in creating appropriate methods for different types of medical data collection, but in promoting an atmosphere conducive to the resultant changes in current medical practices, particularly those of physicians and other medical professionals. This latter is very difficult. It requires overcoming the enormous inertia created by centuries of tradition in a complicated socio-scientific and labor-intensive system, and probably will be best achieved in medical schools and post graduate medical training.

A few other reasons for the evolutionary nature of MIS development are worth noting. First, some people hold that the structure of an information system tends to, and perhaps must, reflect that of the organization in which it is developed, and that it must be recognized that the organization itself changes, structurally and/or functionally, as it adopts and adjusts to new information management capabilities. The initial analysis of a user environment may lead to a computer solution that in fact creates a different user environment. Even in the case of many so-called 'turnkey' products, one often doesn't simply buy an information service, he also reorganizes related functions, and must be prepared to deal with possible resulting behavioral changes of his personnel [8,9]. Not to be ignored are the observations that, within some organizations, there are political and economic effects of attempts to control the development of MIS's that may affect the outcomes in several ways, some good, some bad.[10]

Finally, at all levels, but particularly that of planning and review of projects, there has been the important problem of inaccurate or unrealistic expectation of what can be achieved over short periods of time. We have been warned of this for at least two decades yet, in much of the literature of today, one still sees styles of presentation of state-of-the-art topics and reports of projects that leave many readers with an inappropriate sense of *now.* At least two negative effects of this phenomenon can be observed: on the one hand, overzealous planning and subsequent frustrated expectations and, on the other, a jaded view of the potential benefits that current computing capabilities have to offer.

Conclusion

Experiences in the seventies have revealed that the application of computer technology to medical care, particularly the development of medical care support *systems,* is an evolutionary process. Most systems develop as outgrowths of systems designed primarily for administrative purposes, and provide incomplete and/or inadequate support to the delivery of health care. The closer one draws to direct patient care, the more complex the problems and the fewer the operational systems. For these reasons, and because of our social environment and trends in the organization and delivery of health care, progress in medical computing is likely to continue in a fragmented way, probably for most of the decade.

It seems reasonable to expect an ever greater involvement of data processing services in financial and administrative activities. One important effect will be that providers of medical care, both institutional and individual, will become progressively involved in the collection and documentation of clinical medical information. Inevitably, this will result in increasing demand for more effective and acceptable methods of collecting clinical medical data, and for more understanding of the nature of medical data and medical information requirements in different environments. By environment, I refer not only to small and large groups of medical personnel, institutional teaching centers, health care and preventive medical centers and the like, but to specific clinical settings, such as hospital emergency rooms, gynecology offices, well baby care centers, etc., where the individual physician, nurse or pharmacist, but ultimately the patient, can benefit from the availability of needed clinical information. This would seem to be especially important in the outpatient environment where the bulk of medical care is practiced and where the need for efficient handling of information is the most pressing. It also appears logical that the designs of systems should be addressed primarily to *clinical medical* requirements—most administrative information needs would be a natural resultant benefit.

It follows, then, that the providers of care must become more involved in the processes of planning, development and implementation of these clinical information-handling systems. Such involvement exists, of course, but must be intensified, particularly at the levels of undergraduate and graduate medical education.

Strong arguments can be made that the potential benefits of medical computing outweigh the costs and risks. That there may be occasional negative effects, either physical or moral, is predictable; their magnitude will perhaps be a measure of our capacity to learn by our experiences. As members of a dynamic society, we have no alternative but to adapt continually to change.

References

[1] Symposium on "Technology and Health Care Systems in the 1980's", San Francisco, 1972. DHEW Publ. (HSM)73.3016.

[2] Collen, M. F. Systems Evaluation of MHTS (Chapter 17) in *Multiphasic Health Testing Services,* M. F. Collen, Ed. New York: Wiley, 1978.

[3] Lindberg, D. A. B. *The Growth of Medical Information Systems in the United States.* Lexington, Mass.: Lexington Books, D.C. Heath and Co., 1979.

[4] Barnett, G. O., Boston (Personal communication).

[5] Warner, H., Salt Lake City (personal communication).

[6] Van Brunt, E. The Kaiser-Permanente Hospital Information System, in *Hospital Computer Systems,* M. F. Collen, Ed., New York: Wiley, 1974.

[7] Levy, A. H. Is Informatics a Basic Medical Science? Proceedings: MEDINFO '77, Toronto, 1977.

[8] Dickenson, G. W. and Simmons, J. K. The Behavioral side of MIS, in Trent, R. T. and Wheeler, T. C., *Development in Management Information Systems,* Dakmoore, California, 1974.

[9] Goodfriend, S. T. The Introduction and Utilization of a Computer-Based Program Monitoring System in a Comprehensive Child Welfare Agency. Doctoral dissertation, U. of Calif., Berkeley, 1979.

[10] Spencer, W. A. An Opinion Survey of Computer Applications in 149 Hospitals in the USA, Europe and Japan. *Med. Inform.* **1,** 1976, p. 215.

II. HOSPITAL INFORMATION SYSTEMS

A hospital is a natural environment for an automated system. It is relatively large, labor intensive, information oriented, and dependent upon communications among different operational units. Difficulties associated with the implementation of an automated system include the cost and reliability of computers, the lack of standards among hospitals, and the personal nature of health care delivery. These problems combined to frustrate the early developers. By the early 1980's, however, the cost and reliability of computers was no longer a major factor. Consequently, many vendors now offer products which will support the nonfinancial functions of a hospital. The papers in this section survey these commercial systems, describe several different approaches to an HIS, and consider some of the problems in the selection and installation of a system.

The first chapter in this section, by Ball and Jacobs, contains a summary of Level 1 hospital information systems based on a 1980 survey. A Level 1 system provides Admission-Discharge-Transfer (ADT), order/entry and charge query facilities. (As the name implies, there are Level 2 systems which do even more.) The chapter identifies the vendors and the capabilities provided by their systems. An interesting table illustrates the massive investments that the vendors have made in the products which they market.

The next chapter, by Young and his associates, considers the use of automated data management systems in small to medium sized hospitals (100–300 beds). For these institutions, the installation of an HIS may not be cost-effective. The authors present a five-step growth model for incremental implementation and service. Twelve vendors were surveyed and their capabilities identified at each step.

The following four chapters each describe a system. The systems are quite different in their objectives, architecture, and style. The chapter by Mishelevich and his colleagues describes the installation of a commercial package. The system is the Parkland On-Line Information System (POIS) which was developed using IBM's Patient Care System (PCS) which was originally developed at Duke University Medical Center and then marketed as an IBM Installed

User Product. Put another way, an HIS initially developed for one hospital (Duke) has been modified for use by another (Parkland).

Two classes of problems are encountered when an existing HIS is to be installed in a new hospital. The first is the problem of local definition and training. This is a prerequisite for the integration of the automated system with the hospital's organizational structure and philosophy. A second class of problems relates to the installation process itself; this is generally only a factor where an inhouse development staff is used. In the case of POIS, both classes of problems were confronted. Furthermore, the application chosen as the first to be implemented did not exist already in the basic PCS package. The chapter discusses Parkland's general approach to the installation of an HIS and describes the radiology system they created. The description illustrates the general kinds of user inactions associated with a commercial HIS.

The chapter by Weed and the PROMIS Laboratory, on the other hand, takes a very different view of the HIS function. Where the commercial systems are primarily concerned with management and communication of data and information within the hospital, the PROMIS system has a very different orientation. It is based upon a model of medical knowledge which structures the users' interactions with the system. Rather than providing a tool to manage data, Weed sees the computer as a device for coupling medical knowledge and patient care through use of the information system. In this chapter the concepts are clearly described and an excellent example is provided.

The chapter by Pryor and his associates describes a third orientation. The HELP system provides the functions associated with a typical commercial HIS but with some very significant additions. The primary orientation is towards patient care, and it supports this in two major ways. Through use of a definition language, HELP sectors can be established to differentiate among classes of patients. Using these HELP sectors, it is possible to identify potentially dangerous interactions and thereby initiate an alert. A second application of the HELP sector is to process patient data from the data base in order to generate conditional probilities to evaluate risks and outcomes for a particular patient.

The final descriptive chapter is by Tolchin, Simborg and their coworkers. It presents an alternate approach to HIS implementation. Each of the previous chapters described a single unified system. While multiple computers may have been used, a single data base and set of unifying programs was assumed. In the distributed approach presented here, all hospital functions are fragmented and decentralized. There is a separate ADT system, a separate radiology system, etc. Each of these systems is integrated by a local area network (LAN). As the cost of computers falls and the popularity of microcomputers increases, one will find that new advances will be made in various aspects of hospital management and patient care. Using the central computer model, these advances can be implemented only through reprogramming. With the network model, individual components can be upgraded and new functions can be added. Clearly, the direction of technology favors this approach. But vendors must recognize that the UCSF system is still a demonstration system; many problems remain to be solved.

The final four chapters of this section are concerned with how to install an HIS. The chapter by Kolenaty presents an overview of the planning process. Sobolewski's chapter concentrates on the process of implementation, and Rusnak presents a case study. The final chapter of this section, by Jacobs, contains a survey of facility management services for hospital data processing. The first chapter should be of interest to all institutions which contemplate the installation and operation of an HIS. The next two provide some insights into how an inhouse project can be managed. The final chapter presents some alternatives for institutions which prefer to contract out this function.

In summary, we note that there are many commercial systems to choose from. The marketplace dictates a set of functions which include some which impact patient care. For example, more timely, error free and legible test reports are taken for granted by the health care providers. Yet few of the systems are designed to impact patient care. That we know how to do this (to some degree) is illustrated by the chapters by Weed, Pryor, etc. One can assume that the HIS will continue to focus upon the hospital communication problem until there is a recognition by providers that the HIS can play a more active role in the process. And when that happens, the marketplace will respond.

4

Hospital Information Systems As We Enter the Decade of the 80's*

Marion J. Ball, Ed.D., and Stanley E. Jacobs, Ph.D.

In the June 16, 1980 issue of *Hospitals,* the authors traced the development of hospital information systems, provided a working definition of a Level 1, Second Generation Hospital Information System, and predicted the rapid prolifiration of such systems [1].

The Classical First Generation Level 1 HIS of the 1960's encompassed a basic set of applications for inpatients. These were:

(1) An Admission-Discharge-Transfer (ADT) application with bed status and census reporting.
(2) An order/requisition entry, communication, and charge collection application.
(3) An inquiry application for today's charges for demand bill purposes.

By way of contrast, a Level 2 System is one that has been designed from the start to capture and maintain online the entire patient record. As such the computer can prepare medication schedules, the data bank may be queried for patient medication, profiles, etc.

Second Generation, Level 1 systems now include these same applications, but in addition all vendors offer a variety of additional applications which greatly enhance the usefulness of the HIS to the hospital and its high volume departments. A listing of "enhancement applications" is contained in Table 1 together with indication frequency that each application is offered by a sample of 8 vendors.

The information provided by Table 1 shows that today's Second Generation, Level 1 Systems go way beyond the three applications of First Generation Level 1 HIS. For every one of the 36 applications listed in Table 1 over half of the vendors indicated that they now offered or planned to offer the application. Careful study of the application list will reveal that certain applications require a computerized patient record to be maintained throughout the patient

Table 1. Enhancement Applications (8 Vendors); 1979 Data

Department/Application	Number of Vendors	
	Now Offer	Plan to Offer
Nursing		
Order Set Entry	5	1
Auto-Kardex (Charting)	3	2
Patient Care Level Tracking	3	2
Staffing Requirements	5	2
Nursing Notes	5	0
Care Plan	1	6
Medication Schedules	6	2
Medication Monitoring	3	2
Pharmacy		
Patient Medical Profile	7	1
Drug Precaution/Interaction	2	4
Unit Dose Care Replenishment	5	2
Pharmacy Inventory	3	4
Lab		
Specimen Collection Lists	6	1
Specimen Collection Labels	8	0
Results Reporting[a]	7	0
Cumulative Results	6	1
Interface to Lab Computer	6	1
Radiology		
Scheduling	4	2
Report Normals Only	3	3
Report All Results	2	5
Radiology Index	4	2
Medical Records		
Medical Record Index	5	2
Medical Record Abstract	5	2
Chart Loc. & Delinquency Tr.	3	3
ER/Outpatient		
Registration	8	0
Order Entry	8	0
Demand Billing	7	0
Patient Scheduling	6	2
Miscellaneous		
Preadmitting	8	0
Computer Assigned Patient I.D.	8	0
Time Clocking	5	0
Surgery Scheduling	2	3
Diet List Preparation	7	1
Utilization Review	7	0
Cash Receipts Entry	6	0
Doctors' Registry	6	1

[a]Some report Stats only.

Note: This table is based upon the responses of 8 vendors to a questionnaire listing the enhancement applications and asking whether the vendor now offered or planned to offer the application. The 8 vendors are HBO, SHS, MCAUTO, MCR, Burroughs, DATX, Pentamation, and HDC. In addition to the 36 applications listed, collectively the vendors reported, 24 more applications on a "write-in" basis. All of these were unique to the vendor that reported them.

stay. For instance, a patient medication profile requires that medication data be retained for the entire stay of the patient.

What has happened, of course, is that just as the Level 2 systems have grown over time in the number of functions they accomplish, so have the Second Generation Level 1 systems grown. The survey showed all vendors to be offering dynamic, continually growing systems.

The scope of a vendor's Level 1 system will vary from hospital to hospital. Some hospitals may contract for most or all of a vendor's enhancement applications, while others may be satisfied with little more than a classical Level 1 HIS. Some vendors will even sell an Admitting-Discharge-Transfer system only, perhaps hoping that the hospital which purchases the ADT system will upgrade to a full Level 1 system in the future.

A word of caution is in order. Within an application, the amount of support it provides will differ from vendor to vendor. When choosing between vendors, a hospital must look into each application in depth to decide upon its merits.

The findings from a May 1979 survey of vendors were updated in April 1980 to provide current information in Tables 2–5. These findings are summarized as follows:

(1) Table 2. The fact that 18 vendors now offer Level 1 HIS is testimony to the adaptability of American industry. In addition, the survey found 5 more firms which will probably soon enter the marketplace.

Collectively, more than 500 Second Generation Level 1 Systems have been sold as of Spring, 1980.

Hospitals of all sizes have installed Level 1 HIS's. This is contrary to earlier beliefs that a hospital must have over 250 beds to justify such a system.

The Cathode Ray Tube (CRT) is by far the input device preferred by the industry.

IBM is listed in Table 2 because it does provide software and hardware for a Level 1 HIS. However, the support to get a hospital up and running provided by IBM is somewhat less than by the other vendors. It is up to the hospital to have a relatively sophisticated data processing department to implement the IBM Level 1 system. In other words, turnkey support is not provided. IBM also offers Level 2 HIS software. Nevertheless, the IBM approach is an option that should be considered by hospitals which have in-house, well qualified data processing staffs. The Burroughs Level 1 and 2 offerings are likewise much less a turnkey proposition than are those of the other vendors.

(2) Table 3. Eight of the vendors also offer shared business and financial systems. (Most of the other vendors now offer in-house business and financial systems.)

(3) Table 4. Six of the vendors also offer Level 2 HIS's first and then, perhaps spurred by the success of HBO, brought a Level 1 system to the marketplace.

(4) Table 5. The magnitude of funds expended by vendors for Level 1 and/or Level 2 HIS development, and the funding they have committed for research and development for the next year or two are shown in Table 5. This certainly suggests that these companies are deeply committed to the HIS marketplace. Because of their R&D investments, they can now bring HIS to hospitals at much lesser cost to the hospital than the funds required for a self developed HIS. At the same time they can save the hospital a lot of development time and implementation trauma.

Table 2. Present Vendors of Level 1 HIS

Vendor	No. Installed or in Process—April 1980	No. of Beds Least	No. of Beds Most	Order Entry Terminal[a]
HBO & Company (HBO)	237	100	829	CRT
Shared Medical Systems Corporation (SMS)	100	154	1,725	CRT
McDonnell Douglas Automation Co. (MCAUTO)	53	103	650	CRT
IBM Corporation (IBM)	50 (est.)			CRT
NCR Corporation (NCR)	25	284	641	BCWR or OCR
Burroughs Corporation	20	191	737	CRT
DATX Corporation (DATX)	10	176	534	CARD READER
Pentamation Enterprises	9	280	800	CRT & OCR
Tymshare Medical Systems	5	80	436	CRT
Compucare, Inc.	4	250	440	CRT
Space Age Computer Systems, Inc.	4	349	815	CRT
Hospital Data Center of Virginia (HDC)	4	256	800	BCWR[b]
Spectra Medical Systems, Inc.	2	140	600	CRT
Technicon Medical Information Systems Corporation	3	446	719	CRT
Interpretive Data Systems, Inc. (IDS)	1	250	250	CRT
National Data Communications, Inc. (NADACOM)				CRT
Genitron, Inc.				BCWR
Informational Resource Electronics Corp. (IREC)				CRT (Intelligent)

[a]CRT = Cathode Ray Tube; BCWR = Bar Code Wand Reader; OCR = Optical Character Reader.
[b]HDC also utilized CRT's for inquiry purposes.

Table 3. Vendors of Shared Business and Financial Systems which Offer Level 1 HIS

Vendor	1979 Data No. of Hospitals in Shared System	1980 Data No. of Level 1 HIS Installations
MCAUTO	561	53
SMS	490	100
Tymshare	150	5
Technicon	45	3
Pentamation	31	9
Space Age	26	4
HDC	24	4
HBO (IFAS)[a]	14	237

[a]HBO introduced a new real-time, terminal/network based financial based financial system in 1979. Their new IFAS System directly interfaces inhouse to HBO's Hospital Information System (MEDPRO System).

In these times when hospital administrators must be extremely cost conscious, can they justify the installation of any level of HIS? In his remarks to a congressional committee, Hodge advised that "Level 2 and 3 systems are or could be cost effective. Level 1 systems are adding to hospital costs" [2]. If the latter statement were true, then the sales of Level 1 systems would have come to a screeching halt. But fortunately for the industry and for hospitals, the facts deny Hodge's statement.

For instance, this article documents the fact that over 500 Second Generation Level 1 HIS's have been sold. Most of these had to be cost justified for Health Systems Agency approval. Apparently, hospitals with vendor assistance have been quite successful in proving to HSA's that Level 1 systems are cost effective.

Table 4. Vendors of Level 1 HIS which also Offer Level 2 HIS

Vendor	No. of Hospitals with Level 2 HIS	No. of Hospitals with Level 1 HIS
Spectra	18	2
Technicon	17	3
NDC (NADACOM)	10	0
Burroughs	8	20
MCAUTO	6	53
IREC[a]	4	0
IBM	Not available from IBM (estimate 50)	50 est.

[a]IREC is a relatively new vendor of HIS. Its development site was The Sisters of St. Mary Hospital Group, St. Louis. Its system is modular, and IREC claims that its complete system is a genuine Level 2 HIS. This system together with facilities management therefore is also being marketed by Electronic Data Systems, Dallas, Texas.

Table 5. Vendor Research and Demonstration Budgets for Hospital Information Systems (Survey: M. J. Ball, 1980)[a]

Firm	R & D $ Invested to Date	R & D $ To Be Invested, Next 2 Yrs
Compucare (HIS)	3,500,000	2,000,000
Datacare (PCIS)	4,000,000	1,000,000 (vary by client needs)
HBO and Company	6,000,000 (approx.)	6,000,000 (approx.)
Hospital Data Center of Virginia (HDC)	Not available	250,000
Interpretive Data Systems (IDS)	1,000,000 (approx.)	300,000
McDonnell-Douglas Automation Co.	20,000,000	4,000,000 plus
Meditech	300,000/yr since 1969 (3,300,000)	600,000
Nadacom	15,000 on full line of patient data systems	will continue to make a significant investment in new development
NCR Corporation	6,000,000	3,000,000
Shared Medical Systems	6,200,000 during 1979	no specific data, but in keeping with past efforts
Spectra (Medicus)	25,000,000	"proprietary information"
Technicon Medical Information Systems	24,000,000 (incl. Lockheed)	4,000,000
Tymshare Medical Systems	"well over the million dollar mark"	not available
Burroughs	"does not report R&D expenditures specifically for the health care industry"	"does not report R&D expenditure specifically for the health care industry"
IBM	"not available for public disclosure"	"not available for public disclosure"

[a]The data listed are intended to convey the scope of the aggregate funds currently spent on development activities. This data in no way should be used to compare individual vendors to each other since each of the vendors budgeting and reporting conventions differ. The main point is that HIS development is not a minor expense.

It is beyond the scope of this article to go into the factors by which one justifies a Level 1 HIS. A brief statement on costs must be addressed, however. Let it merely be said that some of the factors may be categorized as "revenue increasing factors," some as "cost reduction factors" and some as "cost avoidance factors." They are precisely the same factors that are used to cost justify all levels of hospital information systems. For many factors the cost benefits from a Level 1 system will be exactly the same as the benefits from the Level 2.

Summing up the above, cost justification is not a problem. The cost benefits to be derived should stimulate sales in the cost conscious posture of hospitals today.

Another stimulus to continued healthy sales of Level 1 systems is the increasing awareness on the part of the hospital administrators that these systems are no longer just "charge collection and message switching systems." They have become a viable alternative to the much more costly Level 2 systems. The Level 2 systems cost range for hardware and software is $4.00 to

$7.50 per patient day; for Level 1 systems the cost range is $1.50 to $2.75 per patient day. Some hospitals are simply not ready for extremely sophisticated systems, and some administrators are not yet ready to risk the Level 2 implementation process.

Prospects for continued healthy sales of Level 1 HIS are predicted for the next five years. By 1985 at least 1500 short term general hospitals will have installed the Level 1 systems, and most of these will have a full spectrum of enhancement applications.

The Level 1 vendors will continue to expand upon the applications they offer. Some will enter the Level 2 HIS marketplace in addition to those that already offer both Level 1 and Level 2. They will be forced to do so to retain their client base as hospitals become desirous in the future of upgrading from Level 1 to Level 2. They will also modify their systems so that they may provide direct usage by physicians, nurses and other professionals.

References

[1] Ball, M. J. and S. E. Jacobs, "Information Systems: The Status of Level 1," *Hospitals,* June 16, 1980, pp. 179–186. (Most of the information provided herein is taken directly from this article.)

[2] Remarks by Melville R. Hodge, Computers in Health Care. Hearings before the subcommittee on domestic and international scientific planning, analysis and cooperation of the Committee on Science and Technology. U.S. House of Representatives, 95th Congress, Second Office, Washington, D.C. (35-763-0).

5

Evaluation of Automated Hospital Data Management Systems*

Edward M. Young, Earl W. Brian, M.D., Dorcas R. Hardy, Alan Kaplan, and James K. Childerston

Introduction

The application of automated technology to the information flow needs of the modern hospital has dramatically increased and expanded in recent years with the introduction of minicomputer technology. It has been claimed that at a fraction of the cost of the large scale, general purpose computers such as the Technicon system, minicomputers are now theoretically capable of performing much of the same work. In economic terms, the import of such a claim is that a moderately sized hospital which heretofore was excluded from using such a tool may now be able to acquire and implement an automated information system.

A recent study, conducted at the University of Southern California School of Medicine Center for Health Services Research, addressed the use of automated data management systems in small to medium sized hospitals (100–300 beds). Among the findings of that study was an assessment of the limits of available technology, and the extent of utilization of that technology by hospitals.

Implementing an information system in a hospital is not necessarily a simple or straightforward effort. Computerized information processing is not merely a matter of installing a machine with certain capabilities; it is, rather, a dynamic situation which combines the functional characteristics of computer hardware and software with the needs of an institution involved in providing health care. The institution's needs are continually growing and changing, as are the characteristics of the state of the art of the computer industry. The result of this dynamic situation is that the automation of a hospital's information handling functions is less a static event than a dynamic process.

While this conclusion may be evident in the development and utilization of large scale, in-house developed systems, it also holds for supplier developed package systems. Need changes and technological advances create a situation where new types of applications and individual applications are developed and to some extent, implemented in hospitals. The extent to which, and the manner by which, such development and implementation occur has been examined to determine whether any patterns exist and, if so, the implications of such patterns for current and future automation in hospital information processing.

Hospitals in the 100–300 bed size range were selected as the target for this study because they are the focus of much of the minicomputer based technology that has been developed in recent years. Numerous questions, addressing issues ranging from the nature of systems appropriate to this class of hospitals to the impact of those systems upon the hospital, have been raised. This paper addresses primarily the former issue. The question of impact is addressed here not from the standpoint of the costs and benefits of the system, but rather with regard to the implications of the manner in which such systems are implemented in hospitals. In order to distinguish such systems from both smaller department specific systems (finance, laboratory, pharmacy, etc.) and larger, more comprehensive automated hospital information systems (AHIS), the term AHDMS for automated hospital data management systems was coined.

Approach

The primary approach which was utilized to identify the systems available to hospitals and the manner in which those systems were implemented consisted of a survey of both potential system suppliers (those which actively marketed an AHDMS) and a select number of hospitals that had implemented an AHDMS. The former consisted of a mail survey with telephone followup to 113 potential AHDMS vendors, of which twelve were ultimately selected [1]. The latter consisted of a mail survey supplemented by site visits to eight hospitals.

To facilitate the collection of comparable information from all subjects, a list of approximately 75 AHDMS Application Descriptions was developed. These descriptions presented computerized applications available to hospitals in this size range in a functional manner, as opposed to utilizing vendor specific titles for either systems as a whole or individual applications. This list comprised the basic framework by which available and utilized systems were identified.

An additional framework was termed the 5-Step AHDMS Growth Model. This model is a series of five groupings of applications which facilitate an identification of the manner by which such systems have been and are being implemented. It was used to identify both the extent of system implementation in a given hospital as well as the points at which equipment resource needs increase.

Information obtained through the surveys was arrayed to present a comparison between the applications available from each of twelve system suppliers and between the systems offered and those actually implemented in the hospitals studied. The growth aspect of the model was utilized to make projections with regard to the resource needs of various sized hospitals within the target category, operating at different occupancy.

Findings

The study identified a type of system, defined in terms of levels of applications implemented, that is applicable to the needs and resource availability of hospitals in the 100–300 bed size range. This type of system may be identified by the fact that it incorporates some level of interdepartmental communication, as opposed to those systems which are department specific. In general, however, they fall short of the more sophisticated AHIS which includes a variety of patient care or clinical applications and are sometimes referred to as medical information systems (MIS).

It is difficult to categorize AHDMS in terms of maximum hardware size or system cost, since even between the time at which the study was designed and the site visits were conducted, initial assumptions regarding these two issues were discarded because of technological advances. Such advances have facilitated the use of powerful minicomputers with an extended range of applications in this class of hospital.

An AHDMS may be viewed as an interim stage in the development of automated hospital information systems. It is interim both from the standpoint of the manner in which systems have been developed for use in hospitals and from the standpoint of a logical growth pattern in the use of types of applications by hospitals. The 5-step AHDMS Growth Model presents a more comprehensive description of this type of system. It affords a picture of the level of resources required to add the more sophisticated groups of applications to a hospital's information processing system and presents implications for optimizing the effective implementation of such systems in hospitals.

The 5-Step AHDMS Growth Model

The 75 application definitions, identified and developed with the assistance of AHDMS vendors, were grouped into five sections, each representing a step in the development of a hospital's automated information processing. Each application in the five steps was calculated in terms of memory, on-line storage and processing time requirements. Additionally, these calculations were made for six cases of hospitals within the 100–300 bed size range, including 100, 175 and 250 bed general acute care facilities operating at 70% and 85% occupancy. The steps represent the sequence by which such systems are normally implemented in hospitals. This sequence, although not totally logically necessary, is the product both of the needs and desires of the hospital combined with certain characteristics of the type of application in each step.

The following brief descriptions of the five steps indicate the type of applications provided in each, and the implications of those applications on the processing resource needs.

Step 1. The first step in the development of an AHIS entails the implementation of a basic financial system, including basic billing and accounting functions, with patient data collection limited to basic patient census information. The resource needs for this step are substantial in that it entails establishing a considerable amount of patient data, primarily focusing on the charges accrued by the patient. Data are required in a format and with the capability of being retrieved for recording and computing charges. Of the twelve applications included in this step, it was found that the majority of vendors offered almost all applications [2] and that the eight site visit hospitals had implemented them as offered. See Table 1 which presents vendor/application and hospital/application matrices.

This step also impacts hospital personnel in that it begins to impose some of the standardization requirements necessitated by the computerization of the financial processes. It also generates an initial awareness of the capabilities of computers in the hospital, facilitating, if not stimulating, the addition of the more complex applications of steps 2–5.

Step 2. The 14 applications included in Step 2 expand on those implemented in Step 1, utilizing much of the same data, which can be processed in a batch mode. Some on-line data entry is added in the area of the admission/discharge/transfer function, along with some real time processing requirements and the addition of terminals at various stations in the hospital. While, in theory, this is not a major addition to the information processing capabilities of the hospital, it raises the critical issue of interface between types of applications.

Many hospitals in the past implemented the type of applications presented in Step 1 with little or no awareness of the possibility or need for interfacing with more sophisticated on-line applications in the future. As possibilities arose for hospitals in the 100–300 bed range with the development of advanced minicomputer technology, many of these hospitals were faced with the reality that advanced applications could not be interfaced with existing systems, and required major changes often including replacing the original Step 1 system with those affording greater compatibility with advanced types of applications. While future developments in the field cannot be predicted with certainty, this experience has implications for the entire process of planning for future system growth.

As Table 1 demonstrates, the availability of applications in vendor developed packages is still relatively consistent with those actually implemented in the site visit hospitals.

Table 1. Vendor and Hospital Matrices; Five Step Growth Model

	AHDMS Vendor/Application Matrix												Hospital/Application Matrix							
Step 1—Initial Implementation (Batch processing and primarily batch entry using no or very few terminals)	Burr.	Datx	HBO	HDS	HDC	IBM	McA	NDC	NCR	Pent.	SMS	SACS	H#1SMS	H#2-McA	H#3-SMS	H#4-HBO/McA	H#5-McA	H#6-NCR	H#7-Burr	H#8-IBM
Inpatient census and census reporting (1)	X	X	X	X	X	X	X	X	X	X	X	X	+	+	+	+	+	+	+	+
Charge compilation (2)	X	X	X	X	X	X	X	X	X	X	X	X	+	+	+	+	+	+	+	+
Inpatient billing (3)	X	X	X	X	X	X	X	X	X	X	X	X	+	+	+	+	+	+	+	+
Inpatient accounting (4)	X	X	X	X	X	X	X	X	X	X	X	X	+	+	+	+	+	+	+	+
Outpatient billing (if service is provided) (5)	X	X	X	X		X	X	X	X	X	X	X	+	+	+	+	+	+	+	
Outpatient accounting (as above) (6)	X	X	X	X		X	X	X	X	X	X	X	+	+	+	+	+	+	+	+
Outpatient visit recording (as above) (7)			X	X		X	X			X	X	X	+	+	+	+	+	I	I	+
Medicare billing (8)	X	X	X	X	X	X	X	X	X	X	X	X	+	+	+	+	+		+	
Insurance billing (9)	X	X	X	X	X	X	X	X	X	X	X	X	+	+	+	+	+	+	+	+
Coordination of insurance benefits (10)	X		X	X	X	X	X		X		X	X	+	+				+	+	+
Checks and registers (payroll) (11)	X	X	X	X	X	X	X	X	X	X	X	X	+	+	+	+	+	+	+	I
Hours and wages reports (12)	X	X	X	X	X	X	X	X	X	X	X	X	+	+	+	+	+	+	+	I

	AHDMS Vendor/Application Matrix												Hospital/Application Matrix							
Step 2—Follow-up Systems and Addition of Patient Data (Limited real-time processing and on-line data entry is added through the addition of a limited number of ternimals)	Burr.	Datx	HBO	HDS	HDC	IBM	McA	NDC	NCR	Pent.	SMS	SACS	H#1SMS	H#2-McA	H#3-SMS	H#4-HBO/McA	H#5-McA	H#6-NCR	H#7-Burr	H#8-IBM
Patient record data collection (1)	X	X	X	X	X	X	X	X	X	X	X	X	+	+	+	+	+	+	+	+
Admission/discharge/transfer notification (2)	X	X	X	X	X	X	X	X	X	X	X	X	+	+	+	+	+	+	+	+
Reigstration reports and form printing (3)	X	X	X	X	X	X	X	X	X	X	X	X	+	+	+	+	+	+	+	+
Patient identification number assignment (4)	X	X	X	X		X	X	X	X	X	X	X	+	+	+		+	+	+	+
Bed assignment and status (5)	X	X	X	X	X	X	X	X	X	X	X	X	+	+	+	+	+	+	+	+
Outpatient assignment and status (6)			X	X		X	X	X		X	X	X		+			+		I	I
Credit and collections (7)	X	X	X	X	X	X	X	X	X	X	X	X	+	+	+	+	+	I	+	+
Bad debt accounting (8)	X	X	X	X	X	X	X	X	X	X	X	X	+	+	+		+		+	+
General ledger interface (9)	X	X	X	X	X	X	X	X	X	X	X	X	+	+	+	+	+	+	+	+
Accounts payable (10)	X	X	X	X	X	X	X	X	X	X	X	X	+	+	+	+	+	+	+	I
Capital equipment accounting (11)		X	X	X	X	X	X	X	X		X	X	+	O	+	O	O	I		+
Capital equipment depreciation (12)		X	X	X	X	X	X	X	X		X	X	+	O	+	O	O	I		+
Labor cost distribution (13)		X	X	X	X	X	X	X	X	X	X	X	+	+	+		+	+		I
Medical staff rosters (14)		X	X	X	X	X	X	X	X	X	X	X	+	+			+	I		+

X = Application available
+ = AHDMS vendor application implemented
I = In-house developed AHDMS application
O = Other vendor supplied application (not interfaced to AHDMS).

Table 1. *(Continued)*

Step 3—On-Line Addition of Data Through Nursing Station and/or Ancillary Department and/or Cashier Terminals	AHDMS Vendor/Application Matrix												Hospital/Application Matrix							
	Burr.	Datx	HBO	HDS	HDC	IBM	McA	NDC	NCR	Pent.	SMS	SACS	H#1SMS	H#2-McA	H#3-SMS	H#4-HBO/McA	H#5-McA	H#6-NCR	H#7-Burr	H#8-IBM
Order transmission (1)	X	X	X	X	X	X	X	X	X	X	X	X	+	+	+	+	+	+		+
Message communication (2)	X	X	X	X		X	X	X	X	X	X		+	+	+	+	+	+		+
Demand billing (3)	X	X	X	X	X	X	X	X	X	X	X	X	+	+	+					+
Cash accounts (4)	X	X	X	X		X	X	X			X	X				+			+	+
Patient number retrieval (5)		X	X	X	X	X	X	X	X	X	X	X	+				+	+		I
Discharge abstract preparation (6)	X	X	X				X	X	X	X	X	X		+			O	+		
Cost allocation by cost center (7)		X	X	X	X	X	X		X	X	X	X		+		+	+	+		I
Salary budget preparation (8)		X	X	X	X		X			X	X	X					O	I		I
Salary budget control (9)		X	X		X		X	X	X	X	X	X		+		+	O	I	I	I
Pharmacy central inventory (10)			X				X	X	X	X	X	X								
Office supplies inventory (11)			X				X	X	X	X	X	X	O				+	I		
Purchasing storeroom inventory (12)			X				X	X	X	X	X	X	O		+		+	I	I	
Sterile supplies inventory (13)			X				X	X	X	X	X	X	O					I	I	
Non-sterile supplies inventory (14)			X				X	X	X	X	X	X		O			O	I	I	

Step 4—Automate Remaining Manual Steps Add Complex System Operations (System begins to use and reuse data which has been stored)	AHDMS Vendor/Application Matrix												Hospital/Application Matrix							
	Burr.	Datx	HBO	HDS	HDC	IBM	McA	NDC	NCR	Pent.	SMS	SACS	H#1SMS	H#2-McA	H#3-SMS	H#4-HBO/McA	H#5-McA	H#6-NCR	H#7-Burr	H#8-IBM
Medical record number assignment (1)			X	X	X		X	X	X	X	X	X		+			+	+		
Discharge analysis and reports (2)	X	X		X			X	X	X	X	X	X						+		I
Supplies budget preparation (3)		X		X	X		X		X	X		X		O			O	I		
Equipment budget preparation (4)		X					X			X		X		O			O			
Budget preparation—Other (5)		X			X		X	X	X	X	X	X		O			O			
Budget/actual expense reports (6)	X	X	X	X	X		X	X	X	X	X	X	+	+		+	O	+	+	
Labor time collection (7)		X		X		X		X	X	X		X						+	+	I
Prescription issue/label (pharmacy) (8)	X	X	X		X		X	X	X	X	X			O			O			
Formulary (pharmacy) (9)			X		X		X	X	X	X	X	X								
Narcotic usage/control (pharmacy) (10)	X				X			X	X	X	X									
Physician BNDD numbers (pharmacy) (11)							X		X			X								
Worksheets (laboratory) (12)		X	X		X	X	X	X	X	X	X	X	+				O	+	+	+
Specimen pick-up schedule (laboratory) (13)	X				X		X	X	X	X		X					O	O	O	
Other work schedules (laboratory) (14)							X	X	X	X							O	+	+	

X = Application available
+ = AHDMS vendor application implemented
I = In-house developed AHDMS application
O = Other vendor supplied application (not interfaced to AHDMS).

Table 1. (*Continued*)

Step 5—Two-Way Data Transmission, Clinical Applications	AHDMS Vendor/Application Matrix												Hospital/Application Matrix							
	Burr.	Datx	HBO	HDS	HDC	IBM	McA	NDC	NCR	Pent.	SMS	SACS	H#1SMS	H#2-McA	H#3-SMS	H#4-HBO/McA	H#5-McA	H#6-NCR	H#7-Burr	H#8-IBM
Results reporting (1)	X		X		X	X	X	X		X	X	X						I		+
Personnel history (2)		X	X		X	X	X	X	X	X	X		+			+		+		I
Utilization review (3)		X	X		X	X	X		X		X	X						+		I
Outpatient scheduling (4)			X				X		X	X	X	X								
Medical chart reports and form printing (5)				X				X	X		X		O			O	O	+		I
Tumor registry (6)						X	X		X	X										
Generic name cross reference (pharmacy) (7)		X			X		X	X	X	X	X	X								
Pharmacy floor stocks inventory (8)			X				X	X	X	X	X									
Solutions (pharmacy) (9)			X				X	X				X								
Drug/drug incompatibility (pharmacy) (10)			X				X		X											
Digital work station input (laboratory) (11)	X	X	X		X		X		X	X		X								
Analog signal acquisition (laboratory) (12)					X		X		X	X		X								

Miscellaneous—Other Applications Occasionally Implemented	AHDMS Vendor/Application Matrix												Hospital/Application Matrix							
	Burr.	Datx	HBO	HDS	HDC	IBM	McA	NDC	NCR	Pent.	SMS	SACS	H#1SMS	H#2-McA	H#3-SMS	H#4-HBO/McA	H#5-McA	H#6-NCR	H#7-Burr	H#8-IBM
Physicians index (1)	X	X		X	X	X	X	X	X		X	X						I		
Medical staff privileges (2)		X	X		X		X	X				X						I		
Professional billing (3)		X	X	X		X	X			X	X	X								I
Professional fee: Revenue and cash reporting (4)		X		X		X	X	X		X		X						I		
Nursing staff rosters (5)		X	X				X	X		X		X						I	I	
Wage and salary studies (6)		X			X	X	X		X	X	X	X						I		
Personnel survey and questionnaires (7)		X		X			X			X									I	
Patient medical history data collection (8)					X	X	X	X		X		X								
Emergency room statistics (9)		X		X			X	X		X								I	I	

X = Application available
+ = AHDMS vendor application implemented
I = In-house developed AHDMS application
O = Other vendor supplied application (not interfaced to AHDMS).

Step 3. Step 3 presents a shift in the capabilities of the system in terms of an increase in applications involving on-line use of data. The focus of this data use shifts from the financial applications to patient care oriented data. Users of the system include patient care professionals, primarily nursing personnel. Prior to this stage of development, the majority of applications may have been handled in-house or through the use of an offsite computer service. At this stage, support for the number of terminals involved in interfacing ancillary departments and nursing stations requires that the central processing unit be located near the facility, if not inhouse.

This change in computer hardware needs represents one of the first system choices which will affect the long range technological limitations of the system. Computer hardware choices made at the initiation of Step 3, in terms of expandability and growth capability, could be limiting factors in subsequent expansions. This is also true of operational software, the computer programs used to control operation of the terminal network, manage the data files and develop new applications. The choices made at this point will be difficult or perhaps impossible to reverse later, and should be made with the end goals of the hospital for the AHDMS in mind.

The workload of the computer begins to increase at this point, a result of the amount of memory being occupied by the operating software to process the more complex data handling requirements. The fact that the two most commonly implemented applications of this stage of development, order transmission and message communication, occur in a real time mode during the busy times in the hospital day introduces the phenomenon of peaking. Operational and hardware capabilities that are sufficient for the average daily workload become overburdened, with resulting increases in computer response time and the possibility of temporary computer shutdowns.

While such occurrances are not critical from the standpoint of the computer's ability to ultimately process required data, they are particularly serious from the user standpoint. Planning for increased automation of information processing, such as the addition of those applications included in Steps 3–5, requires consideration of the amount of data which the computer will have to manage and the points during the day when those demands will be highest. Consideration must be given to the interaction between the number and type of patients in the hospital, the number and type of system applications implemented and the size and efficiency of computer hardware and operating software in order to effectively design a system which will meet the hospital's needs.

Table 1 depicts a change in the availability and use of the Step 3 applications. The study found that, while many of the AHDMS vendors were making most of the Step 3 applications available (less than in Steps 1 and 2, however), few of the 100–300 bed hospitals studied were actually utilizing many of the available applications. Some hospitals which had the applications had developed them in-house, demonstrating a low match between the vendor supplied system and that actually in place in the hospital. This situation is magnified in the following steps.

Step 4. The applications in Step 4 build on those developed in Step 3. The primary significance of Step 4 applications lies in their development costs, rather than in the computer resource costs. The Step 4 applications are relatively complex, more difficult to design, and require more interaction with the daily activities of hospital personnel. The degree of difficulty encountered in Step 4 is directly related to the success achieved in addressing the problems in Step 3. The system, now largely on-line, and making extensive use of the terminal network, will add volume to the data handled by the terminal network and the data management software, compounding the situation described in Step 3.

The fourth step does not require substantial increases in computer resources, but the peak demand phenomenon is applicable to all increases. This creates problems out of proportion to the amount of increased loading. Also, the increased computer utilization will, for the larger hospitals, begin to create some difficulty in scheduling the extra shift work on weekly and less frequent operations which were previously handled in free time, or on weekends. Systems begin to show definite signs of being fully loaded at this point and the remedies adopted in Step 3 will determine system limits.

The phenomenon noted in Step 3, namely that the number of applications offered by vendors is considerably greater than those implemented in the hospitals visited, may indicate that the cost of such applications was considered too great for the benefit which these applications could be expected to afford the hospital. It may also indicate that the direction of development by the vendors is not always directly responsive to the expressed needs of hospitals. An additional possible consideration is that optimal utilization of the more basic applications of Steps 1–3 is a prerequisite for being able to utilize more advanced applications. Given the fact that most of the site visit hospitals had only recently acquired an AHDMS, their general nonutilization of Step 4 and Step 5 applications may merely be a factor of time. Further monitoring is needed to more fully assess this situation.

Step 5. The applications in Step 5 are best characterized by their complexity. Results Transmission requires storage, interpretation and selective routing of diagnostic ancillary results. It presents complex problems of network management, especially in view of the traffic load which must be added to peak demand periods. Outpatient Scheduling is significantly complicated by the matching of resources to requirements and the need for frequent changes and subsequent recalculation. All of the pharmacy operations are complex, with Drug/Drug Incompatibility the most complex from a systems and maintenance viewpoint. These are the applications which will test the technological limits of the system, both in design and in computer resources. The amount of effort and cost involved in the design of these applications requires extensive attention to the calculation of needed and available computer resources.

Application complexity is confined to online applications. The batch processing applications, occurring weekly or less frequently, are not challenging, either in design or in the demand on computer resources. Step 5 applications are highly individualized for each hospital. Even if commercially available packages match overall needs, their adaptability to existing applications and to the computer hardware may present significant implementation difficulties.

Summary

Automation of information processing functions is becomming a reality for a continually increasing number of hospitals. The process of implementing a system requires careful planning, taking into consideration the dynamics of both the individual institution and the characteristics of the computerization process. An assessment of the state of the art of the development of systems in small to medium sized hospitals, those which can be expected to profit most from advances in minicomputer technology, produces a number of conclusions which address the optimal utilization of those systems. Those conclusions may be summarized as follows:

(1) Automated Hospital Data Management Systems (AHDMS) comprise an identifiable class of computerized interdepartmental hospital information systems which have evolved between the class of narrower scope, department specific (e.g., laboratory, business office, medical records, etc.) systems and the broadest scope, hospital-wide Automated Hospital Information Systems (AHIS).

(2) The development of an AHDMS is normally a dynamic proceeding from basic fiscal/administrative applications, through patient data collection applications and, in some cases, limited medical applications.

(3) An AHDMS may be implemented by following any one, or a combination, of available approaches (e.g., turnkey, in-house development, etc.). Successful development of a system of this type necessarily includes long range information system plans which identify what applications will best meet the needs of facilities and when and how they should be implemented. The 5-Step AHDMS Growth Model presents a framework which can be employed in the development of an information system plan.

(4) The initial selection of an approach to AHDMS implementation, and the first group of applications implemented, appears to have a disproportionately large impact on later alternatives for system expansion and development of new applications in the hospital.

(5) AHDMS actually implemented in small to medium sized hospitals differ from AHDMS available from vendors in terms of applications included. Hospitals appear to utilize considerably fewer applications than are poten-

tially available from vendors. This discrepancy should be further assessed from the standpoint of developing systems which may optimally meet the current and future needs of hospitals.

Notes

[1] *AHDMS Vendors:* (Abbreviations also used in Table 1)
Burr. —Burroughs Corporation
Datx —Datx Corporation
HBO —HBO and Company
HDS —Healthcare Data Systems
HDC —Hospital Data Center of Virginia
IBM —International Business Machines Corporation General Systems Division
McA —McDonnell Douglas Automation Company
NDC —National Data Communications
NCR —Corporation
Pent. —Pentamation Enterprises, Inc.
SMS —Shared Medical Systems
SACS—Space Age Computer Systems, Inc.

[2] "Available" entails implemented in at least two independent hospitals at the time of the study.

6

Implementation of the IBM Health Care Support/Patient Care System*

David J. Mishelevich, M.D., Ph.D., Betty G. Hudson, Donald Van Slyke, Elaine I. Mize, Anna L. Robinson, R.N., Helen C. Brieden, R.N., Jack Atkinson, and James Robertson, Jr.

Introduction

The revitalization of the data processing function and the installation of a comprehensive Hospital Information System (HIS) in a relatively short period of time is a momentous, exciting and distinctly challenging task. A succession of fits and starts in data processing characterized the evolution of the Department of Information Services during the period 1963–75.

Thus in 1975, there were archaic, inadequate financial systems. Patient Accounting had no consolidated financial view of the patient. Inpatient patient accounting was done with a computer based system which was different from the Emergency Room patient accounting system and both were written in assembly language. Outpatient billing was done manually. The General Ledger was manually posted on ledger cards.

Types of On-Line Hospital Information Systems

An on-line HIS may take many forms [1]. While various classifications are possible, the most common division is based along broad functional lines; typically a Class I system would be for input only, a Class II system would include feedback of administrative information, a Class III system would include medical and patient care information (primarily in an input mode) in addition to administrative information and a Class IV system would include medical interaction including results reporting and thus allow active participation in the health care process. In order to illustrate what can be done in a Class IV system, a specific example will be described: the installation at the Dallas County Hospital District.

*This chapter has been modified from a version previously published in *Proceedings of the Fourth Annual Symposium on Computer Applications in Medical Care,* IEEE, 1980.

Many factors will influence how well the installation of a Hospital Information System will go [2]. Many techniques found useful at Parkland can be applied beneficially elsewhere [3].

The Parkland On-Line Information System (POIS)

The HIS will reflect the environment in which it is installed. It is thus extremely appropriate to provide a brief description of the hospital and give a sketch of recent data processing history [4].

The Dallas County Hospital District is a 775 bed plus 102 bassinet hospital with 275,000 inpatient days in 1979, along with 220,000 outpatient and 173,000 emergency room visits. The hospital is the major teaching hospital for The University of Texas Southwestern Medical School, a component of The University of Texas Health Science Center at Dallas. In 1975, a plan for a comprehensive Hospital Information System was approved for implementation by the Administrator, C. Jack Price, and Board of Managers. The objective was indeed to "install a health care delivery system with charge capturing as a logical by-product not an end." In 1974, the DCHD and Dallas County data processing groups had been consolidated. In the summer of 1975, the Information Services Department was reconstituted at the hospital. The system selected in 1975 included a combination of IBM and Hospital Financial Systems (HFS) packages running on IBM equipment, initially an IBM 370/145 and currently an IBM 370/168, The HFS Patient Accounting, Financial Management System (general ledger) and Accounts Payble systems were installed. An existing inventory system was converted from DOS to OS.

The IBM HCS (Health Care Support) Registration and Admissions systems were installed in June and July of 1976, respectively, to in part frontend patient accounting. A prime objective was to have an integrated inpatient, outpatient and emergency room patient accounting system. Separate inpatient and emergency room patient accounting systems had previously existed and outpatient billing had been handled manually. In November of 1976, HFS inpatient accounting was placed in production, followed by the Emergency Room in February 1977 and outpatient billing in June 1977.

The Order Communications Component

The keystone of a comprehensive Hospital Information System is the Order Communications System. The Duke Hospital Information System (DHIS) with its new software architecture and facility for user applications interface was selected and Parkland became the validation site (for IBM which now markets DHIS as the Health Care Support/Patient Care System) in the summer of 1977. In August of 1977 the IBM 370/158 (2 megabytes) was moved from Dallas County Data Services to the hospital. In September of 1977 a third megabyte was added, a change to a Model 3 (addition of cache memory) and Integrated Storage Control was made in December of 1977, a fourth megabyte of core added in February of 1978 and a fifth megabyte of core added in July of 1978. In December of 1979, a six megabyte IBM 370/168 Model 3 replaced the IBM 370/158.

Implementation

In December of 1977, Radiology, the first DCHD application of the Parkland version of the Duke Hospital Information System, was installed in two nursing stations. The Parkland Version of DHIS is called the Parkland On-Line Information System (POIS). As of mid-February, 18 nursing stations (each with a CRT and a printer terminal) were using the POIS Radiology Module, including results reporting. All the then existing forty nursing stations were active in late April of 1978. The next application installed was the Central Supply Room (CSR) Module installed at the beginning of March. Pharmacy went into production in July, 1978, Labor and Delivery in September of 1978, part of the Emergency Room in January of 1979, Electrocardiography in February of 1979 and the first Laboratory, the Immunopathology Laboratory in April 1979. In August of 1979, the Medical Records Chart Deficiency and Locater System was installed; in October, the ability to input transfers and discharges from the nursing stations was added. In November 1979, the University of Michigan Clinic Appointment System was put into production for the first clinics, the 10 Obstetrics and Gynecology clinics (the fourth floor of the Clinic Building). In January 1980, the 30 clinics on the second floor of the Clinic Building (Surgery, Surgical Specialties and Allergy) were added. In February, 1980 the General Medicine Clinics on the third floor were added with a new feature. Each of the house staff physicians were scheduled as an individual resource. In March, the rest of the third floor, the Medical Specialties plus Neurology were completed. Some of the faculty physicians in the Medical Specialty Clinics chose also to be scheduled as individual resources. In April the clinics on the ground floor, Ophthalmology, Ear, Nose and Throat and Oral Surgery were added. Psychiatry was also put into production. In March of 1980, the ancillary department of Physical Medicine and Rehabilitation was added.

Implementation Success Factors

In addition to a strong administrative mandate, two user interaction mechanisms have greatly contributed to the ongoing success: the use of the Directions Committee and the concept of the POIS Associates, the full time health professionals working on the system.

To facilitate the development and installation of POIS (the Parkland Online Information System), a Directions Committee was formed whose function is to coordinate activities as well as to underscore that POIS is a hospital project not an Information Services project. The group is chaired by our Associate Administrator and includes representatives from Nursing Service, the Medical Staff, Ancillary Departments (added when appropriate to the phasing in of a given department), Education and Training, Engineering, Administration (including Finance and Patient Accounting), the vendor and Information Services. It meets biweekly and develops policy, coordinates activities and provides a vehicle for information flow. Subcommittees and task forces are formed and dissolved as needed to study and make recommendations on specific issues to optimize the use of the larger Directions Committee time. One of the subcommittees is the Financial Subcommittee which provides an analysis and auditing function of both automated and associated manual procedures to insure all charges are appropriately recorded.

The second user interaction method is the use of the POIS Associates. There are currently five associates in place. The Chief POIS Associate is an R.N. from nursing services. The other four Associates include an R.N. from nursing services, a staff member from Education and Training, a Registered Technologist from the Clinical Laboratories as the Laboratory Associate, and a Radiological Technologist from the Radiology Department representing the nonlaboratory ancillary areas. As discussed further below, these individuals are health care professionals who are performing functions that in the past would have most likely been performed by data processing personnel.

Distribution of Functions Between Programmer/Analysts and POIS Associates

As mentioned above, the POIS Associates are health care professionals who perform many roles which would in the past have been viewed as belonging essentially exclusively to data processing professionals. This in fact is a major strength to the IBM Patient Care System. The distribution of functions between the programmer/analyst and Associate groups is as follows:

(1) Analysis of the old systems is done mainly by the POIS Associates with input from personnel, from user financial, and internal audit areas.
(2) System specifications are done primarily by the programmer/analysts.
(3) Programming is done by the programmer/analysts.
(4) Manual screens are done almost exclusively by the POIS Associates.
(5) Screens generated in an automated manner from machine readable catalogues are performed by the programmer/analysts.
(6) The Data Collection Lists (DCL's) which indicate where to retrieve data that are being displayed on the screen or where the data entered from screen interaction are to be placed in the data base are currently done by the programmer/analysts. In the future this also will be done by the POIS Associates.

(7) User procedures are developed and manuals are produced in machine readible form by the POIS Associates.
(8) User training is performed by the POIS Associates.
(9) The pilot implementation and ongoing production support is provided by both the programmer/analysts and the Associates. Among other things, at least one POIS Associate is available in the hospital 24 hours a day for the first three days of a new system implementation.
(10) The production maintenance support is provided by both the programmer/analysts and Associates.

Clearly one of the advantages of the user of health care professionals is their understanding of the basic problems that they are being dealt with. In addition, however, it is clear that in some cases, the POIS Associates are providing functions that would be difficult to fulfill with data processing personnel because it is currently so difficult to recruit programmer/analysts. In many cases, the salaries may be more.

It is indeed fortunate that the IBM Health Care Support/Patient Care System is in fact a user involved and committed system. It is designed for ease of use by a video display user who may be either a ward clerk or health care professional and also for easy use by a system designer who may be either a health care professional such as one of the POIS Associates or a data professing professional. By use of the IBM Patient Care System, the health care professional may in fact, define procedures, design screens, printer formats, and other reports as well as test the application and respond to the changing requirements with a minimum of data processing assistance. Such a user-driven system is much more likely to succeed because it not only better reflects actual needs but gets people more personally involved and committed.

Implementation Task Forces

Each implementation area has a task force associated with it with personnel from the relevant ancillary department, Information Services and Associate Support from Nursing Services and/or Education and Training. This group analyzes pre- and post-implementation work flow, designs CRT screens and printer formats, performs evaluation, and as appropriate participates in documentation and training.

A workroom for POIS activites is functioning, primarily for development. The major training effort goes on in a classroom/terminal facility with five CRT's and one printer operated by Education and Training.

The Hospital Information System clearly serves the users and thus is *their* system. Thus they must be appropriately involved in the definitions of objectives, design implementation including production of training materials and evaluation stages. User involvement markedly aids in implementation. It preferably will be appropriately directed, however, rather than being an open-ended "What do you want?"

Description and Examples

Registration Admissions

The registration function is the master-patient-index function. The on-line predecessor to our on-line registration replaced almost one-half million index cards in tub files. Data elements allowing for patient identification permitting "connection" of a patient to his or her hospital unit number to make sure that all records of the various types for a given patient are appropriately linked together are of fundamental importance.

First, however, we must find the patient's record if one exists. To accomplish this, we use the NAMQ transaction followed by typing in the patient name. The patients having that name (and perhaps other search characteristics such as date of birth) are displayed in a table on the screen and if the appropriate name is in the list the matching line number is called in and the registration screen for that patient appears. Note if there are more patients of the same last name than can be listed on one screen, one can page forward the next screens until the names of all potentially matching patients are reviewed.

The registration screen can be updated as appropriate. Once the patient is so identified (and updated if necessary) the next step is to go to the Admission process.

The admissions process deals primarily with financial classification, responsible party and admitting diagnosis information. Admissions is the term used whether an inpatient, Emergency Room visit or outpatient visit setting is involved. On the first screen, classification data, guarantor, notification and other items are displayed. On the second admissions screen, third party, next of kin, preadmission and nonresident data are shown and on the third screen in the series, diagnostic and admitting service/physician data are covered with the bed location put in after an appropriate space has been found using a nursing station census screen. The screens can also be updated or added to as appropriate. There will be one set of admission data for each of the incidents of service in which the patient is involved with the hospital.

Now that the patient is "in the system" for the current contact with the hospital, the information elements are present to form the basis for ordering.

The Radiology System

The Radiology Department, as noted above, was our first implementation. In 1979, there were approximately 100,000 inpatient and 40,000 outpatient general diagnostic Radiology procedures, out of the total of 235,000 performed by the department. POIS was initally involved in the inpatient procedures alone, although because of time freed up on inpatient activities with the Radiology Department, an implementation was accomplished in September of 1979 in which the Outpatient Radiology orders are entered within the Radiology Department itself to further consolidate department activities.

```
@@@@@@@@@@@@@@@@@@@@@@@@@@@@@@@@@@@@@@@@@@@@@@@@@@@@@@@@@@@@@@@@@@@@@@@@@@@@@@TOP
                                                                 08/23/78  1241

          PLEASE SIGN ON BY ENTERING YOUR SECURITY ID AND PASSWORD

                  ID:

                  PASSWORD:







--------------------------------------------------------------------------------
DOCSSION
@@@@@@@@@@@@@@@@@@@@@@@@@@@@@@@@@@@@@@@@@@@@@@@@@@@@@@@@@@@@@@@@@@@@@@@@@@@@@@BOT
```

Figure 1. Sign-on screen. ID and password are not echoed for reasons of security.

The best way to give a flavor of the Radiology system as an example is to walk through the placement of an order. The Cathode-Ray Tube (CRT) Terminal (video terminal) used is an IBM 3278 terminal with a light pen which detects light on the screen and creates an interrupt to the terminal when the light pen tip is depressed against the screen where the desired item is located on the screen. The selection process is called "menu selection."

The POIS sign-on screen is shown in Figure 1. The ID and password are individually assigned to system users and are not echoed on the screen for security reasons. The ID and password will determine which screen will come up to a given user first.

If we are a Nursing Services person on one of the forty inpatient nursing stations, the screen in Figure 2 will appear. In the upper right-hand corner, the date and time (in 24-hour time) appears and in the lower right, the screen shows the current POIS user which is encoded in a shorter form within the system. The date, time and user signed on to the terminal are stamped on each transaction that modifies the data base, thus allowing for an effective audit trail. Also please note in the lower left-hand corner there is a nickname for the screen, in this case, "NADMSTR" (for nursing administration master), which allows users to identify the screen to a troubleshooter over the telephone rather than trying to describe the screen itself.

The use of the light pen allows us to accomplish most of the input by menu selection rather than by typing. In this case we need to get the patient for whom the order is to be placed. We therefore probe "NURSING CENSUS ALL" and get the screen shown in Figure 3. The patient we want is on LABOR & DELIVERY so we make that choice, which brings up Figure 4 which is the census for that particular nursing station. If we wish order an x-ray for DORTHY YOUNG, we probe that patient name to make our selection which brings up the screen in Figure 5. Note with respect to this screen, in the upper left-hand corner we have positive identification of the patient we have selected.

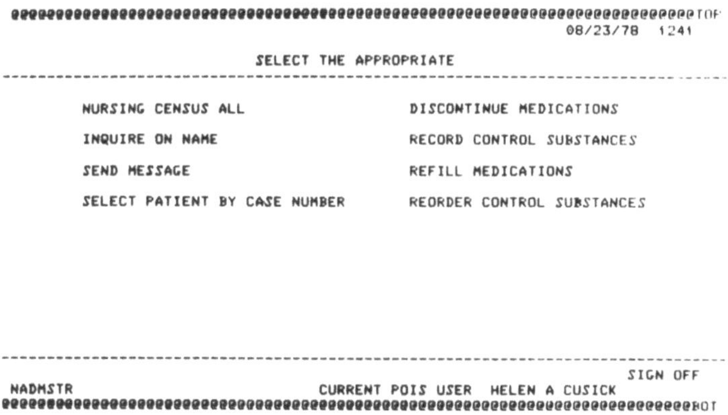

Figure 2. Master screen at a nursing station. NURSING CENSUS ALL is selected in this example as described in the text. Note the date and time (upper right) and the internal code for the individual signed on (displayed in the lower right corner) are stamped on each transaction updating the database, thus providing an audit trail. At the lower left of the screen is the screen mneunonic code (NADMSTR in this case) for screen identification in training and trouble shooting.

We then select "ENTER ORDERS" which brings up Figure 6. If we wish to order a chest film done in the Radiology Department we probe "DEPARTMENTAL" under "RADIOLOGY PROCEDURES." If we have probed "PORTABLE," only the subset of those examinations that could be done portable would appear. Since we wanted the full departmental choice, however, Figure 7 comes into view which allows us to select for chest film the category "ABDOMEN AND CHEST." This selection is positively identified as it appears in the upper left-hand corner of the next screen as shown in Figure 8.

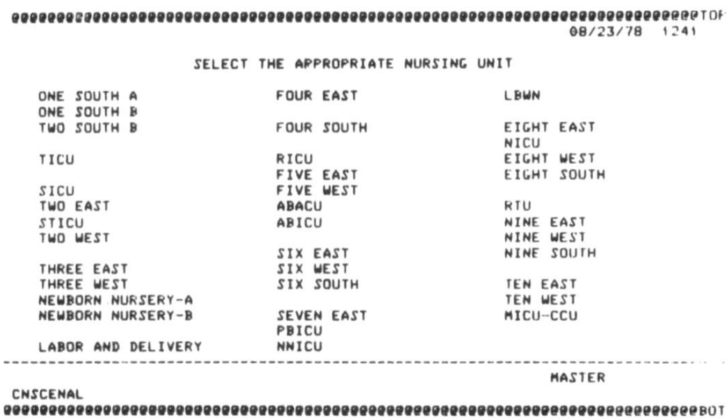

Figure 3. Screen for selection of desired nursing station (LABOR AND DELIVERY in this example).

```
@@@@@@@@@@@@@@@@@@@@@@@@@@@@@@@@@@@@@@@@@@@@@@@@@@@@@@@@@@@@@@@@@@@@@@@@@@@@@@TOP
 CURRENT CENSUS FOR  3S LABOR & DEL SUITE                         08/23/78  1241
  N650
  ROOM BED   PATIENT NAME                    CASE #  SEX STAT SERVICE
  0390-02  HODGES, GLADYS RENEE               444764  F          052
  0391-01  RIBLES, MARY JEAN                  545534  F          010
  0391-03  BIGLARY, ZINAB T.                  557788  F          089
  0391-04  DURTE, ROSA                        408454  F          052
  0391-05  ENGLISH, ROBERTA                   421984  F          051
  0399-01  YOUNG, DOROTHY                     538484  F          010
  0399-04  JAWICKI, SARAH VAUGHN               10101  F          010
  0399-05  GARDNER, HELEN RUTH                444664  F          052
  0399-06  WARD, FLORENCE ODELL               420274  F          051
  0399-08  CURRY, FRANCES PEARL               447084  F          052
  0399-09  SPEED, JANIE MARY                  427424  F          081
  0399-11  FLORES, MATILDE                    421854  F          051
  0399-12  DRAPER, IONA                       444054  F          054



--------------------------------------------------------------------------------
                                                                   RETURN
NSCEN
@@@@@@@@@@@@@@@@@@@@@@@@@@@@@@@@@@@@@@@@@@@@@@@@@@@@@@@@@@@@@@@@@@@@@@@@@@@@@@BOT
```

Figure 4. Screen for selection of desired patient, DOROTHY YOUNG in this example.

```
@@@@@@@@@@@@@@@@@@@@@@@@@@@@@@@@@@@@@@@@@@@@@@@@@@@@@@@@@@@@@@@@@@@@@@@@@@@@@@TOP
 N650 0399-01 YOUNG, DOROTHY                538484            08/23/78  1241
UNIT NO =      667902
                      THESE FUNCTIONS ARE AVAILABLE TO YOU
                              SELECT THE APPROPRIATE
--------------------------------------------------------------------------------

          ENTER ORDERS

          VERIFY/DELETE/CORRECT ORDERS

          CANCEL ORDERS

          DISPLAY ORDERS

          DISPLAY RESULTS

          DISPLAY OCCURENCES

          PATIENT PROFILE

     ......................................................................
--------------------------------------------------------------------------------
               MASTER                     CENSUS
NFUNCSC                                   CURRENT POIS USER  HELEN A CUSICK
@@@@@@@@@@@@@@@@@@@@@@@@@@@@@@@@@@@@@@@@@@@@@@@@@@@@@@@@@@@@@@@@@@@@@@@@@@@@@@BOT
```

Figure 5. Screen for selection of desired function, ENTER ORDERS in this example.

```
                                                                              TOP
 N650 0399-01 YOUNG, DOROTHY                538484            05/10/79  1012
UNIT NO =      667902

                          SELECT THE CATEGORY OF ORDER
--------------------------------------------------------------------------------
     RADIOLOGY PROCEDURES
   PORTABLE       DEPARTMENTAL

   CENTRAL SERVICES
   INDEX 1     INDEX 2

   PHARMACY

   LABORATORY
   SEROLOGY

   EKG


     ......................................................................
--------------------------------------------------------------------------------
               MASTER                      CENSUS                    FUNCTION
NORDSC                                    CURRENT POIS USER PAULA JEAN POISE
@@@@@@@@@@@@@@@@@@@@@@@@@@@@@@@@@@@@@@@@@@@@@@@@@@@@@@@@@@@@@@@@@@@@@@@@@@@@@@BOT
```

Figure 6. Screen for selection of desired order category, in this case RADIOLOGY (DEPARTMENTAL).

```
@@@@@@@@@@@@@@@@@@@@@@@@@@@@@@@@@@@@@@@@@@@@@@@@@@@@@@@@@@@@@@@@@@@@@@@@@@@@@@TOP
 N650 0399-01 YOUNG, DOROTHY                      538484         08/23/78  1242
UNIT NO =     667902
                                    PROBE SELECTION ORDERED
-------------------------------------------------------------------------------
GENERAL PROCEDURES                      GENERAL PROCEDURES
   ABDOMEN AND CHEST                       TOMOGRAMS               TOMOGRAMS
                                           CHEST, SKULL            BONE, SPINE
   ARTHOGRAMS AND BONE SURVEYS
                                        C.T. SCAN
   EXTREMITIES
   A - HI      HU - Z                   NUCLEAR MEDICINE
                                          A - H      I131 - L      M - Z
   GASTROINTESTINAL
   A - C       D - Z                    RADIATION THERAPY

   GENITOURINARY                        RADIOLOGY ULTRASOUND
                                           ECHOGRAMS       OTHER
   HEAD, NECK
   A - P       Q - Z                    SPECIAL PROCEDURES
                                           AORTO & ARTERIOGRAM        OTHER
   SPINE, PELVIS
-------------------------------------------------------------------------------
...............................................                        RETURN
RADINDX                             CURRENT POIS USER HELEN A CUSICK
@@@@@@@@@@@@@@@@@@@@@@@@@@@@@@@@@@@@@@@@@@@@@@@@@@@@@@@@@@@@@@@@@@@@@@@@@@@@@@BOT
```

Figure 7. Screen for selection of desired Radiology general order type, in this case ABDOMEN AND CHEST.

We now select "CHEST, PA & LAT" and this brings up the Transportation and Scheduling screen as shown in Figure 9. Note that the left-hand column has question marks. These question marks, when probed, will change to a greater than sign to let the individual know what selection has been made in case one is interrupted or is considering whether to make a change. If we probe TODAY and then "ST, STRETCHER," we get the next screen, as shown in Figure 10. Note there is no need for question marks to change to greater than signs in the Transportation column because it is the last column going left to right and the screen will automatically change once the selection from this column has been made.

The Nursing History is shown in Figure 10. From the left most two columns, one selects up to four items (for example, "DEAF" and "UNABLE TO MOVE"). ALLERGIES will normally be supplied automatically from the

```
@@@@@@@@@@@@@@@@@@@@@@@@@@@@@@@@@@@@@@@@@@@@@@@@@@@@@@@@@@@@@@@@@@@@@@@@@@@@@@TOP
 N650 0399-01 YOUNG, DOROTHY                      538484         08/23/78  1242
 UNIT NO =     667902
ABDOMEN AND CHEST                  PROBE SELECTION ORDERED
-------------------------------------------------------------------------------
   ABDOMEN, ACUTE SERIES                   CHEST, LORDOCTIC
   INCLUDES, KUB, UPRIGHT, PA OF CHEST
                                           CHEST,PA AND LAT (CANNOT BE DONE
   ABDOMEN, (KUB)                          AS PORTABLE ON ADULT PATIENTS)
                                           CHEST,SINGLE VIEW
   ABDOMEN, MULTIPLE VIEWS                 DONE AS PORTABLE OR ON PATIENTS
   INCLUDES, KUB & UPRIGHT                 UNABLE TO SIT OR STAND ONLY

   CARDIAC SERIES (INCLUDES CHEST)         MAMMOGRAPHY
   MUST BE VERIFIED BY A NURSE             MUST BE VERIFIED BY A NURSE

   CHEST BRONCHOGRAPHY                     PELVIMETRY (SNOWS)
   MUST BE VERIFIED BY A NURSE
                                           PLACENTOGRAPHY
   CHEST, DECUB, LEFT                      RIB SERIES

   CHEST, DECUB, RIGHT                     STERNUM
-------------------------------------------------------------------------------
                                     RETURN RAD INDEX
 RAD01
@@@@@@@@@@@@@@@@@@@@@@@@@@@@@@@@@@@@@@@@@@@@@@@@@@@@@@@@@@@@@@@@@@@@@@@@@@@@@@BOT
```

Figure 8. Screen for selection of desired specific Abdomen and Chest Examination, in this case CHEST, PA, AND LAT.

```
@@@@@@@@@@@@@@@@@@@@@@@@@@@@@@@@@@@@@@@@@@@@@@@@@@@@@@@@@@@@@@@@@@@@@@@@@@@@@@@TOP
 N650 0399-01 YOUNG, DOROTHY                       538484          08/23/78  1242
 UNIT NO =        667902                          CHEST PA AND LATERAL
                         TRANSPORTATION AND SCHEDULING
              SELECT ONE ITEM FROM EACH OF THE FOLLOWING COLUMNS
----------------------------------------------------------------------------------

 SCHEDULING (SELECT ONE)                       TRANSPORTATION (SELECT ONE)
    ?AS SOON AS POSSIBLE                           AB, AMBULATORY

    ?DOWNTIME                                      WC, WHEELCHAIR

    ?TOMORROW A.M.                                 BD, BED

    ?STAT                                          CB, CRIB

    ?TODAY                                         ST, STRETCHER

                                                   PT, PORTABLE


----------------------------------------------------------------------------------
                                                             DELETE ORDER
 RADTRNSH
@@@@@@@@@@@@@@@@@@@@@@@@@@@@@@@@@@@@@@@@@@@@@@@@@@@@@@@@@@@@@@@@@@@@@@@@@@@@@@@BOT
```

Figure 9. Screen for selection of desired Scheduling (TODAY in this case) and Transportation (ST for stretcher in this case).

Patient Profile which is covered below, although additional material can be added, and ADDITIONAL NURSING DATA in the lower right quadrant are added if applicable. Once the ENTER KEY is depressed, the Clinical History screen appears as shown in Figure 11. It is necessary to key something in on this screen for one to be allowed to continue the next screen. If some non-meaningful text is entered, the Radiology Department, when it gets the requisition, will call the physician ordering the examination and get the clinical history before they will in fact perform the procedure. In this case, "POSITIVE TINE TEST" is entered because of a search for possible tuberculosis, and the ENTER KEY is depressed. This brings up the screen shown in Figure 12 in which the name of the ordering physician is keyed into the system. In the future, when the physicians are entering orders and using their own signon

```
@@@@@@@@@@@@@@@@@@@@@@@@@@@@@@@@@@@@@@@@@@@@@@@@@@@@@@@@@@@@@@@@@@@@@@@@@@@@@@@TOP
 N650 0399-01 YOUNG, DOROTHY                        538484          08/23/78  1242
 UNIT NO =        667902                      CHEST PA AND LATERAL
                              NURSING HISTORY
  PROBE APPROPRIATE ITEMS, ONLY 4 MAY BE SELECTED
----------------------------------------------------------------------------------
 ?BLIND              ?CANNOT SIT
                                                      ALLERGIES
 >DEAF               ?SPEAKS NO ENGLISH              (KEY IN TEXT)
                                              PENICILLIN                          *
 ?CHEST TUBE         ?CANNOT STAND                                                *
                                                                                  *
 ?COMATOSE           ?PORTABLE OXYGEN
                                                ADDITIONAL NURSING DATA
 ?DIABETIC           ?PRONE TO SEIZURES              (KEY IN TEXT)
                                              NEEDS ASSISTANCE TO MOVE FROM       *
 ?FOLEY CATH         ?RESTRAINED              STRECHER TO TABLE                   *
                                                                                  *
 ?I.V.S              >UNABLE TO MOVE

 ?COMBATIVE          ?NO PERTINENT NSG HX
                                                 NOW PRESS ENTER
----------------------------------------------------------------------------------
                                                                     DELETE ORDER
 CNSGHX
@@@@@@@@@@@@@@@@@@@@@@@@@@@@@@@@@@@@@@@@@@@@@@@@@@@@@@@@@@@@@@@@@@@@@@@@@@@@@@@BOT
```

Figure 10. Screen for inputting Nursing History. DEAF and UNABLE TO MOVE were selected from Nursing Notes (as indicated by > as opposed to ? at the left of those items), PENICILLIN under Allergies was filled in automatically by the system from Patient Profile (see Figure 21) and Additional Nursing Data were keyed in.

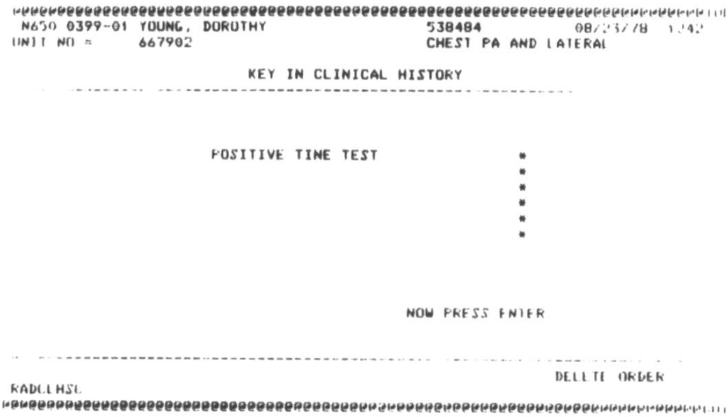

Figure 11. Screen for gathering the Clinic History.

codes, this screen will not be necessary because it will be automatically known who was entering the order. Once the ENTER KEY is depressed on this screen, the Radiology Review Screen appears as shown in Figure 13. This allows one to review the order before either ACCEPTING or REJECTING the order by probing one of those two key words in the lower right-hand corner.

If one wants to change one of the items on the Radiology Review Screen before accepting or rejecting the order, one simply probes that item which will bring the previous screen that is applicable. Once the selection or other entry has been made, one will automatically come back to the Review Screen rather than requiring the whole order to be reentered.

If the order is accepted, two things happen. A Nursing Care Addendum as shown Figure 14 is printed at a speed of 80 characters per second on a quiet printer (IBM 3287) at the nursing station. This includes the information which was on the Radiology Review Screen and is used, among other things, as a receipt for having entered the order. In addition, the Radiology requisition

Figure 12. Screen for gathering the Name of the Ordering Physician.

```
@@@@@@@@@@@@@@@@@@@@@@@@@@@@@@@@@@@@@@@@@@@@@@@@@@@@@@@@@@@@@@@@@@@@@@@@@TOP
 N650 0399-01 YOUNG, DOROTHY                 538484         08/23/78  1244
UNIT NO =     667902
              PLEASE REVIEW THE ORDER, AT THIS POINT YOU MAY
                    CHANGE ONLY THE DETECTABLE ITEMS
----------------------------------------------------------------------------
         ORDER                           FREQUENCY      BGN DT       BGN TM
   CHEST PA AND LATERAL                   TODAY          0823         1300
       CLINICAL HISTORY                   DUR            ORDERING PHYSICIAN
   POSITIVE TINE TEST                     024       DR MILAN ROGERS
   ..............................        TRANSP
   ..............................         ST
   ..............................           REQUISITION :
   ..............................           ........ PROBE TO ENTER DOWNTIME :
   ..............................                    ALLERGIES
                                           PENICILLIN
        PATIENT HISTORY                    ..............................
   DEAF           UNABLE TO MOVE           ..............................
   ..............................            ADDITIONAL NURSING DATA
                                           NEEDS ASSISTANCE TO MOVE FROM
                                           STRECHER TO TABLE
                                           ...... .......................
----------------------------------------------------------------------------
                                          REJECT                   ACCEPT
RADRVIEW
@@@@@@@@@@@@@@@@@@@@@@@@@@@@@@@@@@@@@@@@@@@@@@@@@@@@@@@@@@@@@@@@@@@@@@@@@@@RDI
```

Figure 13. Review Screen. Items can be changed, if desired, and then the order ACCEPTED or REJECTED.

itself is printed out in Radiology. As shown in Figure 15, in the upper portion of the requisition are two "flash cards" that are printed out at the same time and which give the patient identification information which is transferred on to the x-ray film during the film processing. Flash cards of two different sizes are provided because of the use of two different types of x-ray film developing processors within Parkland. One of the flash cards has a hole in it to allow orientation by one of the personnel using that particular processor who happens to be a nonsighted individual.

Once the requisition is received, the computer-based film history (the last two digits of the year that the previous films were taken and one of six different anatomical or functional areas) is updated. This film history is replacing a 3 × 5 card file within the Department of Radiology. The patient is brought to the department by the Radiology Department transport personnel. The procedure is done by the Technologist and/or Radiologist and the results are dictated by the Radiologist as was done previously. The results, however, are not typed onto a multipart form, but are entered into a video terminal. In addition to text processing capabilities, a set of two dozen 3-character codes has been provided to encode normal studies. Examples of normal Radiology Codes are

```
                                                               **CARE ADDENDUM**
N650 0399-01 YOUNG, DOROTHY                      538484     082378          1242
      CLINICAL HISTORY                 FREQ       DUR   BGN DT/TM           DEPT
--------------------------------------------------------------------------------
POSITIVE TINE TEST                     TODAY      024   08-23  1300          RAD
                                        TRANSP:   ST    DR MILAN ROGERS

                                                     ALLERGIES:
                                                 PENICILLIN

     PATIENT HISTORY:
DEAF              UNABLE TO MOVE                     ADDITIONAL NURSING DATA
                                                 NEEDS ASSISTANCE TO MOVE FROM
                                                 STRECHER TO TABLE
                                                   PERSON ENTERING ORDER
     PROCEDURES ORDERED:                            HELEN A CUSICK
CHEST PA AND LATERAL
```

Figure 14. Nursing Care Addendum printed at the Nursing Station.

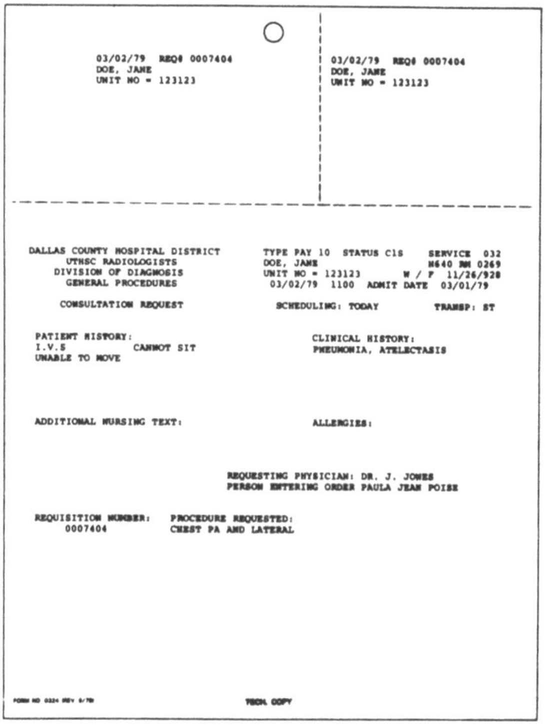

03/02/79 REQ# 0007404
DOE, JANE
UNIT NO = 123123

03/02/79 REQ# 0007404
DOE, JANE
UNIT NO = 123123

DALLAS COUNTY HOSPITAL DISTRICT
UTHSC RADIOLOGISTS
DIVISION OF DIAGNOSIS
GENERAL PROCEDURES

TYPE PAY 10 STATUS C1S SERVICE 032
DOE, JANE N640 RM 0269
UNIT NO = 123123 W / F 11/26/928
03/02/79 1100 ADMIT DATE 03/01/79

CONSULTATION REQUEST

SCHEDULING: TODAY TRANSP: ST

PATIENT HISTORY:
I.V.S CANNOT SIT
UNABLE TO MOVE

CLINICAL HISTORY:
PNEUMONIA, ATELECTASIS

ADDITIONAL NURSING TEXT:

ALLERGIES:

REQUESTING PHYSICIAN: DR. J. JONES
PERSON ENTERING ORDER PAULA JEAN POISE

REQUISITION NUMBER:
0007404

PROCEDURE REQUESTED:
CHEST PA AND LATERAL

TECH. COPY

Figure 15. Requisition immediately printed in Radiology when an order is accepted. The two regions at the top of the form (delineated by the dashed lines representing the perforations) are the Flash Cards used to put the patient indentification information on the films. The two sizes are because one of two types of film processors will be used. The circle is a hole which permits a nonsighted technician on that particular type of processor to orient the Flash Card successfully.

NAS for a normal nasal bone study, FAC for a normal facial bone study, CHS for normal chest film, RIB for normal rib series, and ABD for normal abdominal study.

Let us say one wants to record a result. If one has probed "ENTER RESULT" on a previous screen, the screen shown in Figure 16 is displayed. In this case the requisition number was #9156, and this brings up, after the ENTER KEY is depressed, the screen shown in Figure 17 which allows the keying in of the "CLINICAL IMPRESSION." If this a normal study, all one has to do as shown in Figure 17 to key in the date, a control code, and "CHS," move to the next screen as shown in Figure 18 and key in the name of the reading Radiologist, and the "RADIOLOGY RESULTS REVIEW" screen appears as shown in Figure 19. Note that on this screen the three character code CHS has been expanded to six lines of normal examination result text which has been previously agreed upon by the Chest Radiologist Group in the Radiology Department. At this point one can ACCEPT or REJECT the report.

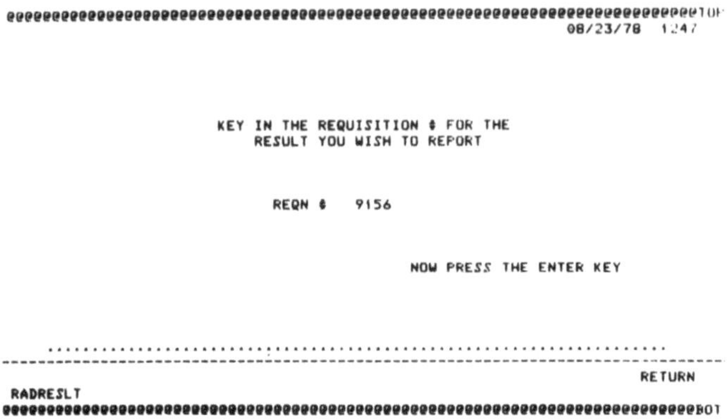

Figure 16. Screen used in the Radiology Department to indicate for which requisition the result report is being input.

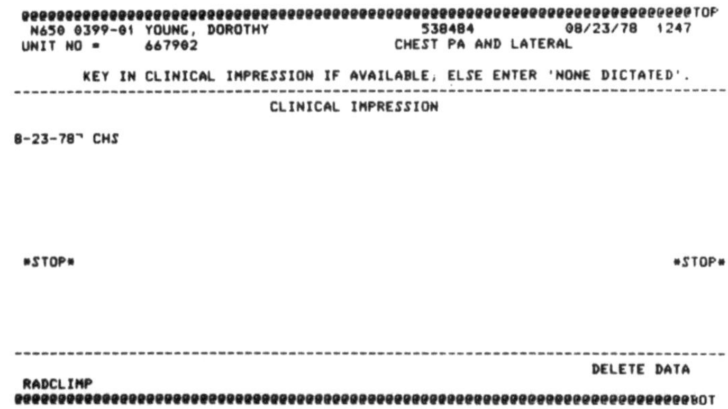

Figure 17. Screen for input of Clinical Impression. This example shows the power of using one of the two dozen 3-character normal codes, in this case CHS. The operator keys in just the date, a control code and CHS, markedly speeding the process.

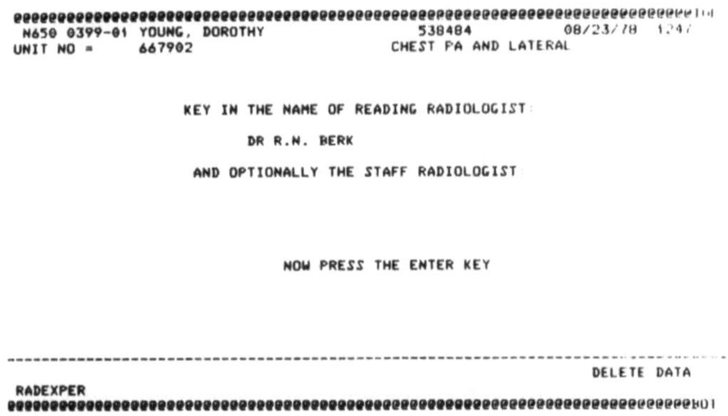

Figure 18. Screen to key in the name of the reading Radiologist.

```
N650 0399-01 YOUNG, DOROTHY                    538484            08/23/78  1:47
UNIT NO        667902          PROCEDURE    CHEST PA AND LATERAL
                      RADIOLOGY RESULTS ENTRY REVIEW
               YOU MAY CHANGE PEN DETECTABLE FIELDS AT THIS TIME
--------------------------------------------------------------------------------
           CLINICAL IMPRESSION
           8-23-78
           CHEST
           THE LUNGS ARE WELL AERATED AND FREE OF CONFLUENT INFILTRATES.
           THE HEART IS NORMAL IN SIZE AND CONTOUR AND THERE IS NOTHING TO
           SUGGEST CONGESTIVE HEART FAILURE. THE TRACHEA IS IN THE MIDLINE
           AND THE MEDIASTINUM IS NORMAL. THERE IS NO EVIDENCE OF PLEURAL
           DISEASE AND THE VISUALIZED OSSEOUS STRUCTURES ARE NORMAL.
           CONCLUSION: NORMAL CHEST.


           READING RADIOLOGIST                         STAFF RADIOLOGIST
           DR R.N. BERK                                ........................

                AFTER VIEWING REPORT PROBE ACCEPT TO PRINT
--------------------------------------------------------------------------------
                                          REJECT                      ACCEPT
RADRESRV
```

Figure 19. Results review screen. The six lines of report text were generated by the use of the 3-character normal code as shown in Figure 17. If the report is ACCEPTED, an immediate result is printed out at the nursing station and the Final Report (figure 20) is printed out in Radiology.

If report is accepted, two things happen. A report is printed out immediately on the nursing station where the patient is currently located (*not* where the order was entered) on paper identifying it as an immediate report, and the final report is printed out within Radiology. Less than three in a thousand reports are changed from the immediate report, so the fact that it has returned immediately to the nursing station, thus saving hours to days is quite helpful. Rapid returns of all types of results will decrease patients' length of stay. The final report that is printed out in the Radiology Department itself (Figure 20) is overread by the Radiologist, signed, and returned to the floor through the normal manual means. It then becomes part of the patient's permanent chart.

The Inquiry Function

On the Nursing Station Function Screen, a number of inquiry functions are available for use with the video terminal, namely DISPLAY ORDERS, DISPLAY RESULTS and DISPLAY OCCURRENCES. After a selection is made, the next screen permits the area(s) of interest, status, and time frame to be selected, with the computer doing the sorting for the user. While currently the majority of terminal activities are performed by Nursing Services personnel, we have begun a pilot program of physician inquiry. Another type of inquiry is that of the Patient Profile covered in the next section.

The Patient Profile

An example of the Patient Profile is shown in Figure 21. This screen has information of generally permanent interest and in the future will be more filled out than in the example given. Note in mid-screen that "PENICILLIN" is given as an allergy and will be filled in on screens where orders are entered where

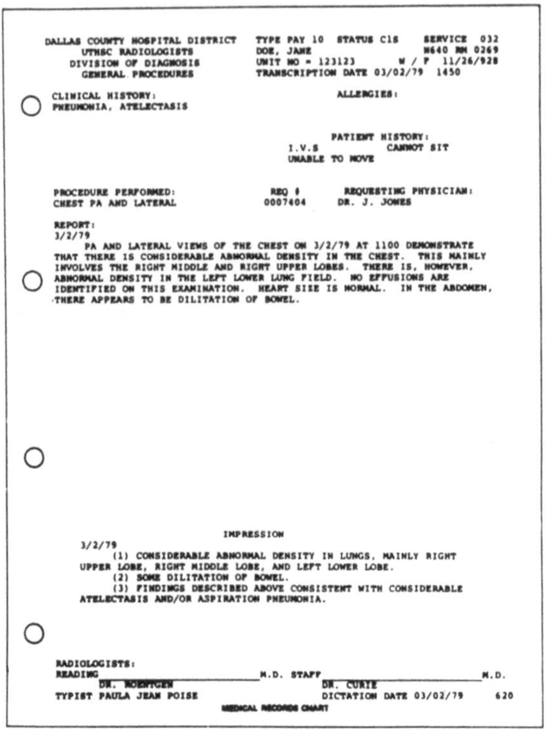

DALLAS COUNTY HOSPITAL DISTRICT
UTHSC RADIOLOGISTS
DIVISION OF DIAGNOSIS
GENERAL PROCEDURES

TYPE PAY 10 STATUS C1S SERVICE 032
DOE, JANE N640 RM 0269
UNIT NO = 123123 W / F 11/26/928
TRANSCRIPTION DATE 03/02/79 1450

CLINICAL HISTORY:
PNEUMONIA, ATELECTASIS

ALLERGIES:

PATIENT HISTORY:
I.V.S CANNOT SIT
UNABLE TO MOVE

PROCEDURE PERFORMED: REQ # REQUESTING PHYSICIAN:
CHEST PA AND LATERAL 0007404 DR. J. JONES

REPORT:
3/2/79
PA AND LATERAL VIEWS OF THE CHEST ON 3/2/79 AT 1100 DEMONSTRATE THAT THERE IS CONSIDERABLE ABNORMAL DENSITY IN THE CHEST. THIS MAINLY INVOLVES THE RIGHT MIDDLE AND RIGHT UPPER LOBES. THERE IS, HOWEVER, ABNORMAL DENSITY IN THE LEFT LOWER LUNG FIELD. NO EFFUSIONS ARE IDENTIFIED ON THIS EXAMINATION. HEART SIZE IS NORMAL. IN THE ABDOMEN, THERE APPEARS TO BE DILITATION OF BOWEL.

IMPRESSION
3/2/79
(1) CONSIDERABLE ABNORMAL DENSITY IN LUNGS, MAINLY RIGHT UPPER LOBE, RIGHT MIDDLE LOBE, AND LEFT LOWER LOBE.
(2) SOME DILITATION OF BOWEL.
(3) FINDINGS DESCRIBED ABOVE CONSISTENT WITH CONSIDERABLE ATELECTASIS AND/OR ASPIRATION PNEUMONIA.

RADIOLOGISTS:
READING ______________ M.D. STAFF ______________ M.D.
DR. ROENTGEN DR. CURIE
TYPIST PAULA JEAN POISE DICTATION DATE 03/02/79 620
MEDICAL RECORDS CHART

Figure 20. Final Radiology report printed out by the system, over-read by the radiologist and then sent back manually.

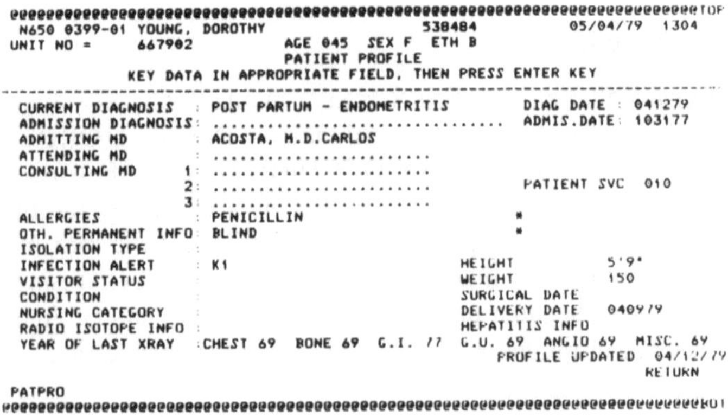

N650 0399-01 YOUNG, DOROTHY 538484 05/04/79 1304
UNIT NO = 667902 AGE 045 SEX F ETH B
PATIENT PROFILE
KEY DATA IN APPROPRIATE FIELD, THEN PRESS ENTER KEY

CURRENT DIAGNOSIS : POST PARTUM - ENDOMETRITIS DIAG DATE : 041279
ADMISSION DIAGNOSIS : .... ADMIS.DATE : 103177
ADMITTING MD : ACOSTA, M.D.CARLOS
ATTENDING MD : ....
CONSULTING MD 1 : ....
2 : .... PATIENT SVC 010
3 : ....
ALLERGIES : PENICILLIN *
OTH. PERMANENT INFO : BLIND *
ISOLATION TYPE :
INFECTION ALERT : K1 HEIGHT 5'9"
VISITOR STATUS : WEIGHT 150
CONDITION : SURGICAL DATE
NURSING CATEGORY : DELIVERY DATE 040979
RADIO ISOTOPE INFO : HEPATITIS INFO
YEAR OF LAST XRAY : CHEST 69 BONE 69 G.I. 77 G.U. 69 ANGIO 69 MISC. 69
PROFILE UPDATED 04/12/79
RETURN
PATPRO

Figure 21. Screen showing Patient Profile. This includes permanent or semi permanent information on the patient. Kl by Infection Alert indicates this sample patient is a Klebsiella Carrier. The year of last X-Ray is the patient's Film History, the last year an examination was performed in one of the six designated categories.

that information would be necessary. The "K1" given next to "INFECTION ALERT" indicates that the patient is a Klebsiella bacteria carrier. At the bottom of the screen is a line captioned to the left as "YEAR OF LAST X-RAY" which is the film history described above.

Benefits of Already-Installed Systems

The initial benefits identified for the radiology application are listed below. Benefits to other applications are included in the original paper [5].

(1) *Eliminated Duplicate Orders.* In the past, nursing did not always know what happened after an order was placed. Therefore, if the service was not performed within (typically) 24 hours, another order was placed by the physician. Eventually, both orders would be performed, even though only one was needed. Today, through audit trails and inquiry into the system *both* Radiology and nursing know the status of an order. Inquiry can show if the order is outstanding, has been performed, or if a result has been reported.

(2) *Detection of Lost Requisitions.* In the past, Radiology had no fool-proof method of detecting a lost requisition.

Currently, with the Radiology work-to-be-done list, and display, Radiology can detect lost or misplaced requisitions.

(3) *Interim Results Reporting.* None produced in the past, only a final copy *after* the radiologist had signed form even though generally only typos and minor modifications would be made in the originally-dictated version in a small percentage of cases. The above plus delays in hospital internal mail meant results were delivered in approximately 1 to 2 days after typing.

Currently, the results are entered into the system by the Results Reporting Secretarial Staff and an *interim* result is immediately printed at the nursing station. This interim copy is used and filed until replaced by the final Radiologist signed copy.

(4) *Tighter Control on Result Reporting.* In the past, some physicians would go to x-ray and read films they had ordered. Occasionally these films would be removed from x-ray and used in teaching and/or consultation, thus becoming "lost" to Radiology, since it was not generally known that these films were being removed. Without the film, the Radiologist had nothing to read and the result could not be reported. In a teaching environment where several physicians/interns may attend a patient, it is especially important that results be entered into the patients chart so that it is available to all those working with the patient. Note that charges for these procedures were not lost since they were captured when the "x-ray" had been taken. Procedures performed but not reported, today, remain on the work-to-be-done list with a "pending result" status. Because of this, the

above situation became apparent and appropriate actions are now being taken to resolve the problem.

(5) Automated functional support such as:
   a. Complete log of patients in x-ray. Gives "time" into department, aids x-ray front desk personnel in answering questions from floor.
   b. Machine printed requisitions and flash cards.
   c. Ability to display all orders and their status for a patient.
   d. Display of all outstanding or unfinished work to be done.
   e. Automated charge capture.
   f. Notification of nursing unit of nonperformance of a Procedure.
   g. More complete audit trail which include times and persons performing the various key elements of the work.
   h. Automated statistical gathering/reporting subsystems. This area was manual and very time consuming in nature.
   i. Ability to collect and display a patients film history. This enables the front desk or radiologist to determine if there are "old films" in the file and if so, have them pulled for reviewing when appropriate.
   j. Results reporting.

Summary

Two factors regarding the implenation of POIS are of particular interest. The first is time, and the second is cost. Table 1 presents the lapsed time necessary for the implementation of the various system components.

The cost benefit analysis is more complex. Because of the large ambulatory-care components, it is not appropirate for our hospital to take the data processing budget, divide it by the number of inpatient days and come up with a cost per inpatient day. Based on volumes of charges from the three service areas (inpatient, outpatient and Emergency Room), a methodology has been developed to distribute all the data processing costs (not just those related directly to patient care) on an incident-of-service allocation [6]. The values for 1979 were $5.81 per inpatient day, $3.64 per Emergency Room visit and $3.05 per outpatient clinic visit.

We have begun looking at the financial aspects of POIS with regard to cost benefit. A prelimary analysis of the Radiology System for the inpatient service

Table 1.

Radiology	3 months
Central Services	3 months
Pharmacy	3 months
Labor and Delivery	3 weeks
Emergency Room Charge Entry	1 month
Physician's Inquiry	1 day

has indicated that the annual savings in this area would approximately $70,000 if all the freedup time would be appropriately captured [7]. Much of the time released has, however, been diverted (primarily due to understaffing) to other worthwhile functions rather than eliminated.

By way of conclusion, we identify the major factors which contributed to the ongoing success of the POIS system are:

(1) A strong and continuing administrative mandate and support;
(2) The use of the Directions Committee as a policy-making group, sounding board for possible project opportunities and information disseminating body; and
(3) The use of POIS Associates, the health care professionals who participate by doing the major portion of the analysis of old systems and the design of new applications from the standpoint of user departments, coding of the vast majority of video terminal screens and printer formats as well as designing report formats and doing user documentation and training materials.

The installation of a comprehensive Hospital Information System can be rewarding experience indeed. It takes a lot of effort and allows the hospital to look at what it is doing if it chooses to do so and make other changes to its operation, sometimes because it is convenient to do so along with the installation of a uniform system (as in essence an excuse), on one hand, and on the other hand, dealing along with problems which are now addressable.

We have been privileged to be working on a living system which has been woven into the fabric of our institution.

References

[1] Mishelevich, D. J. and L. D. Cranfill, "On-line Hospital Information System," *THISS (Texas Hospital Information Systems Society) Installation Planning Guidelines,* pp. 43–54, 1978.
[2] Mishelevich, D. J., Borden, Ruby and J. F. Stay, "Implementation Factors in Hospital Information Systems," In press, *National Computer Conference 1979 Proceedings on Computers in Health Care,* 1979.
[3] Hudson, Betty G., Mishelevich, D. J., Mize, Elaine I., and J. R. Roberts, Jr., "POIS—the Parkland On-line Information System," *Electronic Computing Health Oriented Boondocks* **8,** 1979, pp. 65–116.
[4] Mishelevich, D. J., "Installation of the IBM Patient Care System at the Parkland Memorial Hospital," *Electronic Computing Health Oriented Boondocks*, 7:36–56, 1978.
[5] Mishelevich, D. J., Hudson, Betty G., Van Slyke, D., Cranfill, L. D., Mize, Elaine, Robinson, Anna L., Brieden, Helen C., Atkinson, J., Willis, J. R. and J. Robertson, "The POIS (Parkland On-line Information System) Implementation of the IBM Health Care Support/Patient Care System at the Parkland Memorial

Hospital," *Proceeding of the Fourth Annual Symposium on Computer Applications in Medical Care,* 1980, pp. 19–33.

[6] Mishelevich, D. J., Day, M. W., Gipe, W. G., and L. D. Cranfill, "Distribution of Data Processing Costs for a Hospital Information System on a Cost-Per-Incident-of-Service Basis," Appearing also in this *Proceedings of the Fourth Annual Symposium on Computer Applications in Medical Care,* 1980.

[7] Mishelevich, D. J., Gipe, W. G., Roberts, Jr., J. R., Denney, C., Stem, A. D., and M. W. Day, "Cost-Benefit Analysis in a Computer-Based Hospital Information System," *Proceedings of the Third Annual Symposium on Computer Applications in Medical Care* **3,** 1979, pp. 3339–3347.

7

Representation of Medical Knowledge and PROMIS*

Lawrence L. Weed, M.D., and the Staff of the PROMIS Laboratory†

Divisions of Medical Knowledge

"Representation of medical knowledge" is a very big monster to grapple with. PROMIS divides medical knowledge into two large bodies:

(1) Knowledge concerning the individual patient—knowledge that has traditionally resided in scattered, multiple, and largely unedited, "first draft" medical records; and
(2) Knowledge concerning populations of people or kidneys or bacteria or diseases or drugs or surgical procedures—knowledge that resides in the medical libraries, textbooks, and journals.

The knowledge that exists in medical record rooms should form the basis for much of the medical knowledge that resides in the medical libraries. Unfortunately, the libraries are thought of as the foundation for medical science, whereas the record rooms are thought of as chaotic rooms in the basements of medical establishments. The precise connections between the two bodies of knowledge are not defined and controlled at the present time.

The business of the PROMIS Laboratory has been to house and to couple the two bodies of knowledge to achieve:

(1) Comprehensive care for individuals over a long period of time minimizing confusion, omissions, duplications, and mistakes as the process proceeds;
(2) Documentation of the care provided in a manner whereby the outcomes of the applications of medical knowledge can be studied and the "knowledge" itself updated and corrected; and
(3) A system which is capable of growth and being maintained.

†Editor's Note: A list of the Staff of the PROMIS Laboratory who participated in the preparation of the original paper is contained in reference 10.

For the purposes of achieving the comprehensive care, one must relate all medical knowledge to the needs of individuals in the broadest possible context and in the most coordinated manner possible. Concerning context, from the patient's point of view, medical knowledge can come from many sources. Those sources include: medical textbooks and journals; the patient's own medical records (scattered over offices, hospitals, schools); technical sources of many types that contain information on the toxins in the patient's home and work environment; entertainment opportunities and trends that determine life style; and all the beliefs, facts, and notions in the minds of medical providers and patients that are not recorded anywhere but are an enormous influence as verbal exchanges but are an enormous influence as verbal exchanges abound in the medical care system. The goal of the PROMIS Laboratory is to deal with as much of this information as possible in a coordinated way so that isolated, elegant pieces of medical decision making do not turn out to be trivial, irrelevant, or harmful in the life of a whole individual.

For the purpose of improving the medical knowledge of the PROMIS system by studying the output, we must precisely understand the input and uses of the knowledge in the first place, i.e., the terms must be defined, the logic unambiguously known, the context complete, and the options known at each step. This precludes the direct use of writing or dictating from human memory either to collect basic data or to implement decision pathways. The very tools people use to record their data and do their work must have built into them the parameters of guidance and currency of information to do that work correctly. To the extent that we fulfill the above requirements for the meaningful study of outputs, we go a long way toward establishing the conditions that make automation of the input (i.e., medical actions and decisions) possible.

The first step was the establishment of the POMR system in which the problem-oriented record was the organizing principle. This enabled the precise coupling of action with thought among a group of people working on the problems of a single individual. The POMR was the set of rules that made possible coordination in a broad context over a long period of time.

The next step was taken several years ago when we set out to create a completely electronic POMR—accessible to the patient and all providers through a touch-sensitive CRT. In such a manner, medical knowledge properly represented in computer systems could be displayed almost instantly to any provider in the context of a patient's problems. Direct coupling of patient's problems to the body of library knowledge is possible because of an overall common use of terms and logical approaches to problem-solving and decision-making. The speed and precision of the coupling depend upon the structures that have been created, the codes that have been assigned, and the tools for manipulation of data that have been built.

The details of a patient's problems, the libraries full of medical knowledge, and the individual's values and goals are not rigorously coupled easily or in a single step. The coupling is enabled by structures within structures, proceeding from the largest possible structure required for comprehensive care to the smallest structure that accommodates the final level of detail.

The larger structures that form the foundation of the computerized problem-oriented system are outlined in some detail in previous publications [1,2,3,4]. Inherent in these larger structures are the coordination and context that make comprehensive care to an individual possible. These larger structures are the "algorithm" for making medical decisions. The algorithm is currently "executed" by people as they take action using PROMIS. Varying amounts of guidance are provided depending upon which step is taken and which problem is being pursued. Over time, more of this processing will be done by the computer. In Appendix 4 of a book written for patients (see [3]), the above is particularly well illustrated in the eight steps of a plan for a problem and the accompanying explanations that have been reproduced from the computer displays. It was not feasible to put all the branching logic from those eight steps on paper, but one can see the beginning steps for achieving coordination and context. All the branching logic that is available in the complete electronic system is precisely what makes it possible for the computer system, as opposed to paper systems, to accommodate the many variables and uniqueness of an individual.

The more we have command of the details that uniqueness implies and the broader our understanding of the context of any single problem in an individual, the more we can use those details and the context to guide and simplify a step-wise search for the solutions among the vast possibilities that exist in medical knowledge. The more one knows about a patient, the less chance one takes of falsely categorizing that individual's diagnoses and treatment. There is no question that everyone is unique—the only question is how many variables do you have to look at before that uniqueness is established, and how relevant is that uniqueness to the problem at hand? One x-ray of a femur will tell whether there is a fracture in anybody, regardless of all other characteristics. One x-ray of a heart for size, on the other hand, will not tell us if digitalis is necessary regardless of all other variables. We can, in the latter case, ignore the other variables and make a probability statement and take a therapeutic chance, but we may not need to do that if we look at more variables until we can say with far more certainty who responds to digitalis and who does not. In this regard, some of the most productive areas to explore when caring for an individual are the previous patterns of that individual in manifesting disease and responding to treatments. Although patients are unique, they repeat their patterns of uniqueness. When these patterns are not available to help and we are forced to take chances or use probabilities, we should always remember not to convert what was once a mere probability to an absolute conviction once we have made a decision. We should use probability statements not to deny uniqueness and falsely categorize, but rather to get started in the right direction and be prepared to alter approaches when the patient's uniqueness is revealed by their course. We can get to the bottom of most problems, no matter how unusual (or improbable), if we take small, secure steps in a broad context and keep going until we have reached a secure conclusion (a probability of one as in the fracture) or until we have developed a willingness to tolerate ambiguity and let time and further developments guide us. We can take this latter course if we are prepared to see the new evidence when it appears and if the patient is in no danger.

The specialist has an enormous tendency to try to be certain about what he can diagnose and treat (is there a torn meniscus that can be operated upon in that knee or is there not?) and deny responsibility for all the possibilities outside of his specialty, even though the patient has cast his lot with that specialist at the outset. Computer programmers that use sophisticated calculations of probabilities on some of the causes—not all—for a given problem can act as a specialist in this regard. No patient wants to be tagged with a diagnosis for treatment that is wrong just because it is closer to the specialist's actual diagnosis than all the other wrong choices. The patient should have the answer unique to him no matter how unusual his problem, or he should have the honest advice to tolerate ambiguity as the problem unfolds.

Until better structures for approaching medical records were available, until we recognized the patients' crucial role in guiding their own health care (see book for patients listed as [3]), and until computers for dealing with large numbers of variables came along, we were unable to deal with and to couple the two bodies of medical knowledge as we now can. Indeed, when uncoupled and on paper, both bodies of knowledge are full of holes and misinformation. Coupling of the two bodies forces the quality of each to rise.

Using a portion of the PROMIS system, we can now show how structured and branching logic displays are used, how medical knowledge is represented in those structures and displays, how the structures and displays are used to generate medical records, how the displays are built, and how their maintenance and updating are enabled.

Display Screens of Medical Knowledge

For an illustration, we shall begin with the third phase of medical action, "Initial Plans," which is a choice on a display containing the other three phases of medical action—"Data Base," "Problem List," and "Progress Notes." When one touches "Initial Plans" on the CRT, a display containing eight choices appears. Each of the eight steps then branches to many more choices or options—not just one or two in a chain. And then each of those choices offers many options. No two individuals progress through the series of options in identical ways if the patient's values, unique characteristics, and goals are carefully considered at each step.

As one progresses from the basic structural displays to a finer level of detail, one comes upon choices which are specific medical knowledge such as the causes of a problem or the specific drug used to treat a given disease in a certain age group with a particular set of manifestations. Since one of the main goals of PROMIS deals with coordination and context, ideally I should now progress through each of the eight steps of a plan and show how they work together to achieve meaningful comprehensiveness for an individual—making the specific

```
------------------Initial Plan for Dx Entity or Pathophys. Abn.----------------
-1.  State Aims for problem management    -Exp-

-2.  Check how problem may be contributing to patient's "sickness"    -Exp-

-3.  Check for effects/disabilities produced by problem               -Exp-

-4.  Check function/status of systems that may be involved with problem    -Exp-

-5.  Assess and follow course    -Exp-

-6.  Investigate problem and its etiology                                  -Exp-

-7.  Watch for/prevent complications of problem                            -Exp-

-8.  If indicated, institute and monitor treatment                         -Exp-

-Emergency management *                     -Choose another problem *
```

Figure 1.

medical facts servants to, and not masters of, the approach to an individual. But since that has been done in previous publications, since space is limited, and since many are expecting from this collection "representations of medical knowledge" in the usual sense, I shall take only one of the eight steps of the Plan (Step 6) and progress through it to illustrate a portion of our representation of medical knowledge (see Figure 1).

Display Screen 1. If one touches Step 6 of the above eight steps of a Plan, one immediately progresses to the next display with its seven steps to be considered in the investigation of the cause of a problem. But before touching the choice itself, new users of the system can touch "-Exp-" next to the choice and immediately branch to the explanation seen in Figure 2. (Note that the choices followed by an asterisk (*) are not necessary to our present discussion and will not be explained here.)

Display Screen 2. From the point of view of our present discussion of "representation of medical knowledge," we can see from the above that some of the

```
------------------------------------------------------------------------------
DETERMINE THE CAUSE OF THE PROBLEM:

   Statistics on causes of a given problem in a large population may not be the
   best guide to effectively finding the cause in a given individual.  A
   patient's own data base and problem list are full of clues that help provide
   order to the investigation.  Some clues can force statistically unlikely
   problems to early investigation (for example, absent femoral pulses sug-
   gesting coarctation of the aorta in the hypertensive) and other clues make
   unnecessary investigation of a cause otherwise statistically quite common or
   scientifically quite attractive from other points of view.  The broader the
   knowledge of a patient as a whole, the more efficient and more meaningful
   the investigation of the specialist can be.  Physicians are forced to over-
   emphasize statistical textbook knowledge and under-utilize knowledge of the
   patient himself.  The latter knowledge is not in textbooks and journals - it
   is in good records on a particular patient and in the mind of an old time
   physician in an isolated community who has had sole medical control of a
   patient over a long period of time.
                                                                 -Next Frame-
```

Figure 2.

```
---------------------Investigate problem and its etiology----------

-1.  Check for previous patterns of similar problems

-2.  Characterize problem further (if appropriate)   -Exp-

-3.  Check problem list for possible etiologic relationships

-4.  Check current treatment for possible etiologic relationships

-5.  Investigate causes of initial concern (if appropriate)   -Def-

-6.  Investigate other causes (if appropriate)

-7.  Screen for predisposing problems (if appropriate)
```

Figure 3.

most important medical knowledge to represent and control is the knowledge about individual patients. But this requires a system for generating those records that is far more disciplined than present written and dictated inputs now are. To that end we can now progress to Figure 3 which is the display appearing when Step 6 on Figure 1 is touched.

Display Screen 3. It will be noted that the first four steps above require us to return to the specific knowledge of that patient which in turn can guide us in our interactions with "library" medical knowledge. By touching Step 1 you are led to previous present illnesses and discussions of this problem. In the same touch you are also led to powerful flowsheet programs that reveal trends, patterns, and where and whether data are available within the known structures. In this way, efforts of various providers can be cumulative, the patient's uniqueness can be recognized, and duplications of unproductive and expensive approaches can be avoided.

For implementation of Step 3 above, one merely needs to touch the choice and the problem list appears. One then reviews that list to see which problems are on it which are also present in the textbook lists of causes that are seen when you touch Steps 5 and 6. At the present time this matching is done by the user, but it is apparent that, when all the problems are properly coded and the necessary programs written, the computer can do this matching and immediately confront the user with the causes with which to start the investigation instead of giving the user all the possible causes and a lot of probability data. In the above way, the patient's unique problem list can give priority to investigations and thereby make them uniquely efficient for him as an individual. But that problem list will not be complete and precise if defined data bases are not present, if choices are not made in defined language by all information originators, and if system rules are not followed.

[*Editor's Note:* With the dramatic improvements in computer technology, one further step is now possible: the use of microcomputer systems for coupling the relevant information from the medical literature to the patient's unique situation. Called the "Problem-Knowledge Coupler," it provides tools for building new couplers for specific problems and to structure an information base from the medical literature. The conceptual groundwork for the Problem-Knowedge Coupler was laid during the era of PROMIS, but its present concrete and specific implications came much later. The scope of this paper is restricted to a discussion of the PROMIS system. Readers interested in the Problem-Knowledge Coupler can consult C. C. Weed, Section 2, "The Philosophy, Use, and Interpretation of Knowledge Couplers", of *Manual 1: Understanding and Using Problem-Knowledge Couplers* (PKC Corporation, April, 1983); and L. L. Weed, "Problem-Knowledge Coupling;" presented at the New York Academy of Medicine Symposium on Biomedical Engineering, March 9, 1982.]

Frequently, one immediately sees information on the problem list or treatments that make further investigation of the new problem unnecessary. But if that is not the case, one touches Step 5 and Figure 4 appears. (Since time and space are limited, I shall omit a discussion of Step 2 although it is a powerful step in ordering the investigation of causes on a given individual.)

Display Screen 4. If an investigation of hypertension is necessary and the patient data have not provided clues as to what causes to look for, we have the question: "What causes, of all the possibilities, should be looked for first?," i.e., what are the "causes of initial concern?" For this and many such questions there must be a decision methodology. When deciding on the "causes of initial concern," the builders of the above frame looked at all the possible causes of hypertension and chose those which:

(1) have effective treatments;
(2) progress rapidly; and
(3) may not be easily and eventually revealed in the pursuit of another problem.

The above problem is an illustration of "the representation of medical knowledge" for the purpose of solving the problem of diastolic hypertension. The medical provider is no longer expected to retain this list in his memory. The user may touch one of them—e.g., "Renal Artery Disease" and progress to the next problem-solving step as seen in Figure 5.

```
---------------Diastolic hypertension:  causes of initial concern------

-toxemia                                -essential hypertension
-hypokalemic syndrome                   -pseudoaldosteronism
-renal tumor                            -coarctation of the aorta
-renal artery disease                   -steroid therapy
-renin secreting tumor                  -sympathomimetic therapy
-ovarian tumor                          -monoamine oxidase inhibitor
-primary aldosteronism                  -Cushing's syndrome
```

Figure 4.

```
-----------------------------Investigate for cause------------------

-1.  Screen for cause

-2.  R/O cause

        or

   -report negative screen for cause     -Exp-

   -discontinue investigation for cause    -Exp-

   -report cause as ruled out     -Exp-

   -report cause as present       -Exp-
```

Figure 5.

Display Screen 5. At this point the user is given guidance and is told to first see if the cause is likely from readily available information before plunging into risky and/or expensive "rule-out" procedures. This he does by touching "Screen for Cause" and Figure 6 appears.

Display Screen 6. It will be seen from the above that any reasonably complete data base on a patient will contain the information called for and will guide the user as to whether to proceed. Over and over again a data base can serve many specialists in this way. One can see where a data base reliably coded and available at all times can pay for itself because of its great power in organizing, stopping, and starting more sophisticated investigations by all sorts of providers.

It will be noted that there are columns for sensitivity and confirmation rates. Frequently, we do not have good data in the medical literature for these nor for incidence and prevalence of a problem in a given population. The phrase "given population" is of great importance. It means to say: when you were getting your figures, how many variables did you use to categorize the population that you used in the first place? For example, a given culture medium and level of oxygen may be almost 100% sensitive for septicemia when used in

```
--------------Diastolic hypertension:  cluster to screen for cause--------------

Cause: Renal artery disease
-- Findings that suggest this cause --            -- Sensitiv. -- -- Conf. Rate --
      H/O onset of hypertension < age 30                --              --
      Sudden increase in severity or onset              --              --
        after abdominal trauma
      Epigastric upper quadrant pain                    --              --
      Lateralized bruit, especially in the              --              --
        costovertebral angles                           --              --


-Order screening procedures              -Flowsheet of screening procedures
                                            already in record
-Contingency Plan                        -Recommendations
```

Figure 6.

diabetics with urinary tract infections with E. coli organisms, whereas it may be very poor in finding the problem septicemia in non-diabetic young women who have infections with the organism "bacteroides" after childbirth. Any system, whether it be digital computers that have been programmed in a known way or human computers that have been programmed and which operate in unknown and strange ways, can misuse all sorts of statistical and probability data. How we use data can sometimes be as important as what the actual data are. Up until now in our "representation of medical knowledge" much of these data have not been available because we have not organized and studied most of the "knowledge" on patients. To do that, we need to examine personal records on a day-to-day basis in a serious and disciplined way, in a broad context, and in a manner consistent among many providers so that population studies are possible. We are particularly deficient in our knowledge of the course of various problems, in various contexts, with or without treatment. Both patients and providers should have this knowledge in order to arrive at decisions. We must begin to use structured inputs—gather data, study it, and then slowly the empty columns on the display will be filled out; the numbers will change as populations and diseases change, and the number of variables will increase as new tests appear.

We have probably already arrived at the point where it is not possible for human minds extemporaneously to use these isolated figures at the bedside. Well-though-out guidance is essential just as good, up-to-date maps are useful even to the most sophisticated traveler except over the most traveled and familiar paths.

In the meantime, the present frame extends the human memory and can be used very effectively even before more quantitative data are available and before the inevitable coding and automatic checking are completed.

Medical Decision-Making

At this point we might relate the system thus far described to medical decision-making as described in a general way by many authors. By using defined input and guidance structures, we begin to deal with some of the criticisms of Herrik Wulff in his book on *Rational Diagnosis and Treatment* in which he states that medical students learn to imitate the decisions of the master but do not learn to evaluate those decisions [5]. He also points out that participants in post-graduate courses rarely can agree upon the meanings of terms and that text-book authors have a special responsibility for the standardization of terms. PROMIS believes that the latter can only be achieved when we all use electronic systems which have a common language, interchangeable codes, operate within similar contexts and overall goals, and are capable of corrective feedback loops. Differences in hardware and software will probably not be important when at least this degree of commonality can be achieved. If it can be achieved, we can all grow wiser from our actions, as arbitrary as those original actions may now have been as first approximations.

We can also pause to relook at Figure 4, "Diastolic Hypertension: Causes of Initial Concern," and ask the question: How should a user, and eventually a machine, deal with all the possibilities on that list, and by what calculations and assumptions was the list created in the first place?

Dealing with the last question first, one could say the creators of the list in Figure 4 have merely reviewed standard library medical literature and asked experts for those causes of hypertension that fulfill the PROMIS criteria for "causes of initial concern." Many writers have been critical of that medical literature in that the proper studies and mathematical analyses were not done to support the conclusions in some of those articles; therefore, PROMIS personnel have had to develop rules for selecting data from the literature. After reviewing the best available data, the list was made as a first approximation to be improved with usage and feedback loops.

Dealing with the issue of how one should approach such a list, once it has been created, there are several possibilities:

(1) One could touch each choice, look for the "screen for" data, and match it to the data in the patient's record to see if the fit is sufficient to proceed to the "rule-out" (see Figure 5). Of course, the phrase "is the fit sufficient" is not a trivial one. To answer such a question requires an understanding of probabilities, utilities, sensitivities, and incidences, etc. that the average practitioner does not possess; even if he did, he would not have the time on every problem to apply that understanding and make the necessary calculations. But the user might say he is not dependent upon absolutes—he could see which ones have clusters that fit better than others and pursue them. Indeed, he might say with some validity that he can easily take the time to check each one since much of the time he may never get to this step because the steps of checking patterns and problem lists and drugs, etc. described above will have already given him an answer. Knowing his patient in a broad way does a lot to overcome his mathematical inability to work out the causes based on probabilities starting with a single factor; especially if the computer tells him what to check for on the problem list or actually checks it for him.

(2) Instead of going down the list and doing the above in order, one could start with those diagnoses that, at first glance, would lead one to say such things as "he is not female, so toxemia and ovarian tumor do not need to have their screening clusters even examined" or "he has had the problem a long time and renal tumor is not likely, so I shall skip reading the screening cluster on that," etc.

(3) One could say machines will eventually do that searching and correlating, so in the meantime, let us skip this system altogether and let each provider use his own plans, hoping for reasonable efficiency and correction of poor performance by audit.

(4) One might say medicine is an art, and we should not try to make lists and put numbers and probabilities on everything anyway.
(5) One could say that he is grateful for the list in Figure 4 with all its imperfections because he knows he would not have thought of some of those crucial possibilities; he will go straight to those on the list with an unequivocal "rule-out" (such as a pathognomonic blood level) and just not worry about all these probability problems. For those diagnoses that do not have definite causes and are just patterns of findings to which we have attached a name, he might say he will not worry about mathematically close fits if the treatment may work; just start the treatment and see if the patient can be helped. For those causes where patterns do not work and single causes are not demonstrable and probabilistic reasoning is necessary, he might say he still will not worry about the probabilities too much; he will just keep doing tests on all the possibilities until he comes up with the answer. The individual who reasons in this way might say to himself: "In medical school conferences I may be asked after a presentation what I think the diagnosis is and be judged on an analysis or a guess, but in everyday life I can keep going until I get to the bottom of the matter (even if in retrospect a knowledge of probabilities and methods of analysis would have made the pursuit of the problem more efficient, less dangerous to the patient, and cheaper for society)."

Obviously, what is going on now in medicine is a little of all the above. The belief of PROMIS is that no unaided human mind or collections of minds thinking at the time of action can do what needs to be done. We also recognize that we do not yet have the tools to do what could be done with present knowledge and could be done to improve present knowledge from everyday use in medical care. We are concentrating now on creating overall structures of guidance that assure a broad context for care of a whole individual and then building tools for manipulation of data within those structures in the most rigorous way possible. We can create, and have created, many displays such as Figures 3 and 4 and many programs for the manipulation of data on them. Such displays can raise the overall standard of input of self-corrective systems, even with present limitations. At the same time from that base, we are moving further and further into territory previously under the control of obscure human thought processes. With the proper coding, retrieval mechanisms and tools for mathematical analysis, screening clusters can be matched and the rule-outs ordered in the most efficient way possible. But we shall control this process so that whenever the patient's own value judgments should be introduced to determine alternatives, we shall make every effort to see that it is done. For example, we are promoting the policy that the patient should have a printout of his own record and be free to interact with his electronic record. At each appropriate decision point the patient should be encouraged to look at information in tables on risk factors, complications, manifestations (which include expected course with treatment and expected course without treatment), etc. The more we involve patients, the more they themselves will find ways to be involved and to become more responsible in their own health care. To this end, we have written the book for patients referred to earlier (see [3]).

The patient should be able to have the best of medical science without getting automatically the set values of the person who created the tools for adequate scientific analysis. Unfortunately at the present time, patients are getting the set of values of providers far more than they realize as personal problem-solving techniques are obscurely applied in the name of the "art" of medicine.

We should remind ourselves that all of the last few paragraphs of discussion have grown out of considering one display from Step 5 of seven steps in pursuing causes. These seven steps in turn came from one step, Step 6, of eight more basic steps in the planning for any problem.

In summary, we have been dealing with the issue of the "confidence" with which we make decisions to diagnose or treat. Since total confidence is not possible 100% of the time, we are forced into the position of helping the patients to take chances. We could do this if we could tell the patients the probabilities at each step of doing a new test or starting a new treatment so that they could factor them along with their goals and values into each decision. But frequently even their goals and values are determined by stories they have been told and the opportunities offered which in turn have unanalyzed probabilities. The human need for certainty is so great, the calculation of probabilities sufficiently complex as each new variable is added, and the basic data on probabilities to insert into our calculations so rare, that we all degenerate into quite unrigorous thoughts and actions most of the time. The words "expert," "experience," "art" are frequently used to put the best face possible on this difficult situation. Some investigators have done rigorous calculations using Bayes' rule and other techniques on a few variables in one or two medical situations that lend themselves to reasonable results and have otherwise separated themselves from the millions of everyday decisions in medicine. Statisticians want to help medical people in this difficult situation, but at the very outset they are horrified by the state of the data in everyday medical affairs. Statisticians are absolutely dependent upon consistent use of terms, reliable collections of data, structures and indices for efficiently organizing and accessing that data, and collections on large numbers of individuals so that they can make the most elementary calculations to help medical people bootstrap themselves toward wiser and wiser actions. It will not help to give courses on decision-making to medical students in which we tell them what they could or should do if they had the right data and the time to make the calculations at every decision point. They will not have the time to make the calculations and get their work done anymore than they have the time to build their own car to get to work everyday. Nor will they stumble over the right data to feed into someone else's calculations if we just leave them memorizing facts inadequately, verbally communicating sloppily, and dictating and recording as they please with the terms they personally choose and with the decisions that they generate off the tops of their heads. Never was the following quote of Whitehead's more appropriate to a situation than it is to the situation in which medicine now finds itself.

> It is a profoundly erroneous truism, repeated by all copy-books and by eminent people when they are making speeches, that we should cultivate the habit of thinking about what we are doing. The precise opposite is the case. Civilization advances by extending the number of important operations which we can perform without thinking about them. Operations of thought are like cavalry charges in a battle—they are strictly limited in number, they require fresh horses, and must only be made at decisive moments.[6]

The whole thrust of PROMIS has been first to create the structures and the successive steps such as the eight steps of the Plan described above (and all the other detailed structures described in other publications about the four phases of medical action) that will begin to dictate to us what basic data we need to have to begin to make first approximations, calculate our probabilities as experience accumulates, modify our approximations, and slowly create a wiser and wiser system.

The workers at PROMIS are more than willing to recognize that computers are superb for calculating probabilities, providing structures and guidance, and enabling feedback loops and growth of medical science, but they also recognize their responsibility to base the calculations and growth either on sound data in a broad context or on data from a system that is used consistently and can be updated as insights grow. We do not want to do trivial calculations prematurely in a narrow context just for the sake of "automatic checking," and above all we do not want to deny the patients a great deal of benefit from well-established ideas with a known high probability for success just because the patient was left to the memory and haphazard problem-solving structures of an uncoordinated host of medical providers.

It is perfectly legitimate to take the first approximations for actions at every step from the minds of "experts" and "experienced" people, providing we recognize they are not necessarily valid but just the basis for an initial hypothesis. We cannot let everyone become his own expert using terms in any old way, throw consistency to the winds, and have specialists publish articles on the few little experimental situations they were able to control. If the symphony is going to sound right, then all the players must use the same score and have a common understanding of *standard* musical notation—to say nothing of many years of practice in using it all correctly. Medical providers are likely unicellular organisms that can no longer make it on their own, but who violently resist accepting constraints and limitations that make the multicellular organism possible, with all its homeostatic mechanisms and ruthless demands for conformity (move the body's pH_4 tenths of a unit and the whole, massive, multicellular structure can die).

PROMIS is a multicellular organism whose business is to practice medicine, and the bigger it gets, the stricter the rules for survival of any given cell within that organism gets. To resist enlarging the scope of PROMIS is to resist broadening the context of medical actions; to do the latter is to shatter the hope of patients who come seeking meaningful, comprehensive care of their minds and bodies over a long period of time.

More Display Screens of Medical Knowledge

Having made this digression, we can now continue our step-by-step progression through a very small part of our PROMIS structures. Returning to actual displays from the system, let us consider that the "Screen for" cluster for renal artery disease did match our patient, and we wanted to go further and begin to rule it out with more specific diagnostic measures. We merely have to touch "Renal Artery Disease" as seen on Figure 4, then touch "Rule out cause" as seen on Figure 5, and Figure 7 would appear.

The displays that have appeared thus far (Figures 1–6) have been single frame displays. Some of the figures from this point on will be multiple frame displays. A broken line with the title of that display repeated in the center means that only the information between each broken line is seen at any one time on the CRT in the PROMIS system. A more complete set of examples may be found in [10].

Display Screen 7. At this point the user is given specific guidance. By touching the "-Int-," one can immediately be informed as to how to interpret results as seen in Figure 8.

Display Screen 8. It will be noted that there are numbers and letters after the statements. These designate references to the original articles from which the

```
---------------------------R/O renal artery disease---------------------------

1.  Do            -hypertensive IVP       -Int-

2.  If hypertensive IVP findings are diagnostic,
    then renal artery disease is present.

    If IVP findings are within normal limits but clinical suspicion is high,
    then do            -selective renal arteriogram   -Int-

3.  If selective renal arteriogram is abnormal,
    then do            -renal vein renins              -Int-
---------------------------R/O renal artery disease---------------------(cont.)

4.  If renal vein renins are diagnostic,
    then renal artery disease is present.

-Contingency Plan                         -Recommendations
```

Figure 7.

```
------Procedure interpretation:  renal arteriogram in renal artery disease------

1. Stenosis (occlusion) main renal artery, arteriosclerotic.  /#00F/

2. Branch stenosis, renal artery  /#00G,00H/

3. Fibrosing lesion(s) renal artery  /#00D,00F/

4. Aneurysm, renal artery  /#08H/
```

Figure 8.

information was derived. This leads us to a discussion of how the various lists of causes and procedures, etc., are generated and stored in the system.

When one makes a selection such as "Causes of Initial Concern," the system goes into a table for the particular problem such as hypertension or otitis media, etc., and goes to the appropriate slot in that table. New information about problems or procedures is entered into these tables using pointers to frames. (The frames, therefore, can be used for storage and display.) The use of tables allows one to be sure that a single "up-date" or addition can immediately affect all the points in the branching logic that access the particular slot in the table. One must be able to know these points of access so that information can be used in the right context.

Display Screen 9. Figure 9 is a list of slots in a typical problem table. A builder arrives at such a table easily by progression through a series of displays beginning with Figure 10.

```
------------------------Problem Table Information Access------------------------
-Definition          -"Corr" ther         -Incid/prev          -Unclas rule

-Fol course          -Fol respons         -Predis fact         -Test 1

-Char furthr         -Manifstatns         -Diagnoses           -Test 2

-Cause-init          -Predis prob         -Operat defs         -Updates

-Cause-other         -Rx protocls         -Inst genl           -family members

-R/O entity          -temp-15             -Prev rules          -member of

-Complictns          -temp-16             -ProbsPredTo         -Rad report

-Screen proc         -temp-17                                  -Symptoms

-Prev measur         -Rx rules            -Prob Table Specs    -Pres Illness
------------------------Problem Table Information Access-----------------(cont.)

-To review table content in context of superstructure, touch here and then
  pick appropriate superstructure (no messages generated will store).


-cause cluster

-radiology procedures for this problem
```

Figure 9.

```
---------------------------Table access using alpha list------------------------

 Look up table via Alpha list:
   -Problem table ( > 127/299)              Citation table:    -by title

   -Procedure table ( < 121/5700)                              -by author

      -Drug dosage group table

      -Clinical lab proc table

      -Ward proc table

      -Radiology proc table
        (includes isotope & ultrasound)
```

Figure 10.

Display Screen 10. One merely touches "Problem table", then one types in the first two letters of the problem (e.g., "ot" for otitis media) and Figure 11 is presented.

Display Screen 11. One then touches the specific problem one has in mind, e.g., "otitis media acute," and progresses to Figure 12 which displays slots in a table.

Display Screen 12. The first slot in this table, "Definition," provides the opportunity to develop a common terminology among providers so that communication will be precise and feedback loops and population studies more reliable. It deals with the conclusions of Wulff mentioned earlier. A problem definition is a set of criteria which are fulfilled by all patients who have the problem and by no patients who do not have the problem. The criteria may be terms of clinical findings, clinical chemistry, or tissue pathology. Its purpose is to serve as a basis of deciding on the correct usage of the information in the table.

One notes that selection #4 in this list is the "Causes of Initial Concern" referred to earlier. Another choice is "Rx rules" at the bottom of the second column. If one touches that, he gets Figure 13.

```
                         ||| Type in first 2 chars of problem name/synonym: ot
----------Problem tables beginning with these 2 characters----------------------
-otalgia

-otitis externa

-otitis externa, chro

-otitis externa, with

-otitis media, acute

-otitis media, recurr

-otitis media, serous

-otodynia
```

Figure 11.

------------------------Problem Table Information Access------------------------

-Definition	-"Corr" ther	-Incid/prev	-Unclas rule
-Fol course	-Fol respons	-Predis fact	-Test 1
-Char furthr	-Manifstatns	-Diagnoses	-Test 2
-Cause-init	-Predis prob	-Operat defs	-Updates
-Cause-other	-Rx protocls	-Inst genl	-family members
-R/O entity	-temp-15	-Prev rules	-member of
-Complictns	-temp-16	-ProbsPredTo	-Rad report
-Screen proc	-temp-17		-Symptoms
-Prev measur	-Rx rules	-Prob Table Specs	-Pres Illness

------------------------Problem Table Information Access----------------(cont.)

-To review table content in context of superstructure, touch here and then pick appropriate superstructure (no messages generated will store)

-cause cluster

-radiology procedures for this problem

Figure 12.

----------------------Acute otitis media: treatment rules----------------------

Treatment	Rules	Comments
-ampicillin	-X	
-amoxicillin	-X	
-erythromycin	-X	
-sulfisoxazole	-X	
-penicillin V	-X	
-penicillin G benzathine	-X	
-penicillin G procaine	-X	

----------------------Acute otitis media: treatment rules---------------(cont.)

Treatment	Rules	Comments
-kanamycin	-X	
-aspirin	-X	
-acetominophen	-X	
-codeine	-X	
-actifed	-X	
-phenobarbital	-X	
-chlorpheniramine	-X	

----------------------Acute otitis media: treatment rules---------------(cont.)

Treatment	Rules	Comment
-ENT consult	-X	
-myringotomy	-X	
-co-trimoxazole	X	

Figure 13.

```
-----------------------------Otitis media: treatment----------------------------
If patient greater than 6 weeks or less than or equal to 5 years old,
                                                            /129.805/
   -Ampicillin, P.O., 50-100 mg/kg per day in 4 doses for 10-14 days
      or
   -Amoxicillin, P.O., 20-40 mg/kg per day in 3 doses for 10-14 days

If allergic to penicillins or if local signs or symptoms persist or worsen
  despite ampicillin therapy for 10 days,   /129.685, 129.805, 129.811/
   -Erythromycin, P.O., 30-50 mg/kg per day in 4 doses
      and
   -Sulfisoxazole, P.O., 100-150 mg/kg per day in 4 doses

If patient greater than 5 years old,       /129.810, 129.1082, 129/1093,
                                            129.1094, 129/1098/
   -Penicillin V, 125-250 mg per day or 30-40 mg/kg per day in 4 doses for 10
-----------------------------Otitis media: treatment---------------------(cont.)
If allergic to penicillins,  /129.810, 129.1082/

   -Erythromycin, P.O., 30-50 mg/kg per day in 4 doses for 10 days

If local signs or symptoms persist or worsen despite penicillin therapy for
  10 days,                                              /129.685/
   -Erythromycin, P.O., 30-50 mg/kg per  day in 4 doses for 10 days
      and
   -Sulfisoxazole, P.O., 100-150 mg/kg per day in 4 doses for 10 days

If patient severely ill (temperature greater than 104 degrees F or 40 C,
  vomiting, or diarrhea), give indicated medication I.M. then begin oral
  in 1-2 days.          /129.1093, 129.1094/

   -Ampicillin, I.M.         or              -Penicillin G procaine, I.M.
         or
   -Erythromycin, I.M. (if allergic to penicillin)
-----------------------------Otitis media: treatment---------------------(cont.)

If patient less than or equal to 6 weeks old, use parenteral treatment
  according to sensitivities of aspirate growth     /129.1014/
   and
   -Ampicillin, P.O., 200 mg/kg in 4 doses (unless vomiting, then I.M.)
 Until culture results have returned, in addition to the ampicillin, use
   -Kanamycin, I.M., 15 mg/kg per day in 2 doses

For all patients, the following may be helpful:

   -Analgesics

   -Actifed      /129.805/

If relief of severe pain is needed, or specimen for culture is needed, or
  imminent spontaneous perforation is suspected, then      /129.1098/
Do      -myringotomy
-----------------------------Otitis media: treatment---------------------(cont.)
If H/O or H/O febrile seizures in siblings,        /129.1096/

   -Phenobarbital

If seasonal allergic rhinitis,      /129.805/

   -Chlorpheniramine

If infections recurrent (greater than or equal to 18 days between and greater
  than or equal to 3/year), then update problem to recurrent otitis media.
   -ENT consult (to investigate etiology, particularly to R/O structural
      abnormalities)                    /129.805, 129.1082/
```

Figure 13a.

Display Screen 13. Figure 13 shows a list of treatments for acute otitis media for which there are detailed statements or rules in the literature. To get the rule and the references that support it, one merely has to touch the "X" under the rule column. The rule may be good or bad. The statistics that underlie the conclusions may be faulty, nonexistent, or in too limited a context for transfer to the average clinical situation. At any rate, this is the slot to "represent" that piece of "medical knowledge" with all its virtues and failings.

Decision methodologies will integrate these facts or rules into a series of steps for treatment. Frequently this universe of choices is not known to a reader of a regular article wherein the author may just reference the particular treatments or causes that he uses but fails to show the universe from which his choice was made. An article on coronary type personalities, e.g., may appear scientific and complete, but the reader has no way of knowing about all the articles that come to opposite conclusions. Again, it is the matter of breadth of context that can be so powerful in broadening our perspective and in questioning many of our notions in medicine. Contradictory rules abound in the medical literature on most subjects, and students should be aware of them. Figure 13a, on the other hand, is a step-by-step approach to treatment.

In displays such as shown in Figure 13, the references are directly related to rules in the system which in turn can be referenced to the standard (paper) medical literature.

Let us now review another Rx sequence and note at each step where automatic checking might come in (see Figure 14).

Display Screen 14. In the sequence that follows, we shall begin by choosing Step 8 for the problem of moderate, diastolic hypertension. As we touch it, Figure 15 immediately appears.

Display Screen 15. The first step above reminds us to take advantage of what is already known about a unique individual. The second step reminds us that any treatment we begin has no absolute assurance of performing as predicted

```
------------------Initial Plan for Dx Entity or Pathophys. Abn.-----------------
-1.  State Aims for problem management   -Exp-

-2.  Check how problem may be contributing to patient's "sickness"   -Exp-

-3.  Check for effects/disabilities produced by problem              -Exp-

-4.  Check function/status of systems that may be involved with problem   -Exp-

-5.  Assess and follow course   -Exp-

-6.  Investigate problem and its etiology                                  -Exp-

-7.  Watch for/prevent complications of problem                            -Exp-

-8.  If indicated, institute and monitor treatment                         -Exp-

-Emergency management                      -Choose another problem
```

Figure 14.

```
-------------------------Institute and monitor treatment----------------

-1.  Assess response to treatment (if any currently being done)

-2.  Order procedures for treatment baseline data

 3.  Order treatment:      -supportive/nursing care

                           -symptomatic treatment

                           -"corrective" therapy

-4.  Order procedures to follow response to treatment


-Enter planned treatment protocol
```

Figure 15.

since all the variables in a given patient can never be known. We must be prepared to stop or alter treatments far more than we have been inclined to before serious difficulties arise. It is the immediate "representation of medical knowledge" on unique individuals that is seldom acquired or is ignored in favor of "knowledge" from books and laboratories.

The third step reminds us that there are many aspects to treating a problem

```
----------------Antihypertensive agents - by mechanism of action----------------

 alpha-adrenergic blockers
   -phentolamine (Regitine)            -diuretics, loop acting
                                          ethacrynic acid (Edecrin),
                                          furosemide (Lasix)
 beta-adrenergic blockers
   -propranolol (Inderal)              -diuretics, potassium sparing
                                          spironolactone (Aldactone),
-centrally acting                         triameterene (Dyrenium)
    clonidine
    methyldopa                         -diuretics, thiazide/thiazide-like

-direct arteriole dilators
    diazoxide, (Hyperstat)
    hydralazine (Apresoline)
    nitroprusside (Nipride)
                                                          -Alphabetically
----------------Antihypertensive agents - by mechanism of action---------(cont.)
 antihypertensives - by mechanism of action, cont:

-ganglionic blockers
    mecamylamine (Inversine)
    trimethaphan (Arfonad)


 vasodilator (via effect on cyclic AMP)
     -prazosin (Minipress)

-post-ganglionic agents
      guanethidine (Ismelin)
      reserpine (sandril)


 combination products
     -Dyazide
```

Figure 16.

```
---------------------------Hypertension:  treatment---------------------------
Goal: to maintain BP < 140/90 without development of unacceptable side
   effects.
First start with

   -Thiazide diuretic
      or
   -Furosemide

If BP not controlled after trial of 2-3 weeks, add

   -Methyldopa

   -Propranolol

   -Reserpine

If BP not controlled with acceptable doses of above, add

   -Hydralazine

If BP not controlled by combination of above, may add

   -Guanethidine          or           -Propranolol
```

Figure 17.

and one, without the others, is frequently a futile gesture. However, isolated actions are frequently taken in busy wards and in busy clinics and offices.

Step 4 follows naturally from Step 2.

Display Screens 16 and 17. Let us touch "Corrective therapy" under Step 3 and progress to either Figure 16 or Figure 17 depending on how the system is set up in a given environment. Figure 17 gives much more specific guidance, whereas Figure 16 is a large universe of choices depending on the mechanism of action of a drug. In either case, after choosing a drug, you progress to Figure 18.

Display Screen 18. The above five steps are further opportunities for coordination, understanding context and control in a complex situation. If we touch choice 3, we progress to Figure 19.

```
--------------------General Ordering Methodology for Drugs--------------------
   -Explanation of this frame

-1. Check "Aims for Problem Management" for consistency   -Exp-

-2. Check for duplicate order

-3. Check for findings which may affect use of drug in patient

-4. Discuss use of drug with patient

-5. Check dosage and frequency


-6. Continue to order   -Exp-
```

Figure 18.

```
---------Types of circumstances which may effect the use of a treatment---------

-1. Drug-Problem Interactions   -Exp-     -4. Review mechanism of action

-2. Drug-Drug Interactions      -Exp-     -5. Review absorption, distribution,
                                              metabolism and excretion
-3. Drug-Test Interactions


-Serial Check
```

Figure 19.

Display Screen 19. When each of the above is touched, displays containing the indicated information appear. When a system such as this is complete and all items are coded and quantitative aspects understood, then touching #1 would yield those interactions appropriate to the drug and problem at hand whereas at present one may get a complete list of interactions with all types of problems for a given drug. The same is true for choices 2 and 3. A careful examination of the displays in PROMIS reveals how much information there is that must be integrated with large numbers of facts about individuals. It is easy to write programs for automatic checking that deal with some of these but very difficult to deal with all. The user of the system who knows his patient and the patient who knows his own values and choices about risk should both be given the chance to see all the information; no data or logic should be denied them by oversimplified computer programs that decide risks and values for them. The block at the present time in automatic checking with computers is not any lack of understanding of what it means or how to do it; the block is rather in knowing precisely what to check, how to code it in a consistent fashion as part of a total system for care, how to factor in the patient's values and unique characteristics that so profoundly affect risks and outcomes, how to develop structures and indices and groupings of data so crucial bits of information can be accessed in reasonable amounts of time, and how to assure reliable maintenance and growth of the system itself.

In the present medical care system, which is memory based in large part and in which inputs are poorly controlled, we allow provincial and inadequate decisions to be made and then use computer checking programs in pharmacies and laboratories to point them out. Medical providers make mistakes and often get discouraged about all they have to know and remember. They should resent being but in an impossible system in the first place. Just trying harder with present approaches will not necessarily solve their problems.

Conclusion

PROMIS is attempting to "represent medical knowledge" in a way that is useful in the care of a whole individual. Since each individual is unique, there is almost an infinite number of combinations of variables and personal values.

One cannot create beforehand a fixed series of steps to deal with some textbook definition of a single problem. One must first create a structure that defines the universe of data for the care of a whole individual and which organizes problems for solution. The CPOMR is such a structure, and each basic structural unit such as the "Initial Plans" has content which can become the organizing structure for the next level of content. This cycle is repeated over and over to finer and finer levels of resolution. These structures allow us to start with global questions that are meaningful to the patient (such as "what is the diagnosis or treatment") and then to break them systematically into smaller questions until we reach a level where we can answer with reasonable confidence. In this way we build the larger answer from the smaller answers that the whole framework generates step by step.

The CPOMR structure is used in precise problem formulation. If the user of the system is careful to start the problem at the level for which he has evidence and does not indulge in impressions and guesses, the computer system can guide him to higher levels of abstraction. For example, a single problem, chest pain, can systematically be shown to be a part of a problem stated at a higher level, such as myocardial infarction or pneumonia. One must not only know all the possibilities but the strategies for efficiently ruling them in or out. This really is the process of hypothesis formulation and then pursuit of those hypotheses. The detailed displays of the computer can rapidly provide the user with a broad range of hypotheses suggested by each problem and based on the broad statistics of a population. In contrast, the unaided human mind will tend toward very early hypothesis formation before a full data base is established (i.e., before the full range of important cues is available to him) because he uses hypothesis formation to organize data for problem solution and as "organizing rubrics in working memory" [7]. He will then limit the number of hypotheses and the "problem space" so that it is small enough to contain only the number of variables he can work with effectively. This varies enormously from mind to mind depending on basic ability and experience. Such human "lists of hypotheses and 'problem spaces'" are usually large enough to solve some of the everyday problems (along with nature's help) [8]. But they are not large enough to consistently bring the best of medical science to every problem, simple or complex.

Having generated in the computer a full range of hypotheses for well-defined problems (by use of the literature which in turn uses proper mathematical tools for manipulating prevalence, sensitivities, etc., for specific diagnoses and treatments or by analyzing the data in the computer itself as experience grows), the guidance can then be given for systematically relating that medical knowledge to the unique constellation of data and problems on a given patient. Broad knowledge of a patient is extremely powerful in reordering the pursuit of and safely limiting the number of hypotheses for organized testing. A diminished femoral pulse in the data base, for example, can push coarctation of the aorta from low to high on a list of causes when dealing with the problem of hypertension. A useful analogy is that an automobile map, created beforehand by a skilled cartographer, has on it an almost unlimited number of possible trips, and the user's goals and needs organize the use of it. Although we could not have created the whole map at the time of usage (and we certainly do not memorize maps and/or pass "board exams" before we can travel); we can frequently, at the time of travel, use data on the map to our advantage. Some of our needs become precise and definite as we travel and could only have been presented as a mass of probabilities at the outset of the trip. The principle is that large decisions should be made into as many smaller decisions as possible so that options become manageable and each successive decision can take advantage of and be controlled by the one that preceded it. The more that existing options are clearly delineated on a detailed map, the smoother and more secure the traveler's trip can be between two distant points, choosing the best option at each intersection, and creating a course unique to his needs and resources. The course of a person's health care is calculated decision by decision where each new decision operates on the results of past decisions and influences future alternatives; the options presented at each decision point result from a process which is defined and based on organized data from previous experience. The object of the guidance is to answer a global question, e.g., "What is the diagnosis?" or "What is the best treatment plan?" by a series of small steps and decisions. Given the goal of diagnosis and management and the "conditions at the time," the guidance keeps determining the next logical step and does not try to leap to the final answer. The problem-oriented system requires that the knowledge of those "conditions at the time" be as defined and as broad as possible so that decisions are not irrelevant to the whole individual. The final diagnosis or the final treatment plan is the fallout of this calculus of successive decisions in a broad context; it may be determined early or late in the process depending upon the type of problem and the factors in each patient's situation.

A broken leg can be diagnosed starting with leg pain after one step (x-ray) with 100% confidence; cancer of the pancreas starting with jaundice or vague weight loss or abdominal pain may not be diagnosed until many steps later; one does not go directly to the step of surgery which gives the answer with 100% confidence because the risks and costs are too great and the yield is too small for every case of abdominal pain or jaundice. The patients must understand that how definite we can be and how many steps it will take is determined by the nature of the problem. Furthermore, patients are going to have to begin to appreciate the power of our diagnostic tools and our capacity to grow wiser from experience as we begin to keep track of what we have done and as we continually analyze that experience to learn the safest route and fewest steps leading to answers. Medical providers will always be in trouble if they stick to their present methods for acquiring, recording, and analyzing data and for making decisions. Our present folly grows out of our stubbornness in failing to recognize or admit the limitations of the human mind.

Recognition of limitations of the human mind has led to much of the development of PROMIS. To the extent PROMIS is an extension of the mind for problem solving, just as the automobile is an extension of the muscles for traveling, it may be worth concluding with a discussion of some of the basic characteristics of PROMIS.

The overall integrity of the logical structures of PROMIS is forced by a single characteristic of the system: you can proceed only one step at a time by choosing one display at a time. The vast amount of medical knowledge that has been fed into the computer is available in no other way. (Note: One step can represent the synthesis and application of a great deal of medical knowledge, just as one mathematical formula can represent an extensive understanding of mathematics. For this reason the users' pathways can be efficient, rather than annoying and cumbersome. At the same time, the precise and full documentation for each action can be ascertained.) And no display from which you must proceed can be independent of the structure of the POMR.

The user need not carry in his memory all the options appropriate to his problem or task, but the price he must pay to have at his fingertips such depth of information (literally thousands of displays and hundreds of thousands of choices) is to reveal where he is in his thinking, to relate directly and precisely to the logic pathways built into the computer, and to have the computerized POMR be the complete and sole record of the individual's health. Each of the displays is tightly coupled to the logic of the previous selection.

Contrast this procedure with a library, a lecture, a pharmacy, or a conference, in which there are almost an infinite number of opportunities to absorb information without any obligation to couple it to previous logic or future action. The librarian or journal editor is unable to determine or affect the information that has been extracted, how or whether it is stored or used, and whether it is remembered, forgotten, misinterpreted, or distorted over time. And when a patient is subjected to medical action of any type, it is not unambiguously known what the sources of the information were or what logic led to the actions upon him. The noncomputerized systems of the library and present patient care activities are uncoupled.

Returning to the computerized POMR, not only is a single user tightly coupled to all parts of the system and to sharply-defined logic pathways, but multiple users at multiple terminals are coupled to the same system and, therefore, to one another. Lewis Mumford has stated that one of the great powers of the clock in the early days of technology was that it could synchronize the actions of man; a great power of the computer is that it can coordinate the actions of man and thereby widen our perception of the consequences of what we do [9]. At first these ideas seem dehumanizing, the very antithesis of freedom and art and individualism. Yet consider the symphony as we did earlier in the discussion. How many of us think of it as dehumanizing or the antithesis of art, even though it requires one hundred people playing at the same time to come under the direction of a single individual and to be guided by a score composed by yet another. We do not think of the musicians as slaves to a dead man's wishes whose fulfillment is being demanded by some tyrant.

Paradoxically, the very tool we feared would be dehumanizing enables the relentless pursuit of the pathways that establish both the large and subtle differences in individual requirements. One rapidly goes beyond the capacity of the human mind to retain the logic pathways from which the selections are made or the memory of the pathways chosen for the individuals. Fortunately, because a machine can remember without fail and without distortion, accurate and complete records can be kept on individuals and population studies can be done on their records.

Our desire for maximum information has increasingly revealed the uniqueness among individuals, allowing us to meet their needs in more and more precise ways. But if we let the process of continuous elaboration of facts occur, uncoupled to logical structures and separated from the largest possible universe of a given individual's concerns, we get lost in details, unable to return successfully to the larger purpose for which the pursuit was originally undertaken.

References

[1] Cantrill, S. V., Computerization of the Problem Oriented Record, in MacIntyre, N., Ed., The Problem Oriented Medical Record in Clinical Practice, Edinburgh, Churchill, Livingstone, (in press).

[2] Schultz, Jan R. "Promis: The Problem-Oriented Medical Information System (A medical and technical solution to four problems in medicine)." This paper will be a chapter in a forthcoming book published by the National Science Foundation, Office of Computing Research on Computer Science and Engineering Research. Written June 1977.

[3] Weed, Lawrence L., M. D. *Your Health Care and How to Manage It,* (2nd ed.) Essex Junction, Vermont: Essex Publishing Company, August 1978.

[4] Weed, Lawrence L., M.D. *Medical Records, Medical Education, and Patient Care,* Cleveland: The Press of Case Western Reserve University, 1969. (Distributed by Yearbook Medical Publishers, 35 East Wacker Drive, Chicago, Illinois 60606).

[5] Wulff, Henrik R., M.D. *Rational Diagnosis and Treatment,* London: Blackwell Scientific Publications, 1976.

[6] Whitehead, Alfred N., quoted by Christopher Strachey, "System Analysis in Programming," in *Computers and Computation Readings from Scientific American,* San Francisco: W. H. Freeman and Company, 1971.

[7] Elstein, Arthur S., et. al. *Medical Problem Solving: An Analysis of Clinical Reasoning,* Cambridge, Massachusetts: Harvard U. Press, 1978.

[8] *Ibid.*

[9] Mumford, Lewis. *Technics and Civilization,* Harcourt, Brace, 1934.

[10] PROMIS Laboratory. "Representation of Medical Knowledge" and PROMIS. *Proceedings of the Second Annual Symposium in Computer Applications in Medical Care*, 1978, pp. 368–400.

8

The HELP System*

T. A. Pryor, Ph.D., R. M. Gardner, Ph.D., P. D. Clayton, Ph.D., and H. R. Warner, M.D., Ph.D.

Development of a comprehensive computer system for acquiring medical data and implementing medical decision logic has been ongoing for over 15 years at the University of Utah and the LDS Hospital in Salt Lake City, Utah. This system is known as HELP and is presently operational at LDS Hospital which is a 550-bed tertiary care hospital serving the needs of the intermountain west. This hospital also serves as one of the primary teaching centers for the University of Utah Medical School. Having been developed in this environment, the design of the HELP system was required to meet the administrative, clinical, teaching, and research needs of hospitals, as well as to provide the decision-making capability.

System Objectives

Four major objectives were outlined in the development of HELP. They are:

(1) The system must be able to accommodate an ever-expanding medical data base. That is, the system must easily allow for expanded definition of new medical terms within the internal HELP data base. To facilitate this objective the system had to provide generalized procedures for the acquisition, storage, and review of all data items defined in the medical data base.
(2) The system had to process medical decision logic. The decision logic supported by HELP had to include criteria for diagnostic, therapeutic, and alarm protocols.
(3) The system had to be capable of serving both the medical and administrative needs of the hospital. HELP, therefore, was designed to be a complete hospital information system.

*Reprinted with permission from *Journal of Medical Systems* **7**, Plenum Publishing Corp., © 1983; modified from a version previously published in *Proceedings of the Sixth Annual Symposium on Computer Applications in Medical Care,* IEEE, 1982

(4) The system must provide effective research subsystems that facilitate clinical research on the large data base naturally acquired by the system.

Of the above objectives the most unique and potentially most beneficial is that feature which implements medical decision logic. In developing this capability, certain design constraints were formulated. These included:

(1) The decision logic should be modular with control and interaction of modules determined by medical experts responsible for a given module. Each decision module should be readily added to, modified, deleted, and understood by physicians and other members of the medical community not especially trained in computer science. This constraint lead to the development of a HELP language and programs to write HELP protocols. These programs for developing medical logic were designed to be "user friendly" which allows noncomputer science trained individuals to interact with the HELP protocols after minimal training. The key feature of the HELP language has been the ability to define precise search criteria for patient data within appropriate time windows. Once the appropriate patient data search statements have been written in HELP, the language provides for either deterministic or statistical logic to be easily applied on those data items retrieved to complete the decision criteria.
(2) The processing of decision criteria should be data driven. That is, that the decision criteria once verified by its author, is automatically processed whenever data are acquired which are referenced in a logic sector. For example, if a HELP protocol is written to alarm on certain electrolyte abnormalities, that HELP protocol should automatically be evaluated for a patient whenever clinical laboratory electrolyte data are stored in the patient's medical record. This feature of HELP means that medical decision criteria are evaluated with minimal physician interaction within a timely fashion and conveniently presented to the physician. The physician is not required to enter the data or request evaluation of the decision logic from a terminal before receiving the results of such evaluation. He need only review patient information from either the terminal or hard copy reports. One of the more important pieces of data which can be used to drive the decision protocols is time. Because of this feature, HELP protocols may be written which may be used for hospital-wide patient monitoring at set periodic intervals. For example, in monitoring potassium levels a decision may be written to notify the doctor of the need to order the appropriate laboratory test. Then for a patient who is being administered a potassium sparing diuretic and perhaps a potassium supplement concurrently, the HELP system can check at preset intervals to see if an electrolyte packet is ordered. If not, the physician can be alerted to the need to order such a test to monitor properly the electrolyte levels in the presence of the prescribed drug.

(3) The system has been designed to provide output to a variety of sources. Thus, the designer of HELP protocols may transmit the results of the HELP protocols to several destinations including the patient's medical record, printers located at nursing divisions, back to the departments performing specific tests, special research files set up by the user for other purposes, etc.
(4) The system has been designed to interact with the data acquisition modules of HELP ("ask mode"). This query feature has been added to allow HELP protocols which require that certain items of medical data be present before processing of that module may be completed. After evaluating an appropriate search statement to verify whether or not the data is present, the decision logic may, if the data is not present, instruct the "query" module of the HELP system to alert the appropriate person that specific information should be entered into the system. The information to be entered may refer to a historical or physical finding which would be directly entered through a computer terminal by either the patient, nurse, or physician, or may be a request to order a specific laboratory test. This feature has particular importance in allowing the system to minimize data entry from the physician or nurse since only data necessary to continue evaluation of specific hypotheses will be requested. This feature eliminates the shot-gun approach in which all possible data must be entered by the medical personnel before activation of any decision logic.

History of HELP

The HELP system is the outgrowth of many years of continued development in making the computer a useful tool in the practice of medicine. It had its earliest beginnings in the late 1950's and early 1960's when Dr. Homer R. Warner and associates at LDS Hospital began exploring the use of computers in the diagnosis of congenital heart disease. These early efforts culminated in the creation by the University of Utah of the Department of Biophysics and Bioengineering in 1963. One of the major goals of this new department was to explore ways in which a computer could enhance the decision processes in the medical environment. The early efforts of the department were directed toward the processing of analog signals from patients. A dual CDC 3300 computer system supported by both National Institute of Health and the National Center for Health Services Research was used to develop and implement automated heart catheterization laboratory programs, patient monitoring systems for monitoring both hemodynamic and electrocardiographic parameters, programs for automated interpretation of the electrocardiogram, and programs for capturing and reporting of laboratory values from the clinical laboratory. As this research continued, it became apparent that the integration of the output of these individual systems could be of greater importance than any of the individual system by themselves. Therefore, beginning in about 1970, all clinical systems programs were modified to use a common integrated data base system. The design and development of this data base system was the initial step in the formulation of the current HELP system. The data base was designed for continued enlargement and definition of medical terms as newer systems and data items were added to the HELP system.

With the common data base system in 1975 began the development of a language and a system for processing of medical decisions. This early work defined the requirements necessary to implement the varied medical decisions in a system of sufficient latitude for practical use in a medical environment. Because of the experience which had already been gained in medical computing and noting the similar experience throughout the country, the early design criteria included the need for modularity and the need for data driven processing in order to minimize the data entry required by the medical personnel.

By 1978 terminals had been placed in every nursing division in LDS Hospital, and all of the intensive care wards, surgical suites, laboratories had been instrumented for data acquisition and reporting. It was, however, apparent that the hardware on which the system had been developed was now archaic and there was a need to redevelop and implement the system on modern "state-of-the-art" computers. At this time, the LDS Hospital was also developing plans to meet the administrative needs of the hospital and felt that any subsequent computing system installed in the hospital must satisfy those needs as well as the clinical needs which had been serviced by the HELP system. Therefore, in the conversion and redevelopment of the HELP system, the design criteria were enlarged to include not only the medical goals previously stated, but also the goal of serving the administrative needs of the hospital. A Tandem computer system was chosen as the most logical system for development of the new HELP system since it provided both the capability for redundancy and minimal downtime plus the ability to grow easily to meet increasing needs on the system. The conversion process began in February of 1979. The transfer of administrative and clinical services to the Tandem was initiated in October 1980 with all systems being converted by August of 1981. This conversion included all of the previous clinical functions which had been running on the Control Data equipment together with the newer administrative programs.

The current system consists of a six CPU Tandem system with 770 million bytes of disc storage. They are interfaced to the system 196 terminals and 95 serial printers for entering and reporting at the various nursing stations, intensive care wards and laboratories. Since the Tandem computer system as purchased has capability of only receiving digital data, all analog processing for the system is performed using 14 Data General Nova 3/4's which are interfaced to the Tandem system. The terminals, however, connected to a Nova system have both the capability of calling programs to process on the Nova analog signals from transducers connected to the patient and calling any of the programs on the Tandem central computer system.

Operational Aspects of the HELP System

This section will describe specific applications and capabilities of the HELP system as presently implemented at LDS Hospital.

Clinical Data Base

The data base consists of two elements:

(1) a long term abstract of demographic and clinical information likely to be useful if the patient is readmitted to the hospital; and
(2) a short term comprehensive collection of all data gathered during the current hospital admission.

Table 1 lists the clinical data items which are part of the clinical data base for each individual patient as appropriate.

All data are stored in coded form (as opposed to free text) so that data can

Table 1. Data Items Which Are Currently Part of the Integrated Clinical Data Base

Clinical Laboratory		Chemistry, hematology, urinalysis, microbiology toxicology
Catheterization Laboratory		Oxygen saturation, hemodynamic pressures, valve areas & gradients, shunts, cardiac output, ventricular wall motion.
Pulmonary Laboratory		Blood gases, spirometry, plethysmograph, diffusing capacity, total lung capacity.
Coronary ICU	12 beds	Hemodynamic monitoring, ECG rhythm monitoring, cardiac output, fluid balances, drug monitoring, nutritional monitoring, nurses notes.
Shock Trauma ICU	10 beds	
Neuro ICU	6 beds	
Respiratory ICU	3 beds	
Thoracic Surg. ICU	14 beds	
Surgical ICU	6 beds	
Pathology		Biopsies
Radiology		X-ray findings
Pharmacy		Medications, and TPN administration
Demographic		Allergies, symposis of past hospital admissions, admit diagnoses, discharge diagnosis.
Infectious Disease		Urinary catheter monitoring
Surgery		Hemodynamic monitoring, anesthesiologist report, surgical procedures
Multiphasic Screening		Height, weight, blood pressure, ECG, laboratory
ECG		Rhythm, Morphology, and serial changes

be retrieved and analyzed for use in research and decision logic. Hierarchal codes are defined using a system called PTXT which is basically a computer-based dictionary. Using this program, a user defines the codes for a data item, as well as the associated text to be used for reports and terminal display, and key words associated with the item. Key words enable users who are constructing search items for decision logic sectors to specify easily the codes which should be sought.

For example, to construct a search sector to find all patients who had evidence of pneumothorax by chest radiograph, one would specify that x-ray data was desired then enter the key word "pneu." In response to this entry, 21 different data items are displayed on the screen ranging from pneumonia as a reason for performing an examination, to the procedure labeled pneumoencephalogram. When the term "Radiographic Finding, Pneumothorax" is selected from this list, the corresponding codes for this data item are automatically transferred to the search item logic for the HELP sector or report form. By omitting the specifier limiting the search to x-ray data, all data items defined as part of the clinical data base whose key words began with "pneu . . .", could be reviewed; in this case 59 separate data items are currently displayed for possible selection.

The hierarchal structure of the data base is designed to optimize the speed with which data items can be stored or retrieved. The first level of the hierarchy defines the general data class with which the data item is associated. There are currently 47 defined data classes in the system; one data class for radiology findings, a different data class for pharmacy (medications and allergies), a third for ECG interpretations, etc. There are 5 different data classes for data from the clinical laboratory (chemistry, hematology, urinalysis, microbiology, and drug levels in fluids). The second level in the hierarchal code structure defines the type of data according to the type of format in which the data are stored in the actual record in the computer.

Type zero data are defined according to specific bits or words in a fixed length string. A SMAC 20 result from the clinical laboratory used this type of format and every word in the string represents a value of a particular chemistry value whose units are defined in the PTXT definition.

Type one data codes are preceded by a delimiter which defines the data item which follows. The delimiters specify field codes (general subcategories within the data class), nouns (specific items), adjectives associated with a noun, adverbs (items associated with an adjective), data class modifiers (a modifier which can modify any noun in the data class), field code modifiers (a modifier which can modify any noun in the data class), field code, and various length numeric values. For example, a "PA and lateral chest examination" is a specific noun within the field code of "chest examinations" in the "radiology" data class. "Chest pain" is a field code modifier which could be given as a reason for ordering any of the specific procedures (nouns) in the field code category of "chest examinations." SNOMED SNOP and ICD9 codes are all stored in the Type one data format.

Type three data are the results of previous HELP decision logic which are stored as part of the patient record and give the data class, field code, HELP sector number, value of the sector if appropriate and any appropriate HELP sector modifiers (mild, moderate, severe, etc.).

The data are entered automatically in most instances through digital or analog interfaces (ECG, catherization laboratory, pulmonary function laboratory, and clinical laboratory) or by interaction with a terminal (pharmacy, nurses notes, etc.).

A general question asking program (GQAP) makes it possible for technical personnel without programming skills to construct questionnaires. To develop a new questionnaire for data entry it is necessary to type a free text question, specify the type of answer expected, and define codes which should be associated with the answer so that GQAP can store the entered data in the patient record. Followup questions (depending on the previous answers) and default error detecting logic are built into the system and are routinely used to construct branching type questionnaires for data acquisition.

The data base generated while the patient is in the hospital is generally complete except for physician derived data such as a complete history and physical finding. Research is currently underway to develop ways in which the physician can easily enter this missing data. Because the data base is presently hospital-oriented, physician office visit information and post discharge items are generally not available except in specific research areas.

Computer Based Medical Decision Logic

The most unique aspect of the HELP system is the ability to allow construction of modular sectors which may be used in the medical decision-making process. Sector logic can consist of straight forward deterministic "if . . . then . . ." rules or probabilistic applications based upon Bayes' formula. Sectors can specify a single modular decision, can call additional sectors, or be nested to form a decision tree which gives patient specific results.

HELP allows a user to specify the criteria for a medical decision. These criteria are complied into decision logic sectors which are stored on disc and activated by specific instruction or whenever data referenced by the decision logic are added to the data base.

Each HELP sector (set of criteria for a given decision or diagnosis) may consist of:

(1) a message or title which is displayed when the sector criteria are satisified:
(2) instructions to search for and retrieve specified data items from the patient's computer based medical record:
(3) a list of logical and/or arithmetic statements:
(4) Bayesian probability calculations: and
(5) references to other HELP sector evaluations.

To illustrate how sectors may be written, an example of a sector to predict the presence of pleural fluid by radiograph will be explained (see Table 2). This sector is activated whenever a chest x-ray is ordered for a patient.

Table 2. HELP Sector Logic for Calculating Likelihood of Pleural Fluid in a Patient

		Pleural Fluid Final Evaluation II
A		X-ray, chest, specific exam
B		10
C		If (A EQ 7) then B=524
D		If (A EQ 1) or (A EQ 2)) or (A EQ 3) then B=150
E		If (A EQ 56) then B=79
F		Chest pain (reason)
G	?	Pleural effusion (reason)
H		Pre-op (general reason)
I		Followup known disease (general reason)
J		Chair (transportation)
K		Post op (general reason)
L		Walk (transportation)
M	?	Pneumothorax (reason)
N		Coronary artery disease (reason)
O	?	Pulmonary embolus (reason)
P		Pleural fluid reported in last 2 days

		(Item)	(Prior prob.)	P(S/D)	P(S/D)
Q	Prob binary	F	B	.022	.070
R		G	Q	.193	.029
S		H	R	.009	.150
T		I	S	.210	.146
U		J	T	.546	.420
V		K	U	.395	.255
W		L	V	.004	.084
X		M	W	.059	.021
Y		N	X	.004	.030
Z		O	Y	.053	.018
AA		P	Z	.243	.036
BB	(A) SMA-12, Albumin GM%				
CC	(A) CBC, Hemoglobin GM%				
DD	(A) SMA-12, Calcium MG%				
EE	(A) Age				

		Item	(P)	Min	Max	(1)	(2)	(3)	(4)	(5)	(6)	(7)	(8)
FF	Prob dist	BB	AA	1.1	5.4	6/2	13/4	16/8	23/13	22/19	12/25	6/17	2/12
GG	Prob dist	CC	FF	4.7	18.0	5/2	14/7	18/9	23/11	18/19	13/23	6/16	2/14
HH	Prob dist	DD	GG	6.8	11.0	6/2	12/4	15/10	21/15	18/28	18/27	6/10	3/3
II	Prob dist	EE	HH	0	100	1/1	4/6	5/11	8/11	25/24	34/31	18/13	5/3

Item A causes the computer to search through the patient data base for an ordered chest examination which has not yet been processed by HELP. If such a data string exists, the value of item A is set to a numeric code between 1 and 8 according to the specific type of examination that was ordered. Item B sets the initial *a priori* probability of having pleural fluid (for an inpatient at LDS Hospital for whom a chest x-ray is ordered) at 1% (10/1000). Items C through E change the initial probability (value of item B) according to the specific procedure (value of item A) which was ordered.

After the *a priori* probability of pleural fluid is set according to the type of examination to be performed, the patient record is searched to retrieve other data which affect the likelihood of pleural fluid. Items F through 0 search for specific reasons for ordering the examination, special handling and transportation. These items are used because analysis of the clinical data base showed that they significantly alter the likelihood of pleural fluid. The values for items F through 0 are set to 1 (true), or 0 (false), depending on whether the code for that item is present in the data string which was created when the examination was ordered. These values are used in subsequent probability calculations. Item P is set to 1 (true) if the patient record contains x-ray examinations which showed pleural fluid within the last two days.

Items Q through AA perform binary Bayesian probability calculations based upon the data specified in items F through P. Bayes' formula is an algebraic expression used to calculate conditional probability.

Once the appropriate probability values have been obtained, application of Bayes' formula is straightforward in the binary case, i.e., when the symptom is present or not present. In item Q, the numbers 0.022 and 0.070 reflect the probabilities that, for an inpatient at LDS Hospital with and without pleural fluid by x-ray, chest pain (item F = 1) would be given as the reason for ordering the x-ray. The value of item B contains the prior probability to be used in Bayes' formula when item Q is executed. In the instance of chest pain, the probability of pleural fluid is decreased if chest pain was given as a reason for ordering the examination. If the item upon which the calculation is based (in this case item F, chest pain) is not true, then the posterior probability is simply set to the prior probability. Item R then uses the result of item Q as the prior probability and calculates the conditional posterior probability according to whether pleural fluid was given as the reason for ordering the examination.

In the base of a continuous variable such as a laboratory value, Bayes' formula can still be applied using a distributed probability approach. In this instance the distribution of values of clinical parameter must be known for populations with the disease and without the disease. These probabilities can then be used in the same fashion as binary variables. To calculate distributed Bayesian probabilities using HELP, the parameter distributions are entered at the time the logic is constructed by dividing the distributions into bins and calculating for each bin the probability that patients from the diseased and nonaffected groups would have parameter values which fell into the range for that bin. The HELP logic uses the actual value of the clinical parameter to retrieve the appropriate probabilities for the calculation. Distributed Bayesian calculations are performed in items FF through II using the continuous variables listed in items BB through EE. If they exist, the computer sets items BB through EE equal to the most recent nonzero values for serum albumin (item BB), hemoglobin (item CC), calcium (item DD) and age (item EE). Item FF calculates the probability of pleural fluid based upon the measured albumin value (item BB) for the patient using item AA as the prior probability. The format for the distributed Bayesian calculation specifies the range of possible albumin values (minimum of 1.1 g% to a maximum of 5.4 g%) and, for eight bins, the probabilities that patients with and without pleural fluid would have albumin values within the bin. These are the eight pairs of numbers given at the last entries in item FF. Item GG is also a distributed probability calculation using the value of hemoglobin (item CC) as the variable and the results of the calculation in item FF as the prior probability. If items BB, CC, DD, or EE are not in the patient record, the appropriate item value is set to the prior probability for that line.

The final likelihood of pleural fluid is the value of line II and reflects the influence of laboratory tests, prior radiology findings and reasons for ordering the examination. Only those data which significantly alter the probability are included in the HELP sector logic, and efforts are made to use parameters which are independent.

HELP sectors using similar types of logic have been written for a wide variety of applications. There are currently 2147 operational sectors with clinical application. The most widely used sectors are those involved with ECG interpretation, blood gas interpretation, pharmacy alerts, predicted x-ray findings and clinical laboratory alerts. The tools for constructing the medical logic are now operational, but the number of actual medical logic sectors must be regarded as only a beginning.

Administrative Capabilities

The core of the administrative functions is an admit-discharge-transfer (ADT) system. A patient can be admitted or pre-admitted at several locations in the hospital (admitting, emergency room, radiology, multiphasic screening, clinical laboratory, outpatient area). After the patient's name is entered, the long term file is searched for patients with similar names for verification and selection by admitting clerk. If the address, insurance information, and other financial data are correct, the patient is simply admitted to the active short term file. Otherwise, demographic and financial data are entered and the patient is admitted to the current and long term file.

Pre-admission scheduled patients and pending requests for room transfers generated on the nursing divisions are coordinated by the admitting office which uses the computer to list which beds are expected to be vacated. As soon as a patient is moved to or from a bed this fact is entered into the computer. Using the room trace log generated for each patient and a list of standard room charges, the computer automatically generates room billing.

Ancillary test charges provide an example of the benefit of having an integrated financial and clinical data system. The tests are ordered through the nursing division terminals and the "ready to bill" bit is automatically set whenever results are reported to the clinical data base. Charges for each test are obtained from the PTT file which describes the code associated text and standard cost for each procedure. Each department can make necessary changes to the standard charges for special cases. A transaction file exists to log billable tests or supplies. By using alternate keys, each department can separately reconcile the day's procedures or an up to the minute account may be obtained for a specific patient. Using programs to analyze the transaction file on a daily, weekly or monthly basis, audit statistical summaries for departmental operations are generated.

If the "ready to bill bit" is set, a program which is run at midnight of each day transfers charges to a separate financial billing computer. Via on-line handshaking a "billed" bit is then set in the status word of each entry in the transaction log for those procedures for which charges were transferred.

Research Capabilities

In order to analyze the resultant clinical and administrative data base, a query system called STRATO has been developed. The basic concept of this tool is that a HELP sector can be written to define the query criteria. The STRATO program then goes through the current patient file or predefined list of patients constituting a special study population to see which patients satisfy the criteria listed in the applicable HELP sector. Following this procedure one could, for example, obtain a population (list of patient numbers) of all male patients between 35 and 40 years of age who were admitted to the hospital with suspected appendicitis and received surgery. One could then write further HELP sectors which would define subpopulations of patients whose removed appendix was inflamed, who received certain anti-infective drugs, or who were treated by a specific physician. The STRATO program allows one to combine populations, obtain interactions and unions of populations, etc.

Additional HELP sectors can be written to retrieve and compare variables (patient age, temperature before surgery, length of hospital stay, total hospital cost, etc.) for alternative subpopulations. The variable or variables specified for each patient are stored on disc in a format related to the order of the patient number in the subpopulation. Statistical programs are interfaced to the STRATO system so that one can perform a regression analysis between different variables for the same subpopulation or compare the distributions (parametric or nonparametric tests) of the same variable across two or more subpopulations.

The results obtained from these searches can be used for administrative audits, research, or to obtain the conditional probability values needed in the Bayesian probability calculations of sectors written to make medical decisions.

In summary, STRATO can be used to derive populations or variables which are specified by the logic criteria expressed in specific HELP sectors. Such an approach gives the user ease, flexibility, and rapid results. The statistical packages provide additional tools necessary to analyze the data in a convenient fashion.

Evaluation of HELP Computer System

A survey conducted a few years ago by Battelle Laboratories of Columbus, Ohio, an independent evaluation group, revealed a positive attitude towards the use of the computer. As opposed to the usual negative attitudes and opposition encountered in adopting innovative things especially in hospitals, a positive attitude toward the HELP computer system is prevalent among physicians on the staff of LDS Hospital. A second major finding of the Battelle Study was that the highest proportion of physicians that were very enthusiastic about the system were specialists in surgery and internal medicine. Other specialists have not had as much use of the system. This study evaluated the expectation and utilization of the HELP system by both response to questionnaires and also by individual interviews of key staff physicians. The overwhelmingly favorable response in terms of what physicians expected and received from the system encouraged us to develop further the system decision-making capabilities.

The intensive and critical care units of the hospital have been computerized longer than most other areas of the hospital so evaluation of physician and nurse responses might be representative of what other nursing areas and clinical areas of the hospital after they have had several more years of experience. A report from intensive care physicians as to what they felt were the advantages of the system will give some insight into the system's benefits. These physicians felt:

(1) that computer data is better organized and makes it more easily reviewed for clinical decision making;
(2) that automated computerized decision-making provided valuable data interpretation and alarming functions;
(3) that the computer generated reports and alarms focused attention of the personnel on the treatment and management of most of the problems of the critically ill patient;
(4) that the computer's ability to manipulate and calculate on quantitative data presents patient data and parameters in a form that enhanced patient care;
(5) that the computer accelerated the availability of data because of the on-line communications network interconnecting;
(6) that computerized decision-making provided an excellent basis for teaching logical decision making skills to house-staff and staff physicians;
(7) that the quantitative structure provided by the computer system causes our physicians to think more quantitatively and as a consequence provides an excellent education tool in teaching quantitative and logical medicine;

(8) that having all of the patient data in the computerized data base provided several advantages: It eliminated redundant charting, captured charges for clinical data entry, allowed transfer of patients anywhere in the hospital without a loss of data continuity, or access, and finally allowed data review on an ICU patient from multiple locations.

Evaluation from the ICU nurses of the system showed it to be of advantage in the care of the critically ill patient in the following areas:

(1) All the laboratory data from the clinical laboratory are quickly available for review and where desired a hard copy printout can be generated. These data are available in a legible form as soon as the test is completed in the laboratory.
(2) Blood gas analyses are performed in the pulmonary function laboratory where analysis, computation, and interpretation by the computer, as well as prompt feedback of the results provides a valuable tool in the management of especially ventilator patients in the intensive care units.
(3) The pharmacy module which coupled with the charting function outlined below provides a major advantage to our clinical nursing staff. First, all drug interactions are monitored by the computer, since each drug order is entered into the computer by the pharmacist before the drug is dispensed. Drug–drug interactions, drug contra-interactions because of laboratory data, allergies, and other factors are evaluated on each patient which minimizes drug ordering errors. The computer generates a medication schedule for each patient at the beginning of each shift which provides an easy method for checking what medications are to be given and when they are to be given. All of the drug given information is logged into the computer. This drug logging not only becomes the official record for the patient, but also is used to generate patient charges. The ability to handle drugs and intravenous fluid infusions with the system provides a tremendous advantage for the nurse in keeping track of fluid balance totals as well as charging for the solutions and drugs. Thus, once the nurse has completed the task of a computerized charting, all else that happens to the record is automatic.
(4) The charting function is a very important function for nurses. Currently, the system generates 12 hour "shifts" and seven day reports. These reports are extremely valuable tools not only for the nursing, but also the medical staff in care of the patient.
(5) The fluid intake and output summary provides valuable data as to how many units of blood and what other types of fluid intake and output the patient had during a particular shift or during a particular day.
(6) Nutritional summaries are made available. These are felt to be an extremely valuable tool for both the nurse and physician.
(7) Interpretations of hemodynamic profiles which calculate parameters from cardiac output and pressure measurements are valuable.

The pharmacy module, which has been in operation since 1975, has had wide spread acceptance by the medical staff and the pharmacy. Since the hospital is on a unit/dose system and since the computer provides the surveillance needed, we have been able to decentralize the pharmacy by using movable carts for the pharmacist and placing them with a clinical pharmacist on the nursing division. Having the pharmacist on the nursing division provides several advantages, both for the pharmacist and for the nursing and medical staff. The pharmacist is on the scene; if there are drug questions or problems, they can interact with the medical staff or nursing staff. If there are alerts which come up, these alerts can be quickly followed up with the medical staff. The one-to-one personal communication has dramatically improved the ability of the nurse and the medical staff to order and dispense medication as well as improve the level of communication on drug problems. Ongoing evaluations of alerts from the pharmacy module are performed by review of monthly logs. Table 3 summarizes the monthly logs of drug alerts generated by the system for one year.

The importance of having the computer perform the alert function is clearly demonstrated by the fact that only about 5% of the patients in the hospital have an alert; therefore, there is a large amount of uninteresting data which must be reviewed to get at those few patients and even fewer drugs ordered which represent problems. Those alerts which are truly life threatening and which are not informational or have some subjective reason to be concerned about represent an even smaller fraction of the total alerts. Those alerts which require some physician interaction or changing of a drug or procedure are only about ⅙ of all of the alerts. That is, approximately 1.8 of the patients receive alerts which are truly life threatening. This is considerably lower than reported by other studies which further points out the need to have the computer perform this task. We have now a 94% compliance rate for those truly action-oriented life threatening alerts. Therefore, approximately 50 patients each month are helped in some dramatic way by having the computer monitor the drugs ordered. The pharmacy module has resulted in an overwhelming positive response from the medical staff who make few medication prescribing errors. The computer allows them to conduct their practice with the assurance that the problems will not be a major part of their risk in caring for the patient.

The x-ray decision-making area has also provided some unique advantages. We are currently utilizing the system on daily basis to generate not only the orders and billing, but also to capture the clinical results. We presently have 794 coded examinations, 435 reasons for performing the examinations, and 378

Table 3. Pharmacy Alerting Program

Monthly Averages–July 1981 to June 1982		
Patients (#)	2,684	
Patient generating an alert (#)	227	(5.35%)
Life Threatening alerts (#)	49	(1.8%)
Physician compliance to life threatening alerts		94%

coded findings. Currently, about 62% of all examination results are completely coded and entered into the computer; that is, they require no additional explanatory information to be useful to the medical staff. Using the unique ability of the system to predict from a certain type of examination how the results might turn out, we are then able to generate the results in a very simple and compact form. Approximately 82% of the examinations are now reported merely by checking a box on the computer generated report form. This not only is a great advantage to the radiologists who are reading the x-ray, but is also an advantage to the physicians who are caring for the patients since they can quickly review the information on the computer terminal. The radiologist interpretation data is then available after the report is completed by the radiologist.

As an example of one of the specialized laboratories where the computer has been of great assist has been in the pulmonary function laboratory. During the past year more than 62,000 blood gases were performed in this laboratory and over 1,200 pulmonary function tests. The computer system not only gathers the data, but generates billing information, generates reports which are printed immediately at the appropriate nursing division, and also provides interpretation of the patient data. The system also provides management information for the laboratory as well as quality control information generated as part of the normal course of data gathering. Operation of this laboratory without the computer would be an almost impossible task at this point. The computer provides legible reporting and documentation in a much more timely fashion than what would have been possible with the usual manual methods. The Battelle report showed that the physicians felt that the blood gas interpretation was a particularly effective way of helping the physicians make decisions and diagnosis and adjust therapy. Most of the physicians think that the blood gas interpretations save them time and that they value the service. Physicians had high expectations of the computerized pulmonary/blood gas area and indicated that the system met their expectations 81% of the time. These highly positive evaluation ratings have stimulated further refinement of the computerized pulmonary and blood gas system.

Future Enhancement for HELP

Future enhancement for the HELP system is focused in three areas. The first is a continued effort to add to the medical decision protocols which currently are operational on the system. This is a continuing research effort combining both the clinical data collection and research capabilities of the system in collaboration with clinical researchers to enhance the existing logic and develop new logic. The second area of focus is a continued effort to enhance the speed and throughput of the system. The major redesign in this area will be in distributing the data base and the programs throughout the hospital using "state-of-the-art" microprocessor based systems at each nursing division. In this design the system, while storing the information in its coded form at the central data base for complete HELP processing, will also transmit in a formatted form all data and HELP decisions to the nursing divisions systems where speed of review of this information is provided to the medical personnel at that location. This not only will increase the speed of review for individuals at the respective nursing divisions, but because of decreased load on the central system, also increase the speed with which the HELP processing can be performed at the central system. The third area of enhancement will be to place microprocessor based HELP systems in the doctor's office. These systems will be interfaced to the central data base. Then through communication with central data base and a processing system of similar capabilities for analysis of data recorded within the doctor's office can be provided to users outside the hospital. This will allow continued data entry into the patient data base and allow followup of the patient from the doctor's office.

References

[1] Warner H. R., Olnsted C. M, Rutherford B. D. "HELP - A program for medical decision-making." *Comp & Biomed Res.* **5:** 1972, pp. 65–74.

[2] Warner H. R., Rutherford B. D., Houtchins B. "A sequential bayesean approach to history taking and diagnosis," *Comp & Biomed Res.* **5:** 1972, pp. 256–262.

[3] Pryor T. A., Lindsay A. E., England R. W. "Computer analysis for serial electrocardiograms," *Comp & Biomed Res.* **5,** No. 6, December 1972.

[4] Pryor T. A., Morgan J. D., Clark S., England W., Warner H. R. "HELP - A Computer System for Medical Decision-Making." Presented at National Computer Conference and Exposition, May, 1974.

[5] Clayton P. D., Urie P. M., Marshall H. W., Warner H. R. "A Computerized System for the Cardiovascular Laboratory." *IEEE Proc. of Conference on Computers in Cardiology,* Bethesda, Maryland, October, 1974, p. 97.

[6] Yanowitz F. G., Pryor T. A., Morgan J. D., Warner H. R. The HELP system for medical decision making. Proc. of the 4th Annual Conference of the Society for Computer Medicine. New Orleans, LA., Nov. 7–10, 1974, Ed.: Michael A. Jenkin.

[7] Gardner R. M., Cannon G. H., Morris A. H., Olsen K. R., and Price G. A. "Computerized Blood Gas Interpretation and Reporting System," *IEEE Computer* **8,** Jan. 1975, p. 19.

[8] Gardner R. M. "The Place of Computers in Medicine." Guest Editor's Introduction *IEEE Computer* **8,** Jan. 1975, pp. 39–45.

[9] Warner H. R., Pryor T. A., Morgan J. D., Clark S. J. "Integration of computer support for institutional practice using the HELP system," *Proceedings of SAMS Meeting,* Houston, Texas, 1975.

[10] Yanowitz F. G., Pryor T. A., Frost D. "A Computerized ECG Alarm System for the Coronary Care Unit." *Computers in Cardiology,* J. Cos and P. Hugenholtz, Ed., St. Louis: October 1976.

[11] Hulse R. K., Clark S. J., Jackson J. C., Warner H. R., Gardner R. M., "Computerized Medication Monitoring System," *AM J Hosp Pharm* **33** October 1976, pp. 1061–1064.

[12] Pryor T. A., Yanowitz F. G., "Computer Applications to Exercise Stress Testing," *Computers in Cardiology,* J. Cox and P. Hugenholtz, Ed., St. Louis: October 1976.

[13] Warner H. R., Pryor T. A., Clark S., Morgan J., "Integration of Computer Support for Institutional Practice: The HELP System," *Computer Appl. in Health Care Delivery,* 1976.

[14] Frost D. A., Yanowitz F. G., Pryor T. A., "Evaluation of a Computerized Arrhythmia Alarm System," *Am. J. Card.* **39,** April 1977, pp. 583–587.

[15] Gardner R. M., Scoville D. P., West B. J., Cundick R. M., Jr., Clemmer T. P. "Integrated Computer Systems for Monitoring of the Critically Ill," Proceedings First Annual Symposium on Computer Application in Medical Care, pp. 301–307, October 3–5, 1977, Washington, D. C. IEEE Computer Society.

[16] Gardner R. M., Clemmer T. P., "Computerized Protocols Applied to Acute Patient Care," Advances in Automated Analysis **1,** 1977, pp. 158–163, Mediad Inc. Technicon International Congress, New York.

[17] Pryor T. A., Ho S. J., Stephen R. L., "Computer System for Management and Research in End-Stage Renal Disease," *Current Topics in Critical Care Med.* **3,** 1977, pp. 94–111.

[18] Warner H. R. "First the Electrocardiogram—Then What?," *Am. J. Cardiology,* 1977, pp. 5–8.

[19] Stephen R. L., Pryor T. A., Jacobsen S. C., Janata J., Kablitz C., "A Computerized Closed-Loop Feedback Control of Treatment for Patients on Dialysis," MECO '78, Athens, Greece, June 26–29, 1978.

[20] Pryor T. A., Gardner R. M., Clayton P. D., Warner H. R., "A Distribute Processing System for Patient Management," Proceedings of Computers in Cardiology, September 1978, pp. 325–328.

[21] Gardner R. M., Clemmer T. P. "Computerized Protocols Applied to Acute Care," *Emergency Medical Services,* **7:** November 1978, pp. 90–93.

[22] Gardner R. M., Clemmer, T. P., Larsen K. G., Johnson D. S. "Computerized Alert System Use in Clinical Medicine," *IEEE 1979 Frontiers of Engineering in Health Care.* First Annual Conference, 136–140.

[23] Warner H. R. *"Computer Assisted Medical Decision-Making,"* New York City: Academic Press, Inc., December 1979.

[24] Warner H. R., Clark S. J., Larsen K. G., McNeil F. "The HELP System as a Tool for Monitoring Physician Prescribing Patterns," Proc. of the 13th Hawaii International Conference on System Sciences, Vol. iii, pp. 254–255, Edited by Shriver and Sprague, Western Periodicals Company, January 1980.

[25] Warner H. R. "Computer-Assisted Decision-Making Using Clinical and Paraclinical (Laboratory) Data," Minnesota Symposium on Decision-Making from Laboratory Data, U. of Minnesota Press, June 1980.

[26] Cundick R. M., Jr., Gardner R. M. "Clinical Comparison of Pressure Pulse and Indicator Dilution Cardiac Output Determinations," *Circulation* **62**, August 1980, pp. 371–376.

[27] Pryor T. A., Warner H. R., Gardner R. M. "HELP—A Total Hospital Information-System," Proc. the 4th Annual Symposium on Computer Application in Medicine Care, Washington, D.C., Editor: Joseph T. O'Neill, November 2–5, 1980, pp. 1557–1561.

[28] Warner H. R. "HELP—An Approach to Hospital-Wide Artificial Intelligence,"

Computer-Assisted Decision Making Using Clinical and Paraclinical (Laboratory) Data, Gernard E. Statland and Stanley Bauer, Eds., Tarrytown, New York: Mediad Inc., 1980.

[29] Johnson D. S., Ranzenberger J., Herbert R. D., Gardner R. M., Clemmer T. P., "Computerized Alarm System for Acutely Ill Patients," *J. Nurse Admin.* **10:** 1980, pp. 26–35.

[30] Cannon S. R., Gardner R. M. "Experience with a Computerized Interactive Protocol System Using HELP," *Comp & Biomed Res* **13:** 1980, pp. 399–409.

[31] Sorensen J. B., Morris A. H., Crapo R. O., Gardner R. M. "Selection of the Best Spirometric Values for Interpretation," *Am Rev Respir Dis* **122,** 1980, pp. 802–805.

[32] Clemmer T. P., Gardner R. M., Orme J. R. Jr., "Computer Support in Critical Care Medicine," IEEE Proceedings 4th Annual Symposium on Computer Applications in Medical Care, 1980, p. 1557–1561.

[33] Clayton P. D., Ostler D. B., Gennaro J. L., Beatty S. S., Frederick P. R. "A Radiology Reporting System Based on Most Likely Diagnoses," *Comp & Biomed Res* **13:** 1980, pp. 258–270.

[34] Pryor T. A., Clayton P. D., Gardner R. M., Waki R., and Warner H. R. "HELP—A Hospital-wide System for Computer-based Support for Decision-making," Proceedings of the 14th Annual Hawaii International Conference on Systems Sciences, January 8, 1981.

[35] Crapo R. O., Gardner R. M., Beus M. L., "Lower Limits of Normal for Spirometric Reference Values," 1981 Conference of Computers in Critical Care and Pulmonary Medicine, Norwalk, Conn (in press).

[36] Gardner R. M., Ostler D. V., Crapo R. O. "Signal Processing for Computerized Spirometry," 1981 Conference of Computers in Critical Care and Pulmonary Medicine, Norwalk, Conn., (in press).

[37] Waki R., Clayton P. D., Jensen R. L., Yanowitz F. G., Liddle H. V. "Patient Specific Medical Decision Analysis using the HELP System," Proceedings of the Fifth Annual Symposium on Computer Applications in Medical Care, H. G. Heffernan, Ed., IEEE Computer Society Press, 1981, p. 237.

[38] Clayton P. D., Gray S., Frederick P. R. "A Code Oriented Radiology Reporting System Based upon Pre-examination Prediction of Likely Findings," Proceedings of the Am College of Radiology Seventh Conference on Computer Applications in Radiology, April 1982, pp. 261–272.

[39] Pryor T. A., Clayton P. D., Larsen K. L. "An Integrated Hospital Information System," Computers in Hospital, November 1982.

[40] Pryor T. A., Clayton P. D., Larsen K. G. "Synergistic Relationship between Test Ordering and Medical Decision-making in the HELP Hospital Information System," submitted for publication in Computers in Hospitals, 1982.

[41] Pryor T. A., Gardner R. M., Clayton P. D., Warner H. R. "The HELP System," Proceedings of the 6th Annual Symposium of Computer Applications in Medical Care," B. I. Blum Ed., IEEE Computer Society Press, 1982.

[42] Waki R., Clayton P. D., Jensen F. L., Yanowitz F. G., Liddle H. V. "HELP Based Decision Analysis Applied to Coronary Artery Disease," *Comp & Biomed Res* **15:** 1982, pp. 188–202.

[43] Neeley J. P., Ostler D. B., Frederick P. R., Clayton P. D. "Pre-examination Prediction of Radiographic Findings," *Invest Radiol* **17:** 1982, pp. 310–315.

[44] Gardner R. M., West B. J., Pryor T. A., Orme J. F. Jr., Clemmer T. P. "Computerized Decision-Making applied to Critical Care," *Critical Care Medicine,* December 1982, (in press).

[45] Gardner R. M., Crapo R. O., Morris A. H. "Computerized Decision-Making in the Pulmonary Function Laboratory," *Respiratory Care,* **27:** 1982, pp. 799–808.

[46] Morris A. H., Kanner R. E., Crapo R. O., Gardner R. M., "Clinical Pulmonary Function Testing: A Manual of Uniform Laboratory Procedures for the Intermountain Area," 2nd ed., 1982, (in press).

[47] Gardner R. M., Clemmer T. P., Morris A. H. "Computerized Medical Decision-Making—An Evaluation in Acute Care," *Computers in Critical Care and Pulmonary Medicine,* **2**, 1980, pp. 147–150.

Implementation of a Prototype Generalized Network Technology for Hospitals*†

S. G. Tolchin, Ph.D., R. L. Stewart, S. A. Kahn, E. S. Bergan, G. P. Gafke, D. W. Simborg, M.D., Q. E. Whiting-O'Keefe, M.D., M. G. Chadwick, and G. E. McCue

Introduction

Approaches to a total, or composite hospital information system (HIS) have been proposed, studied and evaluated for the last twenty years. Despite extensive industry efforts to develop packaged HIS products, a solution to the information processing needs of a "typical" hospital that is readily available, comprehensive, functionally integrated and appropriate to the individualized needs of various institutions has not been demonstrated. Thus, development of a generally acceptable and transportable HIS technology remains a challenge.

Recently, distributed data processing (DDP) approaches to the HIS problem have been considered [1–6]. These approaches support modular, distributed HIS development under in-house control as a viable alternative to the costly, centralized mainframe approach. A particularly important feature of the DDP approach is the potential for the institution to build its own HIS as an integrated system composed of existing or planned heterogeneous functional area systems. Each component system may be chosen to best serve the needs of a given hospital area, within an overall system application design that is tailored to the operational needs of the institution. The component systems may be off-the-shelf, developed in-house or developed by a vendor to specifications prepared in-house.

*This project is funded at least in part by Grant No. 1 R18 HSO4196-01 with the National Center for Health Services Research.

†This chapter has been modified from a version previously published in *Proceedings of the Fifth Annual Symposium on Computer Applications in Medical Care*, IEEE, 1981.

Although such component systems are apparently incompatible with respect to hardware, software and data bases, it is feasible to network these systems together inexpensively (along with peripherals and other computing devices) using recently developed local area communications network (LACN) technology [7–10]. This approach is designed to support integration of functions and data, and to facilitate reconfiguration of the physical plant, user processing, controlled growth, and component system upgrade.

A demonstration implementation of a DDP HIS using LACN technology is described below. The University of California, San Francisco (UCSF) Hospitals in collaboration with the Applied Physics Laboratory (APL) of The Johns Hopkins University has begun implementation of a DDP system supported by a highly intelligent LACN to permit integration of heterogeneous minicomputer systems. The implementation site is at UCSF and uses an application level design developed by UCSF and a fiber-optic based, intelligent LACN developed by APL [7].

System Architecture Considerations

Several factors must be considered in developing a system architecture appropriate for HIS service. These issues may be organized into the applications environment, technical performance issues, and network communications services available to users and computers, including physical architecture, communications media, protocol structure and services and network interface issues.

Applications and Data Environment

The applications environment typically will consist of some existing computers serving administrative or ancillary departments, plans for future automated support and a concept for overall integrated systems application design specific to the institution. The following principles may be used to guide this process:

(1) The composite system should provide immediate, recognized benefits in patient care "... by facilitating communication and reducing errors ..." [11];
(2) The autonomy of participating computer subsystems should continue. No aspect of the integration design should preclude control of, or access to local processing resources by local user communities. [1];
(3) The hardware and software modifications to individual host computers, necessary to join the integration network, should be minimal [14];
(4) The network communication subsystem should be low-cost off-the-shelf technology. It should be easy to install and highly reliable.;
(5) Whatever modifications become necessary to interface a computer or

access node to the composite DDP systems, no such change can be allowed to result in a critical dependence upon external resources. Local autonomy, user access priorities, and dynamic reconfiguration would be seriously inhibited if critical nodes are permitted in the architecture.

The requirement for functional autonomy impacts the distributed data base (DDB) organization. Three approaches to DDB organization exist:

(1) each data item has a unique location; or
(2) all data is replicated at every node; or
(3) some data are uniquely located and some data exists in multiple physical copies at multiple nodes. (The only realistic possibility.)

Approach (1) will have high query processing costs, will require modification of subsystem database access programs to recognize virtual data, will substantially increase communications resource utilization, but will also simplify the concurrency and integrity control problems. Approach (2) will incur excessive storage costs and network traffic costs to broadcast and synchronize updates. Approach (3) supports rapid access requirements, functional autonomy, minimal modification to local DB access systems and reasonable storage cost control but presents potential problems in the management of physically replicated data. Our approach is to resolve these problems within the scope of HIS data access integrity and transaction processing requirements. Within this scheme, databases local to each node contain data of two types: locally generated and remotely generated by other processes or users. Retention requirements must be formulated in a globally consistent manner as part of the data administration policies so that locally stored data in a given functional area can support short term, stand alone operation in the event of network or remote node failure.

A major problem that exists at the applications level is the administration and management of the integration process and design. This problem is similar to that of the database administrator, and may admit to a similar solution. The integration activity, and maintenance of the integrated system can become the responsibility of network administrator (NA) (group) who understands the problems existent at all levels, and can mediate conflicting departmental needs for the benefit of the institution. An identification of some potential responsibilities of a NA will be a product of the implementation experience described below. It is clear at this point, however, that the NA should have no control over departmental authority to choose systems best serving their user needs, nor should the NA have authority to establish end-user priorities. NA authority is likely to include granting authorization/access rights to the various DDP resources, establishing resource availabilities, coordinating new applications development and network applications maintenance within a framework of hospital operational policies and procedures established by the chief administrator or a designated group.

Communications System Architecture and Service Issues

For purposes of this discussion, the domain of the communications network subsystem includes all physical and logical communication pathways between distributed resources plus the software and processing resources necessary to manage the communications services provided to user communities connected directly (through direct access nodes such as terminal concentrators) or indirectly (via an intermediary host computer) to the DDP system. The boundary of the communications network subsystem may logically extend into the operating system of each computer node. Data communication protocols, or rules governing data exchanges, are required to structure the communications network subsystem.

The following issues are considered critical to an evaluation of alternative system architectures for a DDP system:

(1) To what extent must component computer-based nodes participate in communications functions?;
(2) Who should be responsible for developing and maintaining local operating system modifications deemed necessary for a given computer system to join the network?;
(3) What networking topology and control structure is most effective for an integrated HIS composed of heterogeneous subsystems?;
(4) How effectively does the system support performance criteria characterized by connectivity, flexibility, responsiveness and growth potential?

There are a number of answers to each question reflecting different sets of objectives. Answering question (3) so as to develop a system with maximum throughput under heavy DDP loads would tend to rule out any flexibility in answering the first two questions. The result might be a very effective network for the hospital at a point in time, but a network that soon began to degrade in performance because of heavy reliance on particular personnel skill levels or version of computer operating system at an ancillary node.

Another approach is to weigh possible answers to the three questions with respect to the component features of a model DDP system. Thus, preservation of subsystem "autonomy" would require a minimum of participation in networking communications protocols at each node and would assign responsibility for the reduced set of operating system modifications to the affected user community. Retaining the ability for dynamic reconfiguration and a high level operating system for coordination of resources under these conditions requires a new entity—a separate network operating system (NOS) [20,21,22]. This NOS functions within the domain of the communications network subsystem but performs all but the simplest communications protocol processing **external** to all subscriber nodes. This appears feasible by selecting a network topology and control structure that incorporate separate computer processing resources and high level communications protocols within the communications network **but external to all host computers and devices.**

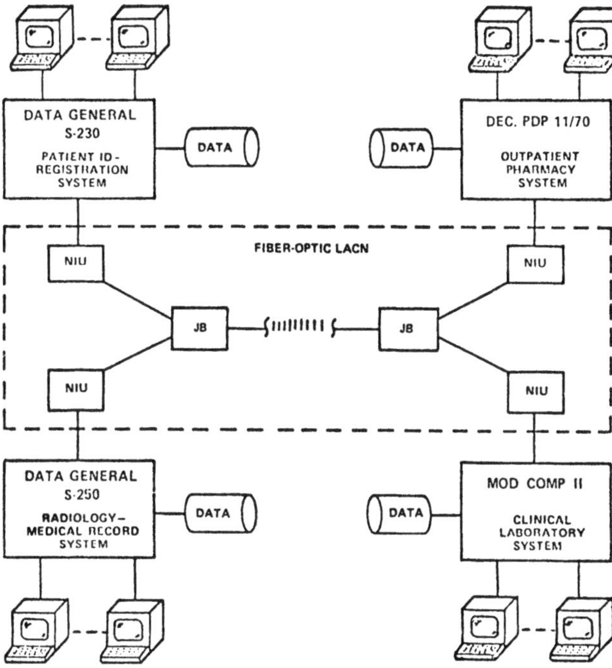

Figure 1. Schematic of UCSF Implementation. Four different computer systems serving different local user communities at the UCSF hospital are being integrated in 1981 via a fiber-optic local area communications network (LACN).

Applications Environment and Design

Introduction

The demonstration project described in this paper involves four minicomputers (hosts) interfaced with LACN support to achieve a level of functional integration that forms an extensible core for a comprehensive HIS. A fifth host computer is interfaced to the network to provide monitoring services for performance analysis and evaluation. Figure 1 identifies the four application hosts and the general system configuration. The system is completely heterogeneous with respect to machine vendor/models, operating systems application, software systems, and local database or file access systems. The geometry of the system is as follows (see Fig. 1). The Patient ID-Registration System and the Radiology-Medical Record System are located in the basement of the UC Clinics/Ambulatory Care Center on the north side of Parnassus Avenue. The Clinical Laboratory System is located on the fifth floor of Moffitt Hospital on the south side of Parnassus. The Outpatient Pharmacy System is located on the twelfth floor of Moffitt Hospital. The distance from the PID system location to the fifth floor of Moffitt is about 400 meters. As indicated in Figure 1, the logical and physical interconnections among the host computers is accomplished by means of an LACN (to be described below). Each host computer system retains its own terminals and its own local database. Although the LACN can support terminals directly, currently all remote access from a given terminal is through that terminal's defined host.

Applications Systems

In this section, we present a brief discussion of the clinical modules involved in the demonstration. The Patient Identification System is the core of all systems in the hospital. Its function is to associate a patient with a medical record number uniquely and consistently at any time or any place in the medical center. Thus, the Patient Identification System file consists of a large number of patients (750,000) with a small amount of information per patient.

The Registration System is the active patient system containing records on patients who have open accounts with the hospital. No patient can receive any services in the hospital unless an account has been opened for the patient. Thus, the Registration System contains a smaller number of patients than the Patient Identification System (approximately 130,000), but considerably more information per patient primarily regarding insurance and payor information.

The Clinical Laboratory System is a general purpose system for managing all laboratory data related to chemistry, hematology, serology and blood bank tests. All laboratory requests are entered into the system and all results are processed through the system. The majority of laboratory results are obtained through direct interfaces with clinical laboratory instruments. The remainder are entered manually. Reports are distributed via hardcopy printed output, as well as through CRT terminal display. Individual patient reports, nursing unit summary reports, and patient cumulative summaries are provided as hardcopy output. Input is generated automatically for the hospital billing system and management statistics are provided.

The Radiology System is a comprehensive radiology department management system. It contains functions for patient login and arrival, x-ray film tracking, patient flow control, management statistics, x-ray reporting, pathology coding, and billing. X-ray reports remain available for on-line display, and retrieval by diagnostic code is available. The coding system utilizes the American College of Radiology coding scheme.

The Outpatient Pharmacy System is to be implemented during 1981. All prescriptions filled at the pharmacy will be entered into the computer. Prescription editing, drug-drug interactions, and allergy checks will occur automatically. Patient drug profiles will be displayed on-line, and prescription refills will be facilitated by the system.

Characteristics of Applications Hosts

Table 1 details the characteristics of the host computers in the application environment, including vendor/model, core capacity, operating system, applica-

Table 1. General Characteristics of Demonstration Host Computers

CPU	Data General Eclipse S/230	Data General Eclipse S/250	Mod Comp IV	PDP 11/70
Core	512 KB	512 KB	512 KB	768 KB
Operating System	AOS	MIIS	MAX IV	IAS
Applications in demonstration	a. Patient Identification b. Patient Registration c. Inpatient Reservation (future) d. Census (future)	a. Radiology b. Medical Records c. Pathology (future) d. Operating Room (future)	Clinical Laboratory	a. Outpatient Pharmacy b. Inpatient Pharmacy (future)
Hours of operation of demo applications	23 hrs/day	23 hrs/day	23 hrs/day	10 hrs/day outpatient 23 hrs/day Inpatient
Uses of CPU other than demo application	None	None	None	Cardiac Cath Lab Research
Location of CPU	A-57	A-57	Moffitt 5	Moffitt 12
No. of application terminals, printers and peripherals	a. 35 OWL CRTs b. 2 Character printers c. 1 Embossing terminal d. 2 196 MB disc drives e. 1 800/1600 BPI tape drive f. 1 line printer	a. 43 OWL CRTs b. 7 Character printers c. 15 Megadata word processing CRTs d. 2 196 MB disc drives e. 1 800/ 1600 BPI tape drive f. 2 line printers	1. 35 CRTs b. 4 Character printers c. 10 Laboratory instruments (including realtime analog) d. 2 256 MB disc drives e. 1 800/1600 BPI tape drive f. 1 line printer	a. 13 CRTs b. 5 Character printers c. 1 Line printer d. 196 MB disc drive
Application Software	UCSF/Fortran	Purchased (CMS)-MIIS	Purchased (CHC)-Fortran	Purchased (Custom)-Fortran

tions, up-time requirements, locations, terminals and peripherals supported and application software source and language. It is important to note that four distinct operating systems are involved, including one MIIS (MUMPS variant) system which is typical in hospital environments. Initially, the interface to each host is a subset of RS 232-C EIA standard running at up to 9,600 baud using only transmit, receive and ground wires. (A parallel interface to the PID system is planned.) The I/O handling for each machine is somewhat different, including input buffer sizes, termination characters, and flow control options.

The AOS is a multiprogramming, multitask system. IAS is multitasking, but each task is a program. MIIS is a timesharing system with each program executing independently. MAX IV is a timesharing system which checks each executing program periodically for status changes. Each system handles communications line inputs differently.

Applications Functional Requirements

The integrated system applications design is implemented by means of transaction exchanges among hosts using services provided by the LACN. Transactions are associated with session types known to the LACN "permanently" or interactively created by some host or network manager. The LACN maintains sessions of transactions among processing nodes. These sessions constitute processing agreements between hosts of differing capacities and functions. A host may honor multiple session requests from remote nodes using a prearranged permanent session structure.

A transaction consists of data and control information in a single message. A session consists of transaction event identifiers and specifications of transaction event interactions for participating entities. Processing attributes of a session may be associated between a host and its network interface unit (NIU) to achieve required processing for a number of applications. For example, a request transaction would contain a "request" identifying code plus data to serve in part as a key or search field. The completion of the transaction session event is specified as a "response" transaction that matches a secondary sequencing field (e.g., patient number). Processing to support this transaction session may be allocated between host computers and their NIUs differently. A host with reduced buffering capacity might require buffering support in the NIU (possibly resulting in reduced throughput). A host with adequate buffering capacity may require processing support from its NIU to time and match returning responses to local requests.

The following transaction types are supported in the demonstration:

(1) **Request:** A request consists of one or more messages containing a request operator and query data. An application may request (patient specific) information from a remote host. The requests in this application are less than 2K bytes and consist of patient unit number check digit, message identifier and from zero to five request modifier fields. For example, a request to the radiology system for x-ray report summaries may specify any (or all) reports for a given patient, all reports for a given date, all reports of a given x-ray type, or a conjunction of these.

(2) **Response.** A response consists of one or more messages semantically and causally associated with a previous request. In this application, responses are paired with requests and are addressed to the requestor. The response to the above request example may be a set of variable length messages. For this application, each such message is less than 2K bytes.

(3) **Unsolicited Data.** This consists of one or more messages that a host wishes to direct to other hosts. Within this application, responses from des-

tination hosts are not required. Updates are transferred by this mechanism. Broadcasts are included in this category, allowing a host to direct a message (stream) to all (or some subset of all) other hosts. New patient data is broadcast by PID/R.

(4) **Control.** These are short messages used to transfer standardized control information, including host non-availability, error conditions and start/restart messages.

Table 2 shows some of the response–request transaction pairs needed to support the applications, along with projected volume by type. Since the request can originate from any of the hosts, twenty-one transaction pairs are represented. The PID/R broadcast will occur about 600 times/day, with a range of 200 to 1,200. The traffic will peak such that 80% of these will occur between 9:00 a.m. and 3:00 p.m.

Transaction processing for the UCSF demonstration requires several kinds of host/LACN processing sessions. Maintenance of sessions will be handled primarily by the LACN. The following session types are supported:

(1) *Permanent virtual circuit.* A number will be provided by the NIU to its parent host to identify a prearranged "circuit" between one of the host's programs and a remote resource. Examples include a general broadcast channel to a subset of hosts or remote processes or an "urgent call" link between two points for conversational traffic. This kind of session provides a permanent assured pathway for independent messages. Options include LACN control of message sequencing and individual message time-out monitoring.

Table 2. R-R Transaction Types

Requests			Expected Volume/Day (Total All Hosts)
To	Requests Request, Modifiers (Optional)	Response	
1. PID/R	None	1. Standard PID/R Record (~ 125 bytes)	50
2. RAD/MR	Date, Procedure #	2. Summary X-ray Report(s)	50
3. RAD/MR	Date	3. Discharge Diagnoses	75
4. RAD/MR	Date, Procedure #	4. Operative Note Diagnoses	10
5. RAD/MR	Date, Procedure #	5. Pathology Diagnoses	10
6. Pharm	Order #, Rx #	6. If no modifiers, entire profile summary. If modified, detailed prescription record	50
7. Lab	Date, Lab #, Procedure #	7. Lab Results	400

(2) *Transaction event monitoring.* Transactions passed through the LACN will be monitored, buffered and processed as permitted by one or more negotiated session control messages. Examples include initiating a timer on numbered request; matching return acknowledgement and remote host receipt; and insuring message sequence and event delimiting/reassembly.
(3) *Unsolicited data transfer.* Any user or process connected to the network may initiate a message to any other network member without prior circuit establishment. Uses include transfer of patient data, conversation, or mail.
(4) *Processing task transfer.* A generic process resource (e.g., accounting) and associated data will be provided to LACN. A servicing process entity will be located and the task will be transferred. Examples include: calculator functions; patient/test scheduling; round-robin data collection.
(5) *File transfer.* A negotiated number of sequenced messages are exchanged between hosts followed by an NIU-to-NIU or Host-to-Host acknowledgement of receipt prior to buffer clearing and additional exchanges. Examples include tape transfer with NIU sequence numbers and program downloads.

Other Requirements

Several issues have been addressed and will be summarized briefly. First, a variety of error conditions (excluding LACN faults) may occur that will prevent normal operations. If a host is temporarily down, messages cannot be delivered and originators must be advised by the LACN. Similarly, a reversal of this condition must be reported. The LACN performs some error detection on messages delivered to it and rejects messages for various errors including length, incorrect transaction or session identifiers, incorrect address, etc.

Flow control is established at various levels in the integrated system, including between the host and its network interface unit. Restore and recovery procedures are defined for handling problems due to network fragmentation, network component failures, and host failures.

To support monitoring and evaluation, the NIU was expanded to support network monitoring software. This documents transaction rates plus errors detected and corrected at all levels of network software. It also records collisions, unresolved errors, and timing events on the network.

Local Area Communications Network

Introduction

The system shown within the dotted lines in Figure 1 is a communications subsystem capable of connecting together heterogeneous computers and computer devices serving different user communities. The communication subsystem provides a range of services to network members. When the networked devices are localized within a geographic area of a few kilometers, such a communications support is referred to as a local area communications network (LACN) [23]. The expected benefits resulting from cooperative use of computer resources by a collection of user communities are substantial. They include reduction of human work load and improved access to and utilization of computing resources.

A network based on an inexpensive approach to local area communications network technology, aimed at providing extensive communications service, has been developed at APL/JHU. This network is called Duplex Bus Communications Service (DBCS) [7]. It uses fiber optics as the communications medium to provide potentially high capacity data communications while permitting ease of installation, physical expansion, and reconfiguration of devices. In addition, fiber optics is relatively immune to electromagnetic interference (e.g., near radiology equipment) and provides enhanced security and reliability. Through the use of microprocessors and comprehensive software support at each network connection, DBCS provides a range of communications services extending from simple message exchange to management of temporary integration of processes in different computers. The physical architecture of the network is reliable with respect to overcoming localized component failures (e.g., link or node failures). The software provides assured and secure data communications. In addition, the DBCS is an "open system" in the sense that new members may be easily added without modification to existing members.

The DBCS networking technology will be capable of supporting a wide range of applications including:

(1) information, data processing, and data acquisition systems integration;
(2) office automation;
(3) electronic mail memorandum routing and retrieval;
(4) remote access to computer graphics, word processing and other special resources or software packages;
(5) distributed database development.

Network Architecture

The DBCS architecture consists of three types of components:

(1) network interface units which permit connection of computers, terminals or other devices to the network;
(2) dual strand fiber optic cable links; and
(3) junction boxes which direct all incoming data on a given fiber (encoded as light pulses) to all other outbound fibers.

The fiber optic links and junction boxes, operating at the same optical frequency, comprise a common communications channel. Figure 2 illustrates a possible network configuration. Computers or other devices such as terminals

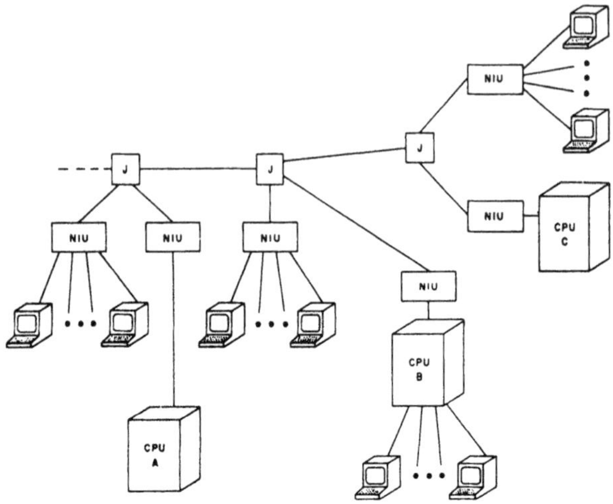

Figure 2. A fiber optic computer network.

or printers are connected to the network by means of the programmable network interface units (NIU). These units are interconnected by means of junction boxes (J) and the fiber optic cables.

A network interface unit transmits data from its host(s) to the receiver of its junction box. The junction box retransmits the data to all of the other outgoing fibers, but not back to the originating source. Since all other junction boxes follow the same rule, data is transmitted to all nodes on the network except back to the sender of the data. Data is passed on the network in discrete packets containing source and destination addresses. Only packets addressed to a given NIU are read by that NIU.

Each network interface unit competes with all others for use of the common communications channel. By listening on the channel until no signal is detected, a network interface unit can determine when to begin transmitting data. Another network interface unit may also attempt to transmit at approximately the same time. This may result in the superposition of two or more messages on the common channel. Such an event, called a collision, is detected by the sending network interface units, which issue a network abort. This abort is achieved by flooding the network with a pulse of light which directs all receiving NIUs to disregard the packet. Each sending NIU then waits a randomly determined delay before attempting retransmission to avoid recurrence of the collision.

Two data packets may exist simultaneously without causing a collision. This situation will exist when two NIUs are sending messages to each other. In this case, the receive lines of these two NIUs do not contain superposed data (however, the receive lines of all other NIUs do contain the superposition of data sent by the two correspondent NIUs). This scheme permits simultaneous exchange, or full-duplex, within the DBCS architecture. (This capability is not used in the UCSF demonstration project.)

In the scheme presented above, each NIU contends with all others for access to the shared communications channel. Thus, DBCS uses a contention bus topology. The combination of listen-while-talk for access and collision detection logic is known as Carrier Sense Multiple Access with Collision Detection (CSMA/CD). The basic CSMA/CD algorithm has been modified in DBCS to permit full-duplex communications. This has been done to improve data transfer efficiency, to permit close coupling between two computers when necessary, and to support real time applications such as feedback control and sensor–processor coupling.

Data communication in DBCS is reliable in two different ways. First, the integrity of bits, bytes, packets, and complete messages transferred across the network is assured. This is accomplished with software embedded in each NIU to perform error detection, correction, and reporting. Second, the basic network is not vulnerable to single-point catastrophic failures, since NIUs on either side of a failed junction box can continue to communicate with all NIUs linked by operative junction boxes. Communications control is distributed among NIUs to avoid or minimize effects of NIU or junction box failures on network members.

Communications functions are performed according to rules known as **protocols.** These functions range from managing the physical transfer of bit streams at the low level through high level functions such as control of integrated and structured data exchanges among cooperating programs in different computers [24–28].

These protocol functions are accomplished by the hardware and software within each NIU. Distributed network control is achieved through use of a common network protocol system design within each NIU. The protocol design is based on the ISO/OSI proposed standard [24].

The design decision to perform most protocol functions within the NIUs rather than within each host computer provides significant advantages, including a resulting simplification of development and maintenance activities. Rather than the need to develop protocols within each computer host for every other host with which it may need to communicate, only one interface (to the network) is needed. Future changes may include protocol revision, changes of computers in various user areas, or changes in the applications designs supported by the network. Rather than implementing new software in each computer on the network to accommodate the new environment, only changes to DBCS software will be required since the protocols are within the NIUs. These changes can be made in a uniform, controlled manner since they will be the responsibility of a network administrator, rather than the responsibility of several distinct user groups needing to cooperate.

Communications Protocols

The protocols are designed to provide various communications services. Simple message exchange without receipt acknowledgement (like a letter dropped in a mailbox) is known as datagram service. The next level of service is called virtual circuit service in which message sequencing and delivery are assured. Finally, to link two computing devices across the network, the content and time reference of messages as well as their associated virtual circuit must be monitored. The administration of such an event-oriented circuit is called session management.

Typically, session management is achieved for a given pair of computers by maintaining separate cooperating programs in each participating computer. The DBCS has been designed to provide session management services via a negotiation process between the initiating computer (or person at a terminal) and its NIU. This NIU then assumes responsibility for establishing and maintaining the required virtual circuits with other NIUs. Permanent sessions may also be defined and implemented in PROM within the NIUs.

Major protocol program modules are depicted in Figure 3 in their relative functional location within an NIU. Briefly, communications protocols functions are performed asynchronously by two separate microcomputer modules. Higher level "management" protocols, such as control of processing sessions between two network hosts and input/output of complete, sequenced messages, are implemented by the host/terminal interface module. Through use of a shared memory for buffering of messages, the network interface module par-

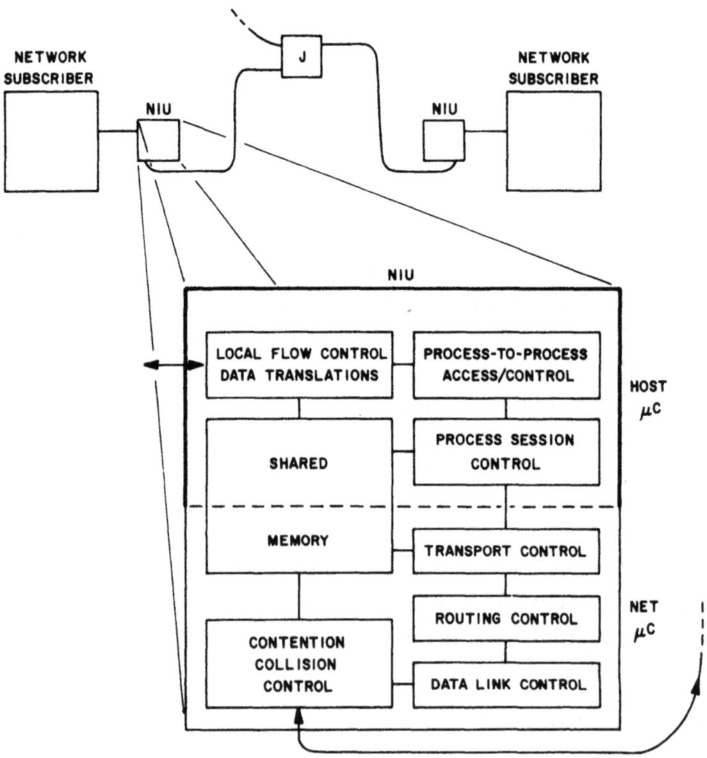

Figure 3. NIU protocol modules.

titions complete messages into smaller, more efficient packets and prepends destination and source addresses. When the network is quiet these packets are then sent (and resent as necessary) to their destination until error-free delivery is acknowledged by a return message.

The following is an example of NIU protocol operation, involving a user at a terminal and a remote computer. Following conversion from the terminal's data code to a single network code, a connection and processing session between the user and the computer is negotiated across the network using the process-to-process and session control protocol modules in the terminal and computer NIUs. Following this establishment phase, data between the terminal and computer are exchanged interactively under the control of the lower level protocol modules. The transport control modules ensure end-to-end accuracy and correct sequencing of messages within a process session. Routing and data link modules ensure that data are properly addressed and accurately delivered to destination NIUs. The contention/collision control modules regulate both the access to the network and the retransmission of packets that collide on the network. Data bit errors occurring during packet transfer between two NIUs are detected by continuous packet collision detection and a standard CRC error detection algorithm. Retransmission of aborted packets is attempted a preset number of times and correction of data packet errors is achieved with a dynamically-weighted random backoff and retransmission algorithm.

Flow control is achieved through a combination of buffer status counters appended to control packets, and an inhibit on further transmission on receipt of a buffer overflow NAK. Unless such a NAK has been received, packets are sent without negotiation for space. Transmit buffers are cleared quickly with priority ACKs. Briefly, during transmission of a packet, all nodes except the destination node are inhibited from acquiring the network at the end of the transmission for a preset time. The destination node sends a priority ACK or NAK during the inhibit time.

To permit network management and tuning, a set of network parameters treated as variables within protocol software have been implemented. These parameters include: abort duration; access inhibit duration; buffer full time-out; resend time-out (data link level); number of permitted resends, etc.

For the initial implementation the operating characteristics were chosen as follows. The total address space for each DBCS is 16 bits. The allocation of this space reserves the first 8 bits for NIUs, thus permitting up to 256 physical address nodes. The remaining 8 bits are allocated as secondary addressing for sessions, ports, and processes. Each NIU may serve up to four subscriber ports, each of which may serve multiple sessions and processes subject to the total address limitation. This implementation of DBCS operates reliably at 0.5 megabits per second in half duplex or simplex modes. This characteristic speed resulted from a decision to use off-the-shelf components and a simple design for the network interface board. Based on the family of fiber optic components chosen the maximum distance between junction boxes may be up to 1 kilometer. The maximum network packet size was chosen as 256 bytes based on an analysis of optimal packet size, and to promote X.25 compatibility. Assured delivery of full-screen transfers (2K bytes) is provided. None of these characteristics reflect fundamental limitations of the DBCS architecture.

Through login negotiations and exchange of control messages as necessary between requesting application (user), local NIU, and destination NIU, the following types of communication services are offered by the initial implementation:

(1) Data transfer;
(2) Broadcast/Multicast (sub-networking);
(3) Virtual circuit;
(4) Encryption/Decryption;
(5) Session control of transactions. Additional services such as remote job entry and file transfer may be available following evaluation of the basic services.

In terms of the integrated hospital system application, Figure 4 is an illustration of the functional relationship of the NIU to the parent host and its input/output handlers, operating system and data calls. Each of the four hospital computers serves its local user community in a dedicated or time sharing mode of operation. For an active process running on the host computer, calls for remote data or other messages to remote hosts are accomplished by building a transaction and writing to the NIU. Transfers are made across a physical and data link path between the NIU and its host under the interface flow con-

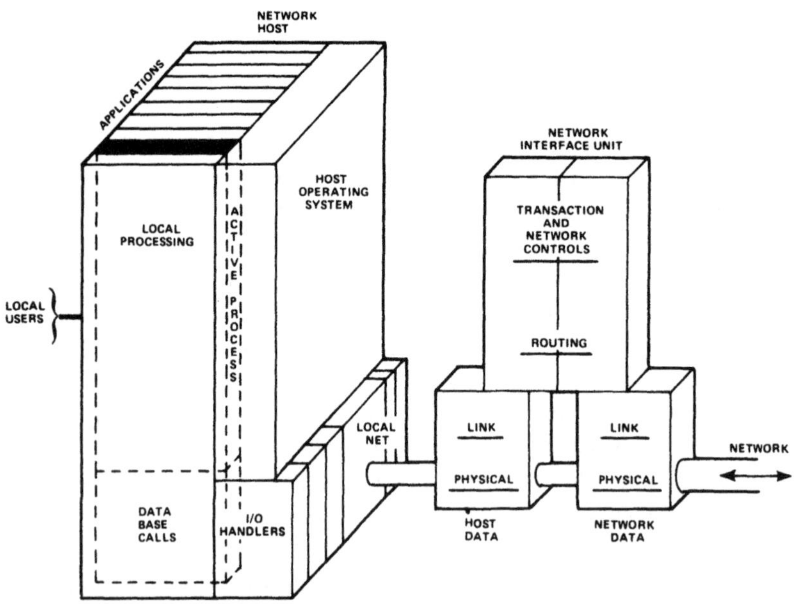

Figure 4. A functional node.

trol protocol for that host. Upon acceptance of the transaction the NIU then assumes responsibility for a proper response or an error message. On the return path, the NIU signals the host I/O handler when a transaction addressed to a resident of the host is available for transfer. When signalled that the host is ready to receive, the NIU makes the necessary code/format changes and transfers the transaction.

The fiber optic LACN described above has been implemented at UCSF to support a core distributed HIS as described under Application Environment. Five NIUs (one for the monitoring node) two junction boxes and approximately 1,000 meters of dual fiber optic cabling are used. The cost of this implementation of the LACN, including NIU interfaces, NIU software for transaction processing, installation and checkout is on the order of $100,000.

Summary

An applications design and generalized local area communications network technology are being implemented to achieve a flexible distributed hospital system capability. The approach integrates heterogeneous stand-alone computer systems within an expandable, reconfigurable architecture while maintaining the autonomy of the individual systems. The applications design, LACN physical architecture, fiber optic communications medium, communications protocols and network interface units have been reviewed. Currently, the application integrates a Patient ID/Registration System, Outpatient Pharmacy System, Clinical Laboratory System and Radiology/Medical Records System. Planned extensions of this initial core implementation, and an evaluation of the project will be reported in future publications.

References

[1] Tolchin, S. G. and R. L. Stewart, The Distributed Processing Approach to Hospital Information Processing, *14th Hawaii International Conference on Systems Sciences,* January 1981.

[2] Tolchin, S. G. and R. L. Stewart, The Distributed Processing Approach to Hospital Information Processing, *Journal of Medical Systems* (to be published).

[3] Stewart, R. L. and S. G. Tolchin, A Distributed Processing/Fiber-Optic Hospital Information System, *Fourth Annual Symposium on Computers in Medical Care,* November 1980.

[4] Tolchin, S. G., B. I. Blum and M. A. Butterfield, A Systems Analysis Methodology for a Decentralized HIS, *Fourth Annual Symposium on Computers in Medical Care,* November 1980.

[5] Zeichner, M. L., P. J. Brusil and S. G. Tolchin, Distributed Processing Architecture for a Hospital Information System, *Third Annual Symposium on Computers in Medical Care,* October 1979.

[6] Whiting-O'Keefe, Q. E., et. al, The Argument for Modular Distributed Hospital Information Systems, *Fifth Annual Symposium on Computers in Medical Care,* November 1981.

[7] Kahn, S. A., R. L. Stewart, S. G. Tolchin and S. J. Healy, Functional and Logical Description of a New Fiber-Optic Contention Bus Network, *COMPCON Fall 80,* September 1980.

[8] Wood, D. C., et. al., "A Cable-Bus Protocol Architecture," *Sixth Data Communications Symposium,* November 1979, pp 137–146.

[9] Carpenter, R. J., et. al., "A Microprocessor-Based Local Network Node", *Proceedings IEEE COMPCON,* September 1978, pp 104–109.

[10] Metcalfe, R. M. and D. R. Boggs, "Ethernet: Distributed Packet Switching for Local Computer Networks", *Comm. ACM,* July 1976, pp 395–404.

[11] "Policy Implications of Medical Information Systems", Office of Technology Assessment, U.S. Congress, Government Printing Office, Stock No. 052-003-00496-8, Washington, D.C., 1977.

[12] Komaroff, A. L., "The Variability and Inaccuracy of Medical Data", *Proc. IEEE,* V. 67, No. 9 Sept 1979, pp 1196–1207.

[13] Payne, L. C. and P. T. S. Brown, *An Introduction to Medical Automation,* Second Edition, J. B. Lippincott 1975.

[14] Giebink, G. A. and L. L. Hurst, *Computer Projects in Health Care,* Health Administration Press, 1975.

[15] Baker, R. L. and W. A. Spencer, "Considerations for Evaluating Transferable Automated Information Systems", *Health Information System Evaluation,* Colorado Associated University Press, 1974.

[16] Lewis, J. W., "Commentary: Clinical Laboratory Information Systems", *Proceedings of the IEEE,* Vol. 67, No. 9, September 1979, pp 1299–1300.

[17] Zachman, J. A., "The Information Systems Management System: A Framework for Planning", *Data Base,* Winter 1978, pp 8–13.

[18] Bernstein, P. A., J. B. Rothnic and D. W. Shipman, "Tutorial: Distributed Data Base Management", IEEE Cat No. EHO 141-2, IEEE Computer Society, Long Beach, CA, 1978.

[19] Chu, W. W. and P. P. Chen, "Centralized and Distributed Data Base Systems", IEEE Cat. No. EHO 154-5, IEEE Computer Society, Long Beach, CA, 1979.

[20] Forsdick, H. C., et. al., "Operating Systems for Computer Networks", *Computer,* January 1978, pp 48–57.

[21] Donnelley, Jed, "Components of a Network Operating System", *Fourth Conference on Local Computer Networks,* October 1979, pp 1–12.

[22] Poulos, J. C. and A. N. Beavers, "Network Operating Systems: A Concept and Protocol", *Proceedings IEEE Trends and Applications:* 1980, pp 64–69.

[23] Clark, D., et al., "An Introduction to Local Area Networks", *Proceedings of the IEEE,* Vol. 66, No. 11, November 1978, pp 1497–1516.

[24] "Data Processing—Open Systems Interconnection—Basic Reference Model", ISO/TC 97/SC16/N537 (DP 7498), American National Standards Inst; New York, N.Y.

[25] McQuillan, J. M. and V. G. Cerf, "Tutorial: A Practical View of Computer Communications Protocols", *IEEE Computer Society 1978.*

[26] Stack, T. R. and K. A. Dillencourt, "Protocols for Local Area Networks", *Proceedings IEEE Trends & Applications: 1980,* May 1980, pp 83–93.

[27] Pouzin, L. and H. Zimmerman, "A Tutorial on Protocols", *Proceedings IEEE,* Vol. 66, No. 11, November 1978, pp 1346–1370.

[28] Sproull, R. F., D. Cohen, "High-Level Protocols", *Proceedings of the IEEE,* Vol. 66, No. 11, November 1980, pp 1371–1385.

10

Hospital Information Systems Planning*

George Kolenaty

Introduction

One of the most important support functions of any goal-oriented organization is its information system. This can be viewed as consisting of the organizational data base, an information distribution net, and an information processing facility. In hospitals, such information systems have traditionally been implemented through manual methods of information processing. A manual information system typically utilizes paper as its information transfer media; mechanical files as a physical representation of the data base; and telephones, pneumatic tube, or messenger services as methods of information distribution. People are used as information processors, performing a variety of tasks such as data recording, storing, sorting, transcribing, duplicating, retrieving, and computing.

An information system based on a manual mode of operation, when contrasted to other available alternatives, may have the following drawbacks:

It may be too costly on a "per-transaction" basis.
It is inherently slow.
It can have aspects of redundancy. This characteristic is eventually reflected in the unit cost of processing.
Because of the high dependence on the human element, it is inherently error-prone.

In the last decade, a general trend of steady decreases in average period of hospitalization developed. This resulted in a rather sharp increase in "intensity of care," as the required medical procedures had to be compressed into increasingly shorter time spans.

As the intensity of services continues to increase and the pace of organiza-

*This chapter has been modified from a version previously published in *Proceedings of the Fourth Annual Symposium on Computer Applications in Medical Care,* IEEE, 1980.

tional interactions between various units accelerates (in acute care type hospitals, for example), the demand on information availability and quality increases proportionally. Manual methods of information processing may thus at some point actually impede the growth of organizational effectiveness. To alleviate the problems inherent in manual systems and to meet the increased information demand, hospitals have been for some time evaluating and implementing EDP-based methods of information processing. Over time, the term Hospital Information Systems or HIS has become synonymous with the use of EDP methods in hospitals. It will be used throughout this chapter in such context.

Categories of HIS

While it is possible to classify HIS using many different criteria, the three classification categories below appear suitable from the planning viewpoint as each calls for a different level and type of planning effort.

Category 1: This includes stand-alone, usually single-purpose systems that-may support one or several primary functions, for example, Blood Analysis or Pharmacy, plus some additional auxiliary functions such as Admissions or Patient Registration. Such systems are often supplied as "turnkey" systems by specialized vendors.

Independent, stand-alone operation of several such systems may present the problems of processing redundancies and inconsistencies in data definitions.

Integrated operation, on the other hand, may require incremental investments in interfaces to accommodate hardware, software and application design differences of various suppliers. Such expense can be minimized or avoided through advance planning.

Category 2: This includes highly modular, component-oriented designs, developed with the integration capability in mind. Depending on hospital needs, the individual components can either operate independently or can be assembled into larger agglomerations. Offerings of this type are available on "ready-made" basis, from several major vendors. In such case, many of the HIS planning decisions have already been done by the vendor. Alternatively, the hospital can acquire some of the components through internal development and combine these with specialized, "ready-made" applications from Category 1. This approach requires that the decisions on the overall HIS acquisition strategy be done well in advance.

Category 3: This includes architecturally robust, monolithic, and simple batch-oriented processing. The planning and implementation of such systems most likely involved interactions of EDP personnel with the limited number of staff from the controller's and administrator's office. The usual method of acquisition typically involved selection and some customization of a ready-made software package, with the resulting EDP application serving the information needs of a single group of users, mostly from accounting and general administration areas.

With most hospitals presently either completing or having completed the automation of their business functions, the 1980s will be characterized by a large number of hospitals venturing into the area of automating the information processing functions that directly support the patient care delivery, such as those shown in Table 1, Sector I. Because of the logical precedence relationships inherent in the flow of information, automation of functions in Sector I should precede those in Sectors II, III and IV. Through proper planning, functions in Sectors II and IV can be obtained at relatively small incremental cost as a byproduct of automating the Sector I.

Such strategies, however, will be somewhat more demanding in terms of the

Table 1. Overview of Hospital Information Processing Functions

	Production-Oriented Functions	Management or Decision-Oriented Functions
Patient Care Services	I. Retrieval of patient medical history; Communication of diagnostic/ therapeutic order information from patient care process to support services; Communications of diagnostic test results from support services to patient care process; Order status monitoring; Patient monitoring and general communication of patient status; Patient medical history update	II. Patient management; Quality assurance-patient care services; Medical decision-making; Research and development of medical knowledge
Support Services	III. Financial transaction processing	IV. Resource planning, allocation, and utilization monitoring: Manpower Facilities and equipment supplies; Operations management; Financial management; Quality assurance-support services

amount and type of planning as they will require that a number of various choices be evaluated in advance.

Technically, this development calls for utilization of date communication and data base management systems. These two technologies alone require a significant amount of planning as the incorrect choice of common support software systems may adversely affect the operational characteristics of an information processing system for years.

The situation is even more perplexing from the functional viewpoint. The patient care delivery process involves chains of physical processes densely interwoven with clinical decision points. The clinical and operational effectiveness of the overall health care delivery process depends heavily on a timely supply of information to these decision points. There are much more stringent requirements on system's availability and general reliability than was the case with the business-oriented systems. Such requirements are demanded by a wide variety of professional users (M.D.'s, nurses, technicians) that perform a broad repertoir of highly diverse operations directly supported by the HIS. These users have generally little tolerance to deficiencies in data quality, particularly if it means reduced level of service as compared to existing, manual or EDP systems.

HIS Planning Concepts

In the majority of hospitals the manual methods are still a predominant mode of information processing in the areas of patient care. For reasons stated earlier, this area can be considered the primary, general HIS target in the 1980s. In order to successfully serve the information needs of hospital users, certain characteristics and properties of the existing (manual) information systems must be *defined* and *analyzed* before they are *replaced* by an alternative (EDP) system. The objective is to ensure that the replacement system will provide at least equal and hopefully better level of service at less cost than the existing one. Exploratory analyses of this type are typically performed during the HIS planning phase.

In order to identify some of the planning issues of strategic importance to HIS implementation, let us examine what is actually involved in installing a computer-based information system in a hospital environment. The actual implementation of an HIS involving multiple functions is most practically done in semisequential fashion, where individual HIS components are installed and integrated one or several at a time. By contrast, the HIS planning process must consider all information processing functions globally. The primary reason is the need to ensure a high degree of integration and compatibility between individual functions and to ensure efficient utilization of available HIS project resources during the implementation phase.

To illustrate this concept, consider the following situation. During the *HIS planning* process, we identify through operational analysis a set of functions F, currently performed through a method A. Some of the functions identified in set F can also be performed by an alternative method B, yielding various types of benefits in the process. If we construct an objective function O which assigns different relative measures of importance to various benefit categories yielded by method B, we can then evaluate F through O, and identify a subset of functions F', which will "optimize" the overall objective criteria as expressed by O. This process will result in elimination of the marginal functions from further consideration for automation.

During the *HIS implementation,* the existing methods of processing A in subset F' are replaced through methods B, which are in effect "implanted" into the overall body of organizational processes and "linked" with the remaining, manually supported functions. Depending on whether the hospital has already implemented some EDP systems, the replacing methods may also have to be integrated with the existing technological environment.

Assuming that an HIS objective function has been defined, the two problems that must eventually be addressed by the planners are:

Identification of the set F and selection of the optimal subset F' ("automatable" functions).

Identification of available HIS acquisition alternatives and selections of the "best" HIS replacement (method B) for F'.

The primary objective of HIS planning then is to research and resolve the various issues associated with this replacement process in order to select the best replacement strategy that maximizes the overall objective.

If we visualize the process of HIS implementation as a route starting at the point A and ending at point B, we can identify a series of decision points with alternative paths or "intersections" at which the HIS project manager must decide which branch to take. As each possible "route" will eventually result in different cost/benefit and success/failure outcomes before reaching the "finish," it may be desirable to make such a trip in advance, and use a small group of planners to quickly map the territory before sending out the entire implementation team. The movement of the implementation team can then be guided on strategic intersections by information concerning the correct path selection.

This is what HIS planning essentially entails. In this chapter, planning is viewed simply as a process that generates strategic type information on which major implementation decisions can be rationally based. Through availability of such data, the risks of HIS project failure hopefully are reduced and opportunities for success increased.

The HIS planning efforts should be justified as any other investment—on their own merits. An organization should be better off with the HIS plans than without them or, in other words, the planning activity itself should be cost effective. Studies of life-cycle costs for large EDP projects indicate that there is indeed a very good case for doing advance planning studies. Figure 1 illustrates the cost of correcting a design or implementation error over the life cycle of an information system. As indicated by Curve A, the longer in the system's life cycle an error is detected, the more it costs to correct it. At the same time,

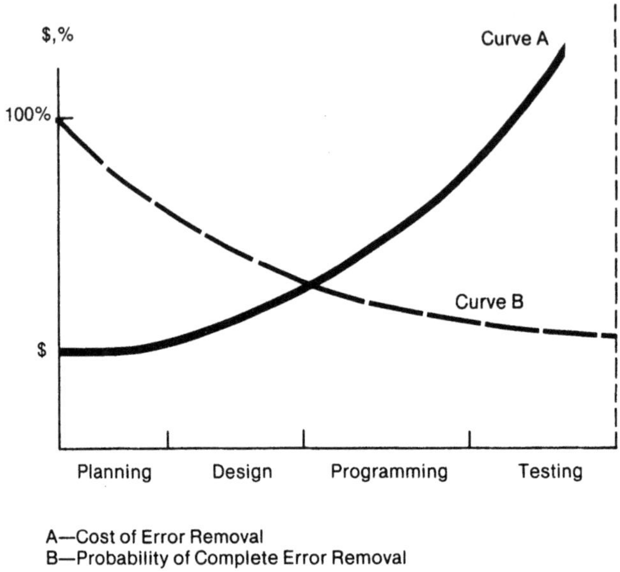

Figure 1. Error correction during information system life cycle.

it is less likely that the error will be corrected properly (Curve B). Also, once a "critical mass" of design errors has been accumulated, it may be impossible to recover.

Objectives of HIS Planning

Our chances for success in developing good quality HIS plans can be increased if we specify up front what we are about to do, or in other words, if we define the objectives of the planning process itself. The following statements are offered as suggested areas of focus of an HIS planning process. They should be, however, reviewed and correspondingly tailored to the needs and situations of individual institutions.

Statement of the HIS Planning Objectives (Example)

The purpose of the HIS planning activities is to outline a *long-term strategy* for the improvement and development of information systems in the area of patient care, teaching, and research as well as in the administration of these areas. The specific objectives of the HIS planning process are:

To identify those information processing functions and processes that would benefit from methods improvement through the use of EDP.

To define the information processing functions in modular units in order to facilitate concurrent design and implementation.

To identify the general requirements of each module so as to ensure a high degree of integration regardless of the method of systems acquisition (internal/external).

To quantitatively and qualitatively predict the outcome of automation projects through milestones within the systems acquisition process so as to have a measure of achievement and to limit the risks in misuse of resources.

To ensure information systems compatibility with planned developments in medical and administrative areas.

To achieve top management review, and general management participation, in the design and implementation process.

To provide cost-benefit analyses for decision making and budgetary schedules for funding.

To plan for acquisition of number and type of human resources required to support the HIS implementation.

To develop long- and short-term schedules for the general management of internal or external acquisition of EDP capabilities.

HIS Planning Methodology

The process of developing an HIS long-range automation plan summarized in Figure 2 is based on a general methodology for problem solving. Important considerations, relevant planning techniques and tools applicable to individual elements of the process are discussed below.

Definition of HIS Objectives and Selection Criteria

This step is one of the most crucial if not the most significant step in both planning and implementation of the HIS as it addresses two fundamental question:

Why do we want to install an HIS? and
Which areas do we want to automate?

The selected objectives will eventually direct resource allocation and guide potential conflict resolution between various organizational units, competing for limited HIS resources throughout the HIS planning and implementation. There are several important considerations related to this step.

A well-organized, consensus-making entity, i.e., HIS planning committee, with representation at all levels and specialties is crucial to any meaningful progress. An example of one such arrangement is shown in Figure 3. The HIS objectives should be agreed to and set by such entity. The functions of the planning committee itself must be well defined and understood. One principal function of the planning committee should be to oversee, review and approve the work done by the planning team.

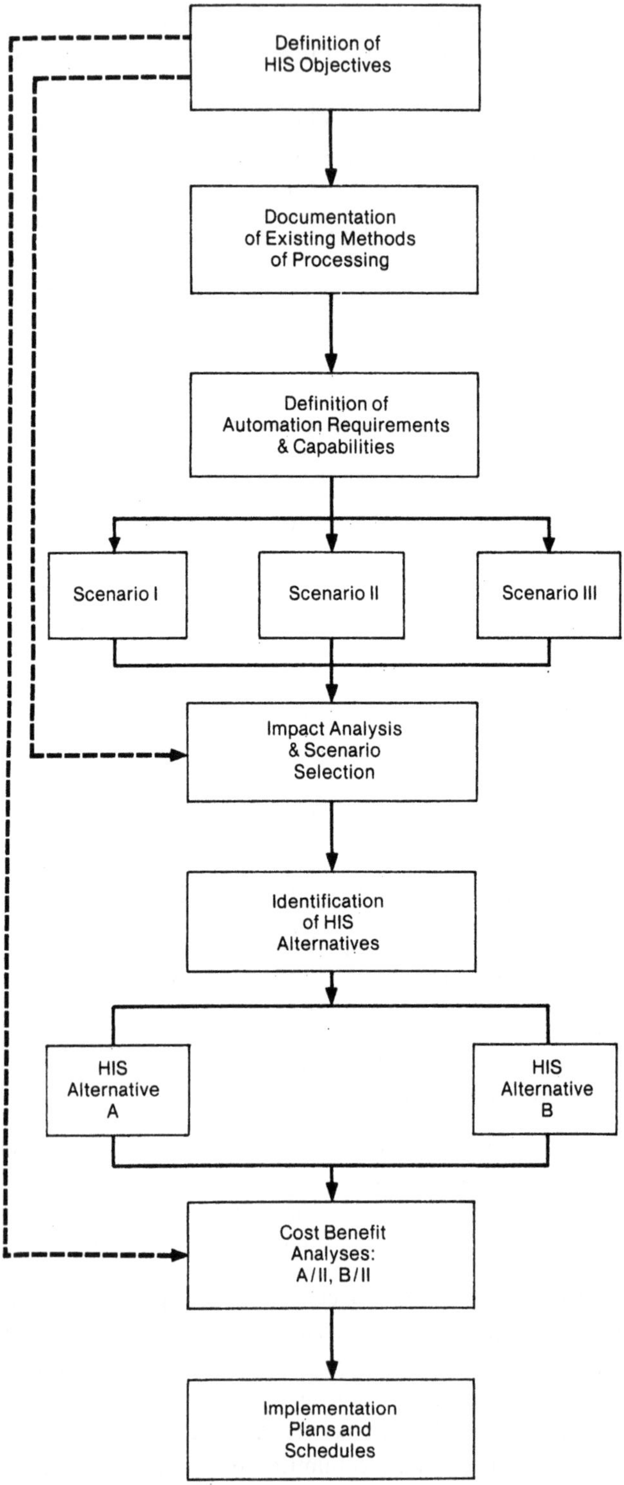

Figure 2. HIS planning methodology—overview.

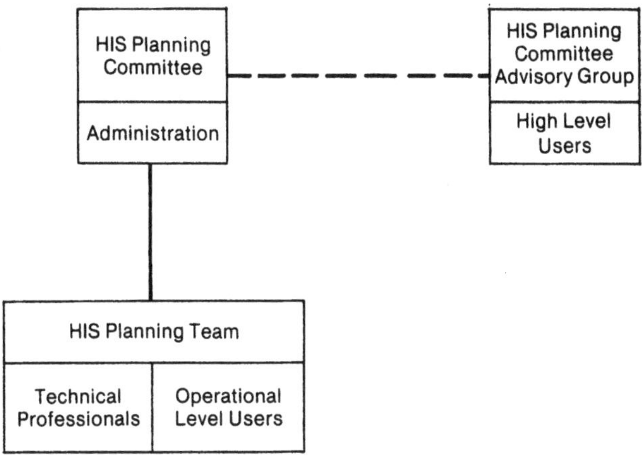

Figure 3. Organization of HIS planning process.

Agreed-upon objectives need to be formalized and expressed in quantitative terms. For example, " . . . improve the utilization of the Operating Room through better scheduling methods by 10% within one year . . ."

To do such a job effectively, the goals and objectives set by the planning committee should be based on and validated against actual operating characteristics. For example, an objective to improve error rate on medical prescriptions should be related to the actual, measured error rates.

If multiple objectives are formulated, relative weights or priorities must be assigned to facilitate the process of final selection of those functional areas most desirable for automation.

In addition to controllable objectives, the administration and the planning committee need to consider other factors, for example:

Logical precedence relationship between individual information processing functions.

Complexity of a specific information function, availability of funds and skilled resources, and a probability of successful project completion.

Availability of ready-made systems and resulting ease of implementation.

As these additional factors may influence the final selection of automation target functions, their relative weight should be formally expressed in the overall selection criteria.

Qualitative and Quantitative Documentation of Existing Information Processing Functions

The replacement HIS should not only provide additional new capabilities, operationally or economically not feasible with the manual methods, but it should also provide the capabilities currently performed, automatically or manually. The documentation of functionally comprehensive designs. A system of this type may actually contain more functions than a hospital may need at any specific point in time. Nevertheless, systems of this type must be installed in its entirety, with majority of functions installed on day one. This HIS type requires the most complex planning.

Benefits of HIS

Most health care institutions are interested in HIS implementation primarily for:

Economic benefits. These are attributable to the proven concept of substituting ongoing labor expenses through capital investments—in this case in EDP technology. HIS can replace labor-intensive information processing functions and allow transfer of human resources from these manual processes to other tasks. Depending on hospital policy, the freed resources can be either entirely removed from the production process, or "plowed back" to patient care.

Improvement in information quality and related improvements in quality of care. In specific situations, the improvements in quality or availability of information may actually improve the quality of patient care. Illustrative situations may involve, for example, automatic validation of diagnostic test results against normal value ranges or patient vital signs monitoring.

Improvement in management control and organizational effectiveness. Improved management control is made possible by the availability of "real time," control-type information. To affect productive organizational activities, correct decisions must be made consistently at various organizational levels. Correct decisions, business or patient care-oriented, depend in turn on the quality of information available at the time a decision must be made. A properly designed and properly utilized management information system can thus contribute to organizational productivity gains, far outweighing those attributable to method changes alone.

Since most health care delivery processes are interrelated and highly information-intensive, it is possible, through careful analysis and design, to fulfill a multitude of objectives at a relatively small incremental cost over the cost of meeting a single objective. One of the reasons for developing the long range HIS plan is to maximize this opportunity.

Information Processing in Hospitals

Before we begin to discuss the issues and processes of HIS planning, let us briefly review the principal hospital information functions and information-intensive processes that can benefit most from EDP methods. Since the major hospital function is the production of *patient care services,* or health care delivery, it is appropriate to begin by examining a simplified model of such a process. Several major steps, each heavily supported through information processing activities can be identified:

First, it is customary for a health care practitioner to retrieve and review a patient's medical history in order to determine possible links of the current disorder with those from the past. At the same time, current disorder is documented for future reference, using symptom-oriented descriptors.

Drawing on personal knowledge and past experience, the health care practitioner formulates a hypothesis about the most probable cause of current disorder (diagnosis). Various diagnostic services are then ordered to confirm or to reject the hypothesis.

Having properly diagnosed the disorder, the practitioner formulates a treatment program, which may consist of predetermined sequences of: (1) Therapeutic services; (2) Diagnostic services to periodically validate the effectiveness of the therapy program.

At the point when the disorder is eliminated and the treatment terminated, it is beneficial to update the patient medical history and document in detail all medical decisions and actions taken during the period of hospitalization. This accumulated data provides a valuable learning base for improved disorder diagnosis and therapy in future episodes.

Other hospital functions also heavily supported by information processing are *research and teaching*. The data base generated by the patient care process contains relationships between actual diagnoses and therapies accumulated across a large patient population and time period. It is one of the primary and most valuable resources utilized by researchers to identify and calibrate new medical knowledge. It is also used by medical students to absorb the accumulated medical knowledge across many medical cases. As the EDP technology can actually broaden the scope of access to medical data bases to both researchers and students, it may also accelerate the process of new knowledge generation.

Additionally, the process of patient care delivery, and to some extent teaching and research, require that a variety of support services such as engineering, laboratories, dietary, etc., be provided on demand in required quality and quantity. *Hospital administration* therefore involves monitoring service demand, determining resource requirements and allocating resources to meet fluctuating demand, ongoing monitoring of operational efficiencies and ensuring the financial viability of the entire organization. Information systems play an essential role in support of these critical management functions.

Overview of HIS Functions

Information processing activities constitute an integral part of overall hospital operations. Upon closer examination we can identify two principally distinct types of information processing:

Processing to perform specific tasks, for example, update a temperature chart or update an accounting ledger. This type of processing will be referred to as *production-oriented.*

Processing to supply data to a decision branch, either in a physical or in an information process which itself could be production-oriented. We will refer to this type of processing as management or *decision-oriented.*

Applying this classification, we can develop the taxonomy of major information processing functions in a typical hospital environment and use it to review past automation efforts and to formulate future HIS implementation strategies.

The four sectors shown in Table 1 represent the major target opportunities in automation of hospital information processing. The actual HIS implementation strategy of a particular hospital will depend on (1) the existing level and type of hospital automation and (2) hospital's HIS objectives. To provide the necessary background for the formulation of HIS acquisition strategy, the past, present and future HIS developments will be briefly reviewed.

Past, Present and Future Developments in HIS

Computer-supported methods of information processing, storage, and distribution have been applied in hospitals throughout the late 1960s and 1970s without much formal planning. The initial automation efforts involved mostly the administrative and business functions represented in Sector III, Table 1. This development has occurred because of several factors. First, there exists a reasonably high degree of functional similarity between, for example, hospital payroll, accounts payable, and general ledger systems and those found in other industries. As other sectors of the economy installed and operationally tested such systems earlier, hospitals in general had very little to risk. Second, these functions typically contained a high concentration of clerical labor, repetitive in nature, and were, therefore, highly attractive targets for cost reduction programs through automation.

From the EDP planning viewpoint, these functions are positioned off the mainstream of hospital operations, with most of the patient-related data processing activity performed after discharge. The most common EDP method used in these systems is relatively existing information processing flows and its subsequent analysis is therefore important from several aspects:

It provides additional HIS objectives validation and calibration feedback to the planning committee.

It provides a comparison baseline for "before/after" analyses.

It provides an authoritative, common source of data for HIS requirements definition process and general communication.

During this step, information processing flows are documented by means of operational flow charting techniques. Both physical processes and information

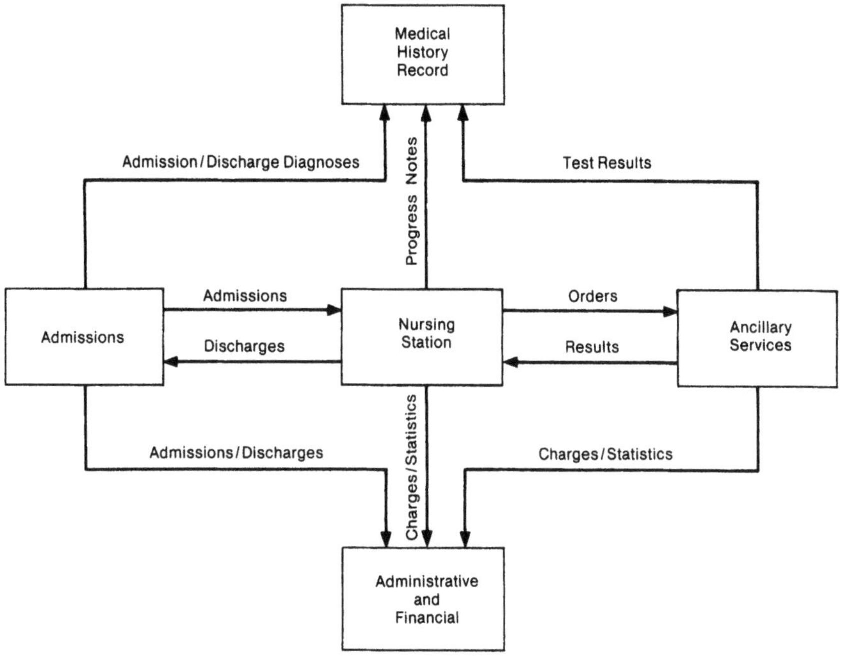

Figure 4. Simplified model of hospital information flows.

flows should be documented, preferably on the same chart for later analysis. Processing steps, timing aspects, frequencies of occurrence, transfer distances, processing delays and information elements processed are indicated on individual process flow charts.

Additionally, a macro-level diagram should be developed through synthesis of developed data to provide a global model of overall information flows throughout the organization. A simplified example of such a model is shown in Figure 4.

Various indicators can guide planning groups' decisions on how to best allocate their own resources to various functional areas during the documentation phase. For example, if the stated objective is labor cost reduction, a Payroll report on distribution of labor costs by type and functional area is a quick source of data that can be used for this purpose. Such indicators should be used, however, with caution and only for preliminary selection of high-potential functions.

Requirements Definition

Using the techniques of methods analysis and the documentation of existing methods of information processing, the planning team can develop a statement of functions, automation of which is desirable based on stated objectives and selection criteria. This process involves identification of "clusters" of "automatable" information processing functions using developed documentation as a base, and draws on familiarity of planners with operational capabilities of EDP technology. By varying the design factors, the planning team can conceptualize new methods of overall hospital information processing. Several such "scenarios" can be prepared for consideration by the planning committee.

Impact Analysis

Using the several scenarios developed in the previous step and documentation of the existing processing methods, the planning team can now proceed with the impact analysis. This process involves a quantitative and qualitative assessment of projected operational performance achieved under different scenarios against stated objectives and current operating characteristics. Using basic decision analysis techniques, the most preferred scenario selection can be derived either on a basis of maximum expected payoff, minimum expected risk, or other criteria. There should be sufficient "contrast" between the projected scenario performance and stated objectives to allow for estimation and projection errors. Also, if the projections for several alternatives appear mutually "fuzzy," additional data may have to be collected to improve the resolution between the alternatives. After review and approval of the recommended design by the planning committee, the next step involves identification and evaluation of alternatives for acquiring the EDP capabilities required to support the selected scenario.

Identification and Evaluation of Alternative Methods of HIS Acquisition

Numerous alternatives of acquiring the specified EDP capabilities are available to the planning team. During this process, the *total investment* outlay and the *annual investment flow* will most likely be the important determinants in planning the HIS acquisition strategy. The total, economically warranted investment level can be determined initially through the use of payback method, and then fine-tuned using the Return on Investment and Discounted Cash Flow analyses (note: A conservative "rule-of-thumb" for estimation of the total, economically warranted HIS investment is to multiply the difference in annual operating costs (Before/After) by a factor of three (3).

The total warranted investment level can be used, along with other criteria, to initially eliminate those alternatives that fall outside of the acceptable range.

The annual investment flow will probably vary with each hospital, depending on the hospital's ability to finance the HIS acquisition program, either through internally generated cash flows (savings), or through borrowing. As there is almost always a requirement for some "seed money," well-laid plans are often instrumental in obtaining required financing from the commercial sector or third parties.

The identification and evaluation of commercially available HIS products and services is a tedious and time consuming task, particularly when an HIS scenario must be integrated into existing technological environment. Various consulting services specialize in this field to assist hospitals in the process of HIS selection and acquisition from commercial sources.

Cost-Benefit Analysis

Once the HIS alternatives have been reasonably narrowed down, it will be necessary to undertake detailed analyses of costs, risks, and benefits for each alternative. If internal systems development is preferred, allowances for the learning curve effect should be made in cost and benefit estimates. Operational cost savings can be estimated using predetermined or measured time standards applied to "before/after" scenarios of information processing. If the future HIS is to be integrated into an existing EDP environment, allowances should be made for cost of integration and development of interfaces.

Implementation Plans and Schedules

This step addresses two additional strategic planning issues:

How fast can or should an HIS plan be implemented?
In what sequence should the implementation take place?

Using the most preferred HIS scenario as a base, development of specific implementation plans, paths and schedules depends, besides other factors, on the annual feasible rate of investment flow. In addition to *selection* of HIS functions, consideration should be given to *sequence of implementation* of the various functional modules. This depends on several additional factors such as:

Logical precedence relationship between functions implied by the natural flow of information and physical processes
Availability of ready-made HIS components and skilled personnel
Existing EDP environment and level of automation
Probability of success based on relative complexity of information processing functions and selected mode of HIS acquisition.

As each implementation sequence will affect differently the rate of accrual of projected benefits and at the same time will generate different requirements in terms of investment spending levels, individual schedule variations should be subjected to extensive quantitative analyses of the total strategy effect on cost/benefit characteristics.

Regardless of the method of HIS acquisition, the overall risks of an HIS implementation project can be minimized by using the "component-oriented" approach. The "monolithic" designs carry probably the highest relative risk. There may be, however, other offsetting factors that may make such designs an attractive alternative for some hospitals.

To minimize the misuse of resources during the implementation process, critical-path type of scheduling method is essential. Specifically defined milestones and interim deliverables should be generously interspersed throughout schedules to facilitate progress monitoring and control. Milestones should be accomplishment rather than "percent completed"-oriented.

GO/NO GO decision points with predefined conditions should be inserted at strategic points, i.e., prior to major investment expenditures, to enforce ongoing validation of implementation progress against stated objectives.

HIS Planning Documents

The HIS Automation Plan is the final product of the planning process. It documents the various analyses performed by the planning team, findings and recommendations made to the planning committee, and the resulting decisions.

Overall, the HIS Automation Plan can be utilized for the following purposes:

Budgetary planning/acquisition of funds
Organizational planning and control vehicle for the HIS implementation process
General communication source and a design document

HIS Automation Plan outlines the selected strategy for achieving the hospital's objectives. This includes restatement of the objectives and purposes of the HIS project, discussion of priorities and criteria of evaluation, final decisions on what functional areas and in what sequence will be automated, conceptual designs of future information processing based on EDP methods, specification of EDP environment, means and methods of acquisition of required EDP capabilities, cost/benefits and operational impact analyses, user attitudes survey, resource requirements, schedules, milestones, deliverables and expected responsibilities.

Because of the volume and variety of use of such information, it is sometimes beneficial to organize the plan in several *separate documents,* serving specific audiences and purposes. An example of one such possible organization is shown in Figure 5.

The individual subsets of the plan could include, for example, the following information:

Information System Plan—Functional Section

Documentation and analysis of selected operating characteristics of the existing information system of the organization including the identification of problem areas.
Identification and definition of functional areas, that are most likely to benefit from automation, considering the stated objectives and present operational problems. Determination of priorities for implementation.
Description of proposed alternative approaches to automation in specified area,

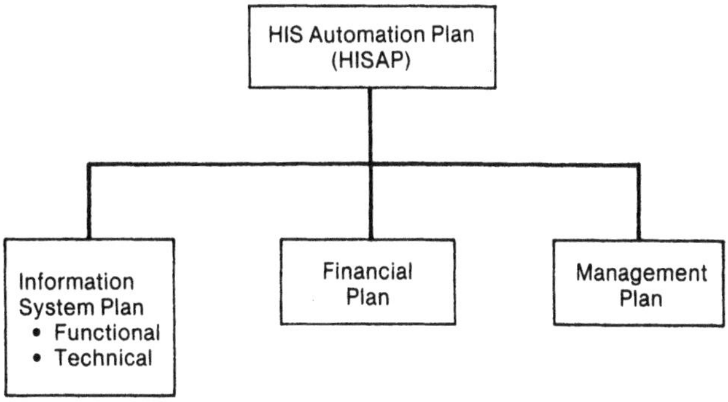

Figure 5. Organization of HISAP planning documents.

along with the evaluations of total payoff for each "scenario" in terms of stated objectives.

Cost/benefits and operational impact analyses.

Selected implementation strategy and reasons for selection.

Overall implementation schedules, budgets.

Information System Plan–Technical Section

Projection of processing workload by type and phase of HIS implementation strategy. Capacity requirements and acquisition plans. Buy, lease, rent types of evaluations.

Initial hardware and software configurations. Final configuration; year-by-year migration paths and upgrades.

Data communication network plan. Network geography, supported locations and type of equipment. Data traffic projections by transaction type and routing pattern.

Data Base Architecture. Overall macro-level diagram of data element types and its logical relationship. Storage requirements projections. Security and Integrity Plans.

EDP Standards and Procedures supporting the selected HIS strategy.

Financial and Management Plans

Overall budget with detailed schedules of expenditures and description of corresponding deliverables.

Overall critical path method diagram of the HIS implementation including the progress monitoring mechanism.

Manpower hiring schedule. Type and number of required personnel, required hire points, skills and qualifications.

Equipment acquisition schedule. Includes completion dates for selection, order placement, delivery, inspection, testing, and installation along with assigned responsibilities for these activities.

Organizational policies and procedures. Includes required or anticipated changes in policies, procedures, organizational structures and their expected effective dates.

Responsibilities, deliverables, acceptance procedures and performance specifications.

Depending on the selected method of HIS acquisition, some of the items above may either be not applicable or may need to be expanded.

Concluding Remarks

The 1980s will be a critical test period for further advances of HIS technologies, particularly in the area of patient care-oriented information processing.

Patient care functions, being highly decision and information intensive, depend on and affect a large number of secondary and tertiary support processes.

To support future information demands in this area, the EDP technology must be carefully integrated into the overall process of health care delivery as success or failure of the integration process will be felt throughout the entire organization, multiplied many times over.

Planning methods and techniques outlined in this and the following chapters are being increasingly utilized by hospitals to manage the transition process on systematic and cost-effective basis.

Towards Implementation of Successful Medical Computer Applications*†

J. S. Sobolewski, Ph.D.

Introduction

Over the last ten years there has been an increase in the number and diversity of computer applications in the field of medical research, administration, education and patient care. Many of these applications have been successful, especially those in areas where the objectives and functional characteristics can be clearly described.

In medical research, for example, the tools for management of clinical data and statistical analysis are used to discover unknown relationships or to verify suspected relationships among clinical variables or events. Although the number of patients and variables in such studies may be very large, and there may be some problems in the appropriate selection, collection and error control of clinical data over long periods of time, established techniques in data management and statistical analysis have, by and large, been successful. Funds for such studies have usually been available when the questions have been deemed to be sufficiently important and the methods used to collect and analyze the data to answer the questions have been judged to be valid.

Similarly, in medical administration, computers can be used to collect and process data which is required for inventory control, personnel records, cost accounting, performance analysis and future planning. These are all recognized requirements having counterparts in the other sectors of the business community. Consequently, their benefits have been recognized and they have received appropriate financial support for their development.

The least progress in applying computers has been in the area of patient

*This work was partially supported by PHS Health Resources Administration grant no. MB 00184.

care. One reason for this is that the needs of this area are far less clearly defined and, in fact, the needs are generally viewed differently depending on whether one is a medical administrator, a practicing physician, a public health official, or a medical research investigator. Other reasons contributing to problems include failure to adequately:

(1) Integrate objectives and evaluation criteria into the analysis and design phases of system implementation.
(2) Obtain commitments from administrators controlling funds, personnel and computer resources.
(3) Communicate with the end users, involve them more in the design process and evaluate the impact of the computer application on their environment.

The purpose of this paper is to examine in greater detail some of these problems and to focus attention to items that should be carefully considered as the medical application evolves through the analysis, design, and implementation phases to the operational and post-operational phases. It should be emphasized here that although this list is by no means exhaustive, it does not imply an order of priority and some items considered may apply to some applications and not others, it should at least help increase the odds of developing a better and more successful application.

It should also be noted that the items presented are completely independent of the programming and design methodology used for implementation. Proper use of structured programming, top-down, bottom-up or other approaches for design methodology may further enhance the success of the application. It is claimed, for example, that use of structured programming methods can result in:

(1) Lower program implementation costs
(2) Improved program documentation
(3) Greater program reliability
(4) Easier program maintenance
(5) Systems in which it is easier to incorporate future changes and improvements.

Discussion of these methods and their relative merits is outside the scope of this paper. For further information, the reader is referred to [1-5].

Although there are many characterizations of the total system design process [1], we assume three main phases—the analysis and design, development, operation and evaluation phases. This has been done for the sake of simplicity and because it provides a good basis for discussion. The problems commonly associated with each of these phases, together with the items that should be considered in order to avoid the problems, are discussed in the first three sections of the paper. The fourth section suggests some methods of dealing with human problems that may result from the introduction of new computer applications.

Analysis and Design Phase

During this phase, the nature of the needs must be analyzed, specifications or functional descriptions of the software must be developed and any constraints should be explicitly mentioned. This information is then used to develop a representation of the system at one or more levels of detail. Problems that may arise during this phase include the following:

(1) The application problem has not been clearly understood or is ill-defined.
(2) Objectives and acceptable criteria for judgment of the success of the application have not been adequately developed or are ill-defined.
(3) Users of the proposed system are not entirely convinced of the value of the computer application.
(4) Full endorsement of the medical application from administrators controlling the funding for the resources needed has not been secured.
(5) The proposed application has not been discussed with those who control the computer resources to ensure that adequate resources exist and can be made available for the application.
(6) Technical, medical and administrative project personnel fail to communicate and plan together.

Problem Definition

As mentioned previously, the needs in the area of patient care are far less clearly defined than in other areas such as medical research or administration. Furthermore, the needs may be viewed differently by medical administrators, public health officials, physicians or medical investigators. It is, therefore, most important to obtain concurrence on the problem definition, both through feedback and documentation, from all concerned, including the end-users (nurses, physicians, lab technicians, etc.) of the proposed system. Without a clear understanding and unambiguous description of the problem, the odds of achieving a successful solution are very small indeed. The process of problem definition is so basic to the system implementation process that it must be considered as the first step in the design process.

Statement of Objectives

Once the problem has been clearly understood, the next step is to develop a set of objectives. The computer application should directly or indirectly improve at least one aspect of patient care. Examples may be improved quality of patient data, faster access to patient data or cost containment of health care delivery. These objectives should be stated clearly, together with any constraints that have to be imposed. Such constraints could include limits in total implementation costs or limits in availability of computer resources. For some applications, it may be necessary to state explicitly what the application should not do, or that it does exactly what is specified and no more, no matter how tempting it may be to include additional features.

Development of Evaluation and Acceptance Criteria

The evaluation criteria serve as a yardstick to measure the extent to which the objectives are met and the degree to which the application is accepted by the users. Evaluation attributes that can be applied include, but are not necessarily limited to, the following:

Accessibility	Confidentiality of data
Accuracy	Reliability, integrity
Conciseness	Timeliness
Usability	Completeness, adequacy
Availability	Applicability, usability
Flexibility	Cost (initial and recurring)

Other important factors include acceptability, portability, and adaptability. The eventual acceptance of a computer system into routine usage is often dependent upon a number of psychological factors and failure to consider these may lead to a system which will be rejected by those for whom it was intended. Some of these psychological factors are outlined in the final section. Portability and adaptability refer to programs written in a sufficiently flexible way to allow them to be run effectively on machines of several different types and to fit different system constraints respectively [6].

In order to obtain valid answers regarding the evaluation attributes, techniques should be employed to obtain both subjective and objective data. It is most important to obtain agreement on the evaluation and acceptance criteria to be used to form the basis for judging the success or failure of the application before the design proceeds further in order to avoid future misunderstandings and disappointments.

User Involvement

The users of the proposed application must be convinced that it be of direct or indirect benefit not only to them but also to the patient. Formal and informal discussion with the potential users will help to develop a rapport and understanding of each person's problems and expectation.

Involvement of Medical Professionals

Physicians, nurses, paramedics and other medical professionals directly or even indirectly connected with the proposed application must be involved and committed before there is a chance of success.

Administrative Involvement

Administrators controlling space, funds for equipment and personnel must be involved and committed to ensure adequate resources necessary to develop the application. The application must also be discussed in detail and a commitment be obtained from the administrators controlling the computer resources available.

Project Coordination

Problem definition and development of objectives, evaluation, and acceptance criteria are the essential ingredients to the preparation of general specifications which form the basis for detailed design and eventual implementation of the application. It is, therefore, essential that users, medical professionals, administrators and technical personnel involved in the early decision making process communicate and plan together in order to have the commitments necessary to help ensure success.

Development and Implementation Phase

The general specifications serve as a reference point from which a detailed and complete design can be developed and implemented. During the development and implementation phase, the possible alternatives, cost, response time, reliability, ease of use, and flexibility of the system must be determined and implemented from the statement of objectives and evaluation criteria. Possible problems that may occur during this phase include:

(1) Failure to consider various alternatives.
(2) Excessive cost.
(3) Low reliability and/or availability because of poor software design or unreliable computer system.
(4) Poor man-machine interface making the application difficult to use or understand.
(5) Availability of insufficient data to initiate project.
(6) Difficulties in data collection, patient identification or data retrieval.
(7) Software implementation is initiated before detailed design is complete.
(8) Ad hoc rather than methodical approach to software implementation is used.

Consideration of Alternatives

Most changes, especially those dealing with man-machine interaction, are difficult to accept and hence any new solution must be superior to the old. Various implementation alternatives, including those not using computers, should be considered in detail and the degree to which they meet the previously established objectives, evaluation and acceptance criteria should be determined. Both objective and subjective data should be used. This can be done by performing user satisfaction studies, observations, interviews and including facilities within the application for automatic acquisition of usage statistics once it is operational.

Estimation of Cost

Good cost accounting practices must be used in order to address the cost attribute of the application. The cost should be lower than other alternatives or there must be a substantial justification involving other trade-offs to which one can attach a cost. This, of course, implies that the costs of other alternatives must be evaluated as well. It should be emphasized that the costs must include costs for development, implementation, debugging, documentation, maintenance, periodical enhancements to the system, special equipment purchases and the on-going computer costs once the application is fully operational. Another important point is that for a well designed computer application involving an interactive system, the actual computer costs will in general be a small fraction of the personnel costs involved in system maintenance and the collection, input, processing and retrieval of the data. Exphasis should, therefore, be given, whenever appropriate, to minimizing personnel costs rather than computer costs.

Data Collection

Most medical applications involving interactive computing require data relating to patient encounters, laboratory test results, diagnoses, medications, treatments, surgical procedures, follow-up encounters, etc. Such information must be collected in a timely manner and entered accurately into the system. Attributes of data collection such as availability, accuracy, timeliness, completeness, adequacy, security and integrity should be evaluated to ensure they meet the system objectives and acceptance criteria.

Data Entry and Retrieval

The data collected should be capable of being entered into the system quickly and accurately. Similarly, the system should respond quickly to a query or to a request for data retrieval. Since the response time requirements of all users are not the same, they must be measured against the acceptance criteria but it must be ascertained that these criteria meet realistic goals. The system should also include adequate security measures to ensure confidentiality and integrity of the data during update and retrieval operations. Checks should be made to protect data from unauthorized access while authorized users should have access only to data they need to know.

Reliability and Availability

A medical computer system should be available and working reliably whenever the user wants to use it. Reliability and availability, therefore, should never be compromised against some other design factor. Since reliability to the user means that the system performs as desired whenever needed, not only must the hardware and software work correctly, but the software should be tolerant of user errors. In particular, it is important to ensure that lost data can be easily recreated in case of loss due to hardware, software, or user errors. Failure statistics should be maintained not only for evaluation purposes to ensure that objectives are met but also to help isolate system problems during implementation and operation.

Human Engineering Aspects

Several operational characteristics are important to user acceptance of the system. Some of these have already been discussed and include adequate response time, reliability and tolerance to human errors. Since every computer application will require at least some changes in the user's mode of operation, it is important that the effect of this change be carefully evaluated. The user should not try to circumvent the computer system or improvise in order to use it. The user should be trained in the use of the system, its use should be natural, and user documentation should be available when needed. Before any design changes are made, no matter how slight, their impact on the user should always be evaluated. The last section will discuss some of these points in greater detail.

Software Implementation

No coding or programming of the application should be done until the design has been finalized. It must be completed to a level of detail that enables its quality to be evaluated and any remaining problems to be identified. It should show the various modules, their interfaces and the operation to be performed within each module—all expressed in an unambiguous way. This should then be compared with the specifications to ensure that the original goals and objectives are met. In some cases this detailed design points out potential problems that may be encountered during implementation and it may be necessary to modify the specifications if it is determined that the design cannot be feasibly implemented [7]. Only when the detailed design is reviewed and completed can it serve as the basis to write the program.

The approach to program implementation should be methodical to ensure an acceptable product. Much has been written about programming style and methodologies [4,5,8] but the concept of structured programming [9] is perhaps the most widely accepted. Some of the advantages claimed for this method include:

(1) More readable programs.
(2) Faster implementation.
(3) Lower implementation costs.
(4) Easier program debugging, testing and maintenance.
(5) Greater program reliability.
(6) Changes and improvements easier to implement.
(7) Improved documentation.
(8) Greater portability and adaptability.

The actual coding or software implementation occupies a relatively small percentage of the total process of design and development. Whereas a good deal of creativity is possible in the detailed design phase, the coding stage is much more rigidly constrained and is relatively straightforward given a good detailed design, and a programming language well suited for the intended application.

Operational Phase

Once the application has been implemented, debugged, tested, evaluated, and found to operate as per specifications, it should enter the operational phase. The most common problems that can occur at this time include:

(1) Premature release of the application from the implementation to the operational phases.
(2) Inadequate user training.
(3) Failure to conduct continuous evaluation to ensure that the application is justifiable and continues to meet with user satisfaction under actual operating conditions.
(4) Lack of flexibility. Application is not easily changed to overcome unexpected user problems.
(5) Lack of adequate system and user documentation.
(6) Lack of adequate backup procedures, in case of system problems.
(7) Underutilization of the system due to overestimation of its need and/or to a combination of any of the above problems.

Premature Release

Since first impressions are sometimes difficult to overcome, every effort should be made to ensure that the application performs satisfactorily before it is made operational. Not only is it necessary to ensure that the programs implemented are fully debugged and are working correctly, but it is necessary to ensure that the application does solve the actual problem and meets all of its stated goals and objectives.

User Training

Although the system should be simple and natural to use, there should be adequate user training and documentation. If it takes two years to train an x-ray technician, it is totally inadequate, for example, to spend a few hours teaching nurses how to use a new computerized patient monitoring system which is a more complex piece of equipment.

System Evaluation

Not only should the system be evaluated before it is made operational, but the evaluation activity should continue at least during the early operational stages to ensure it can be justified and meets the original objectives, evaluation and acceptance criteria in an actual working environment. Continuous system evaluation helps uncover unexpected operational problems or possible improvements as the practice of medicine changes.

Flexibility Problems

No matter how much care has been taken in the entire design process, it is more than likely that at least during the early operational phase some unexpected user problems will be discovered. Furthermore, improvements will be needed to keep pace with changes in patient care. The application must, therefore, be flexible to allow the problems to be overcome and system enhancements to be easily implemented. Flexibility should be part of the evaluation and acceptance criteria and can be measured as the elapsed time between agreement as to what modification is needed and the availability of the modification to the user.

Backup System

Once it enters the operational phase, a well designed application should include backup systems. These may range from a dual redundant computer system for critical applications to the use of a fully manual mode of operation. Transitions between the primary operational system and the secondary backup system should be "smooth".

Documentation

In any major computer application, a large amount of effort is expended in communication among users, medical, administrative and technical personnel. Since verbal communication is not only subject to misrepresentation, but is also easily forgotten, good documentation must be maintained through all phases of the project. Detailed design cannot be done in the absence of good written specifications and evaluation criteria. Documentation is especially important during debugging, testing, validation and at all other times whenever changes are made to overcome existing problems or to add new needed features to the system.

Justification

Irrespective of whether the application is in medical research, teaching, training or patient care, it must be justified in terms of its value in order to rank highly in terms of priorities of the organization using it. Underutilization can be due to an overestimation of the application's need or to a failure to carefully consider all of the relevant items from problem definition all the way to implementation and operation. One must be prepared to abandon or at least to modify the application if it is underutilized, the cost is excessive or users are negative. The product should be marketable to users at all times.

Human Problems

A new medical application almost always implies some changes in the user's method of operation. During all phases, it is therefore essential for analysts and designers to evaluate the environmental impact. In particular, they must understand and consider the possible causes of employee resistance to these changes. Failure to consider these may result in a system that is rejected by the users for whom it was developed. Factors affecting user attitudes are their general previous disposition towards computers and the manner in which system design and introduction was undertaken. Other factors include:

(1) Fear of losing one's position.
(2) Loss of status.
(3) Inability to cope with the new work.
(4) Reduced interest in the work due to the computer application.
(5) Suspicions of the motives behind the proposed change.
(6) Resentment of change because it is assumed to be a personal criticism.
(7) Social disruption caused by rearranging work relationships and patterns.
(8) Application is not human engineered.

Some effective methods in overcoming such resistance include:

(1) Informing all affected persons in advance of the reasons and benefits of the changes being contemplated.
(2) Encouraging and providing all employees and users with an opportunity to participate in all phases of the design rather than imposing the system upon them.
(3) Whenever possible, employees and users should be provided with some personal reward associated with the change.

(4) Adequate training and user documentation should be available to ensure that employees are fully prepared to cope with the changes associated with the new application. Ideally, the system should provide on-line assistance if requested.
(5) The change to an operational system should be scheduled and timed carefully. Some changes should be introduced quickly, while others should be introduced gradually. Whenever possible, changes should be scheduled at times of low activity to allow users to get "acclimatized" at a more leisurely pace.
(6) Cultivating the habit of change. Employees should be helped to see the need for change in a dynamic organization and should be shown how change and new technology can bring them personal rewards.
(7) Provide models or examples, whenever possible, of how the contemplated changes have been beneficial and successful elsewhere.
(8) Minimizing the effects of the change by incorporating user oriented features into the human-computer dialog. These include ease of learning and remembering how to use the system, flexibility of usage for skilled and unskilled users, error detection and graceful recovery from errors and on-line assistance to allow the user to interact with the computer in a natural way without having to learn cryptic codes and error messages or type lengthy amounts of text in a precise and unforgiving format.

Awareness of needs of people is highly important to system analysts and designers since system changes must coincide as much as possible with employee benefits as well as the overall organizational objectives. While there are times when analysts must spend an appreciable amount of time in relative isolation and concentrate on non-human aspects of the system such as reports, organizational charts, controls, and other paperwork to show "progress" to management, this should not lead to a situation where they do not consider the most important needs—people needs.

Conclusion

The nature of many potential medical computer applications is such that the needs are not clearly defined and, in fact, the needs may be viewed differently by users, administrators and medical professionals. Furthermore, the majority of such applications involve human-machine interactions which can cause changes in the environment and in people's work patterns. A key toward more successful implementation of such applications is the careful definition of the problem and development of evaluation and acceptance criteria which will be used to judge the success or failure of the application. Since the problem definition and evaluation criteria form the basis for general specifications without which further design should not be undertaken, it is essential to obtain concurrence on these items from all parties concerned including users, medical professionals, technical personnel and administrators who control all the necessary resources for project implementation. Throughout all phases, the user needs and the impact of the application on the user environment should be carefully considered since in the long run it is the users who will determine the success or failure of the application.

For the sake of simplicity, the design process has been divided into three phases—the analysis and design, development and operation phases. Problems commonly associated with each phase and the items to be considered in order to avoid those problems have been discussed. A special section has been devoted to human problems to emphasize their importance. The list of items to be considered is not exhaustive, nor is it in order of priority. Although following the guidelines suggested will not guarantee success, failure to follow any of them will almost certainly guarantee disaster. It is hoped that careful consideration of the guidelines will at least improve the odds of implementing a more successful application.

References

[1] D. Teichroew, "Improvements in the System Life Cycle," Proc. IFIP Congress, 1974, pp. 972–978.
[2] H. B. Liskov, "A Design Methodology for Reliable Software Systems," 1972 Fall Joint Computer Conference, AFIPS Conference Proceedings, Vol. 41, Montvale, N.J., AFIPS Press, 1972, pp. 191–199.
[3] W. D. Stevens, G. J. Meyers, L. L. Constantine, "Structured Design," IBM Systems Journal, Vol. 13, No. 2, 1972, pp. 115.
[4] "Tutorial on Software Design Techniques," P. Freeman, A. I. Wasserman, Ed., IEEE Press, 1976.
[5] "Structured Programming Tutorial," V. R. Basili, T. Baker, Ed., IEEE Press, 1975.
[6] P. C. Poole, W. M. Waite, "Portability and Adaptability" in Advanced Course on Software Engineering, M. Beckman et al, Ed., Springer-Verlag, 1973, pp. 183–277.
[7] P. Freeman, "Towards Improved Review of Software Design," Proc. AFIPS 1975, National Computer Conf., Vol. 44, 1975, pp. 329–334.
[8] B. Kernigham, P. Planger, "The Elements of Programming Style," New York, McGraw-Hill, 1974.
[9] F. T. Bahri, "Structured Programming in a Production Programming Environment." Proc. 1975 Int. Conf. on Reliable Software, 1975, pp. 172–183.

12

The Conduct of a Health Care Systems Project*

James E. Rusnak

Introduction

This chapter is concerned with the procedures used in the acquisition and installation of computer systems for health care applications. As many hospitals are not capable of sustaining the resources necessary for developing computer systems within one facility, they must rely on the availability of systems in the marketplace and/or the use or systems developed at other institutions. I will describe the procedures used at Washington Hospital, a 265 bed facility located in Fremont, California, to develop the requirements for a data collection/order entry system, and the documents prepared during the conduct of the project. The methods used in this project were based principally upon the work by Philip W. Metzger, described in his book, *Managing A Programming Project,* used in conjunction with numerous sources of material which I have read on the subject of project management[1]. The methods described have been proven to be effective in the management of a complex computer system installation.

History

Hospitals have tended to rely upon the use of outside systems and programming professionals in the conduct of systems development projects. Many hospitals have in the past relied on the use of service bureaus to satisfy their data processing needs. Others have used equipment manufacturers and/or contract personnel to perform the systems development functions, including the installation of health care application systems offered by the vendor or by other insti-

*This chapter has beeen modified from a version previously published in *Proceedings of the Fourth Annual Symposium on Computer Applications in Medical Care,* IEEE, 1980.

tutions. The management of the detail project activities has been usually performed by personnel not on the hospital's staff. Many projects have been characterized by the lack of involvement of hospital staff, especially in the initial project phases. In many cases, the hospital staff has participated in these projects in a limited capacity and under the direction of personnel outside the hospital. When projects are conducted in this manner, it results in:

Inadequate attention to the policies and procedures of the hospital;
Continued dependence upon an outside, and often single, resource for data processing expertise;
Problems in selecting and/or adapting the computer application systems to the specific hospital's requirements;
A lack of a definition of the comprehensive requirements to be met by a health care application system developed outside the hospital's environment;
The presence of system characteristics not favorable to the hospital's operations;
The incomplete and/or erroneous utilization of the system's features;
A lack of systems organization in the development, installation, and integration of multiple computer applications within the hospital.

Recognizing the presence of these problems, Washington Hospital authorized the creation of the Data Services Department. The purpose of establishing the department was to provide the necessary internal systems resources needed for the successful conduct of computer systems projects in addition to the existing data processing operations. The remainder of the chapter describes the processes used in the performance of a data collection/order entry system project, the first project for the department. The project demonstrated that, using internal resources and appropriate project management techniques, the system requirements can be better defined prior to the system acquisition process and permit the hospital to improve its chances for successful system installation and subsequent operations.

Project Procedures

A data processing project can be characterized as a series of milestones. Each milestone represents the completion of a set of activities associated with a project phase. The project phases are the logical steps taken in solving a systems problem. The project phases and the deliverable documents and/or program products associated with each of the project phases are defined in the Project Plan, the first document prepared in the project. These deliverable end products of the project phase provide a positive means of knowing when the phase is completed.

The following phases were identified at the beginning of the data collection/order entry system project and communicated to the Hospital's Administration and the Board of Directors:

Background Analysis Phase

The Background Analysis Phase provided a foundation for the subsequent systems work in the remaining project phases. During this phase, the environment in which the system is to be installed was defined and evaluated in general terms. The historical information relating to the problem area was gathered, reviewed, and organized into working files. The information gathered during this phase provided the analyst with an understanding of the organization goals, administrative policy, procedures of the hospital in use in the affected departments, and other aspects of the hospital's operations which related to the system application.

The Background Analysis was performed with participation by several members of the hospital's staff, and provided a means of involving the hospital personnel during the early part of the project.

Problem Definition Phase

The Problem Definition Phase consisted of a preliminary analysis of the existing outpatient and inpatient systems. During this phase, intermediate changes were made to the current manual processes. These changes were limited in scope, and provided relief from some of the problems that were experienced when the project was initiated. A Problem Specification was prepared at the end of the phase. The document was reviewed with each of the departments to verify the findings of the analysis, and to permit continued involvement of the departments in the development of the project requirements.

Functional Analysis Phase

The Functional Analysis Phase was concerned with a detail systems analysis of the pre-admission, admission, outpatient registration, ordering of services, charging, and account collection activities. Each function which was performed in receiving patients, collecting and distributing information, scheduling patient care, collecting payments, discharging the patient, and performing other inpatient and outpatient services was evaluated. As the requirements for each function were identified, the information was recorded in working papers, functional flow charts, decision tables, and other forms of documentation. These documents were reviewed and approved by all affected Hospital departments. The approved documents served as the basis for the preparation of the Request for Proposal.

General System Design Phase

Based upon the results of the Functional Analysis Phase, a general system design was developed in the next phase. The design encompassed automated and manual procedures—it dealt with all aspects of the application area. During the general system design, various alternatives were analyzed in terms of estimated cost and capabilities. These alternatives were reviewed in terms of the data collection/order entry functional requirements, and the long-term data processing hardware and software goals of the hospital. The approach taken on the project would significantly influence the approach taken in other future applications. A Request For Proposal was prepared to convey the findings and recommendations of the general system analysis to Administration. A separate section was prepared for each department. The Request For Proposal sections were separated and distributed. Each section was given an intensive internal review in each of the affected departments. Each department had a personal stake in the successful outcome of the project by having the opportunity to determine the requirements that the system was to satisfy in their department. The Request For Proposal provided a detail list of the system features to be satisfied and each feature's priority, as well as the guidelines to be used by the Hospital in the selection of the appropriate system. The specifications were organized by department, so that each department would be able to evaluate its particular requirements as written in the Proposal. The Request For Proposal was used by each vendor to prepare their response. This permitted the Hospital to control the format, organization, and information content of each vendor's bid.

System Proposal Review Phase

The System Proposals received from the vendor's bidding on the data collection/order entry system Request For Proposal were reviewed by each of the Hospital's departments affected by the system. The bids by each vendor were separated by department and each section was given to the affected department(s). Each section was also evaluated by the Data Services Department, the architect of the overall system design requirements. Meetings were then held with each department to compare the results of the evaluation. Where differences existed between the department's and Data Services' evaluation, a compromise was negotiated. Numeric point values were used to score the response of the vendor to each item in the Request For Proposal.

The Systems Proposals received were evaluated in terms of:

features supported relative to the Hospital's priorities,
ability to support future systems growth at Washington Hospital,
cost effectiveness of software and hardware proposed,
systems support capabilities,
compliance of conditions in the Request for Proposal.

Prior to the final selection, approval for the system expenditure was sought from the Department of Health, State of California, under the provisions of State Law AB40001. During the system selection process, it was necessary for various key hospital personnel to visit the installations of several of the competing vendors in order to evaluate the operational characteristics of the systems being proposed. Telephone calls were also made to hospitals in other parts of the country to obtain a broader base of system information. A survey form was used in recording this information. Upon completing the evaluation, a report was prepared for Administration and the Board of Directors which presented a summary of each system bid including detail financial information, and a recommended system selection. The final selection was then approved by the Board of Directors following a detail presentation for Hospital staff and a visit to a hospital to see an installed system.

Contract Negotiations

Once the successful vendor (HBO & Company) was selected, a series of meetings was held to negotiate the contract between the Hospital and HBO for the system installation. The contract negotiations are the most critical phase of the project, in that the ultimate success of the system installation depends upon how well the vendor's and hospital's responsibilities are defined, and the providing of adequate remedies for those occasions when a deficiency arises. The negotiating team consisted of the Director of Financial Services and myself. The Hospital's attorney worked closely with us during these sessions. In this particular project, a separate hardware maintenance contract was negotiated with Four-Phase Systems. In both cases the time spent in this activity was compensated for many times over later in the project when problems arose (as occurs in all projects). The contract provided the means to reach equitable settlements and to avoid a serious impact upon the progress of the project.

Detail Systems Design Phase

During the initial stages of the data collection/order entry system installation, a detail systems analysis and design was performed. The purpose of this phase was to:

develop CRT screen formats,
define system input functions,
design the system's data base,
design output forms and data formats,
define the system processes for each type of system function,
develop an Installation Plan for the remaining activities of the Installation Phase.

The detail system design was constrained by the functional limitations of the selected system. Meetings were held with the Hospital departments to develop the detail designs and specifications. Each department approved the system functions which affected its operations before the documents were released for programming. This phase resulted in the preparation of a Detail Systems Design Specification. This set of documents served as the vehicle for communicating to HBO the programming requirements for tailoring their MEDPRO System to meet the needs of Washington Hospital.

Physical Planning

The Physical Planning Phase is concerned with the design and specification of the physical changes to the Hospital's building to accommodate the installation of the computer main frame and its peripherals and the placement of the CRT and printer terminals throughout the hospital. A number of modifications were required to permit this equipment to be properly utilized in each location, including modifying work stations, adding electrical outlets, and the routing of communications cables.

System Installation Phase

Once the detail system requirements were specified in the Detail Systems Design Specification, the System Installation Phase was begun. Programming was performed by HBO & Co. to implement the required processes in the MEDPRO System programs. The output forms were ordered per the detail system design specifications. A significant amount of time was spent in this phase in the training of all affected hospital personnel. It was necessary that all personnel who came in contact with the MEDPRO System be proficient in its use. The system was initially installed in the "test state," where the system was used in a non-operational capacity to allow the users to gain "hands on" experience and to permit the testing of various functions in the system. Revisions to the programs were made as operational problems were encountered. Once the system reached a point where it performed the required functions, a formal test of the system was made. During the time that the programming of the system was conducted, test scripts were prepared by the Data Services Department, assisted by key personnel from Admitting and Outpatient Registration. CRT screen format pages prepared in the Detail System Design Phase were copied and used to record the detail test input data.

Detail field information for each test case was then coded on these copies, based upon the detail specifications. The test consisted of three "test days," with a variety of input data specified for each day. The test results were reviewed by key Hospital personnel. The discrepancies were recorded in a log and then turned over to HBO & Co. for correction. After corrections to the system were installed, subsequent tests were performed until the system performed in conformance with the Detail System Design Specification. A formal acceptance letter was then sent to HBO and the implementation was begun.

Implementation Phase

A phased implementation approach was used in this project. The first phase implemented the Emergency Room registration and ordering functions. This approach permitted a large number of the system's functions to be observed in operation while limiting the number of transactions involved. When errors did occur, their impact was more limited and more easily corrected. In implementing the system at the nursing stations, one function (such as Lab orders) was implemented in a phase. Although many hospitals use the "pilot nursing station" approach, where all functions are implemented at one station, we felt that our approach provided a more complete check-out of each function's operations. This method avoided the problems in the ancillary departments of supporting two systems simultaneously during the pilot period.

Acceptance Phase

Once the system was fully operational, additional time was spent in each department to validate all of the system's functions. Due to the impact of the system upon each department's procedures, and the complexity of the system's various functions, some of the processes required further modification. These modifications were due to problems and/or deficiencies not visible during the System Installation Phase. These changes, however, were limited in scope and represented minor changes to the operating programs. This phase was concluded three months after the system was completely installed.

Post Installation Audit Phase

The Post Installation Audit Phase was the final project phase, where the entire project was reviewed. Attention was given to the completeness of the new MEDPRO System's procedure documentation, forms design, system's computer operations procedures and characteristics, and the presence of problems not resolved in the preceding phases. Additional training was required in specific areas where operational problems were experienced. The goals of the project, including cost benefits, were examined in terms of the actual system performance. Additional program changes were identified and scheduled for installation. Procedures and forms were established to permit future changes and/or correction of problems to be made with the appropriate controls to insure the integrity of the system functions. Once this phase was completed, any additional problems and/or requirements were handled as routine maintenance requirements.

Project Documents

Project Plan

The Project Plan is the key to the successful execution of a data processing project. The Project Plan used for the data collection/order entry system contained the following sections:

(1) *Overview.* (a) Project Plan Objectives—a description of the purposes of the project plan. (b) Project Objectives—the scope of the project.
(2) *Phase Plan.* A general description of the phases to be executed during the project, the expected time schedules to be met, and products to be produced by each project phase.

(3) *Functional Responsibilities Matrix.* This section of the project plan provided a brief summary of the responsibilities for each of the Hospital's functional areas concerned with the project and is detailed by project phase. For each project phase a summary list of activities is presented in a sequence that they are to be performed. Referring to the attached example of a Responsibilities Matrix, one can see that even the Hospital Administrator and the Board of Directors are accounted for in the various phases of the project. When a project is initiated, each area of responsibility should be defined for at least the first two or three project phases. As the project proceeds, additional phases are added to the Matrix along with the responsibilities for each functional area. The functional areas are equivalent in most cases to major departments within the Hospital's organization. By using this approach, each area of responsibility is aware of what it is to do and what it is to expect from the other areas participating in the project. Detailed project activity task lists provide detail information for the activities summarized in the Matrix (see Table 1).

(4) *Documentation.* A brief description of the format and content of each type of project document to be prepared during the project.

As each document was prepared, an organized method of defining the systems requirements was performed. With each phase, a more detailed definition of the Hospital's information system needs was realized. The review of the project documents with the departments resulted in a more complete definition of the system requirements for each operating area of the hospital. These reviews resulted in the Hospital staff being conversant and knowledgeable of the functionality of the system and of the various aspects of the project, and insured a greater participation by the staff in the successful installation of the MEDPRO System.

Problem Specifications

Two problem specifications were prepared during the Problem Definition Phase. The first document, the Intermediate Problem Specification, was concerned with the problems initially identified in the Problem Definition Phase, and to offer recommendations for short-term corrections to these problems. These recommendations were in the area of procedural changes, forms modifications, and other manual changes which would have limited impact on the Hospital's long-term project goals. A second document, the Final Problem Specification, provided a more complete definition of the problems associated with the collection and distribution of patient registration and order information. This document set the stage for the preparation of a Detail System Specification for a data collection and order management system.

Table 1. Data Collection System Project Plan Functional Responsibilities Matrix
Project No: 1
Document No: PD001

Project Phase	Functional Positions: Systems Analysis	Admissions/ Registration	O/P Services Nursing	Patient Accounting	Director of Financial Svc.	Administrator	Board of Directors
Background Analysis	1. Establish project files, procedures and standards. 2. Collect information and forms. 3. Prepare project plan. 4. Prepare status reports and other documents.	1. Support analysis. 2. Review current system procedures.	1. Support analysis. 2. Review current system procedures.	1. Support analysis 2. Review current system procedures.	1. Review working papers, plans, etc. 2. Monitor procedures. 3. Approve conduct of project phase 4. Approve project plan. 5. Approve start of next phase.	1. Determine scope of project. 2. Review of progress. 3. Approve project plan.	1. Authorize project.
Problem Definition	1. Prepare problem specification. 2. Update project plan. 3. Prepare recommendations for intermediate actions. 4. Prepare status reports and other documents.	1. Update current procedures. 2. Define problem conditions. 3. Implement intermediate changes.	1. Specify needs. 2. Define problem conditions. 3. Support implementation of intermediate changes.	1. Update current procedures. 2. Define problem conditions. 3. Implement intermediate changes.	1. Review and approve problem specification. 2. Monitor progress. 3. Schedules management review and approval. 4. Present problem specification to Administrator. 5. Approve start of next phase.	1. Review and approval of problem specification. 2. Presentation to Board. 3. Coordinate Board requested changes. 4. Report to Board on status.	1. Review of status report. 2. Approve problem specification.

Request For Proposal

The Request For Proposal was a comprehensive systems document which specified:

Purpose and organization of the document.
Guidelines for submitting quotations for systems proposals.
General conditions to be satisfied in the system proposals.
Bidding specifications.
System specifications, including:
- system features with a relative priority value assigned to each feature,
- equipment requirements,
- selected hospital operating statistics, and
- examples of the current hospital forms used in the collection of patient information.

The System Specifications were so designed to provide a means of receiving consistent responses from the various vendors solicitated. In describing the detail requirements to be satisfied by the system, the selection of a system was based upon an objective selection process. As most systems in the marketplace were designed for general use, the ability of the system to be tailored to satisfy those characteristics which are associated with the requirements of our Hospital were accurately measured. To insure that the Hospital's system requirements were met, the systems design was defined, in general terms, prior to the evaluation of the software products submitted for consideration.

Summary

Although the project management procedures described in this paper are in use in some form in many industrial applications, and have been described in similar forms in various technical publications, they are seldom seen in the hospital data processing environment. The methods employed in the conduct of this project were tailored to the environment of a medium-size hospital. Though the nature of the project phases, and the resultant documents, may vary between hospitals, the basic concept of an organized project activity is valid in all cases, regardless of the size or nature of the hospital. Though the availability of an internal computer systems resource may not be practical for all hospitals, the use of appropriate project management procedures and documents by Hospital staff members can result in improved systems development activities, where the interests of the hospital can be properly controlled during all phases of the project. Through improved methods in conducting data processing projects, the hospitals can realize improved results in the systems they depend upon for patient care and financial processing.

Reference

[1] Metzger, Philip W., *Managing A Programming Project,* Englewood Cliffs: Prentice-Hall, Inc., 1973.

13

Facilities Management of Hospital Data Processing*

Stanley E. Jacobs, Ph.D.

Introduction

A 1974 article by Glenn Anderson *et al.* reported on the status of facilities management (FM) of hospital data processing departments[1]. The article:

(1) Described the many variations in arrangements between the hospital and the FM firm that were possible so as to best meet the needs of the hospital. These variations included such things as whether the hospital or the FM firm provided the computer, the data entry personnel, etc.
(2) Identified the vendors of FM for hospitals and the hospitals they serviced.
(3) Discussed the reasons why hospitals have found it to be advantageous to opt for FM.
(4) Provided guidance for the selection of an FM firm and the negotiation of a contract.

In a full-blown FM arrangement the FM firm supplies all of another organization's data processing personnel from the DP Manager on down. The FM firm maintains old software, develops new software, and operates the computer. However, as pointed out in the Anderson article, many FM arrangements in industry and with hospitals do not go that far. Therefore, Anderson defined FM as follows: "FM is an information processing service in which an outside supplier assumes some level of line responsibility for the operation of a client's computer facility."

Controversy Over Merits and Future of FM

A number of potential dangers of FM were cited in the previous article[1]. Among them were:

> Management may abdicate its responsibilities to set data processing goals and to insure that they are being achieved.
>
> The profit motive may cause the FM firm to cut corners and fail to develop optimal systems.

During the summer of 1978, *Computerworld* published a three part interview with Dick Brandon, president of the prestigeous Brandon Consulting Group[2].

Brandon had had the opportunity to investigate in detail an ill-fated FM contract under which a prominent FM firm provided services for a brewery firm. Brandon's conclusions, as reported in the interviews, painted anything but a rosy picture of FM and its future. Among his charges were the following:

> A facilities management firm is interested in maximizing its profit, which often means supplying the minimum of threshold service provided for in its contract.
>
> The company made the mistake of not having an intelligent contract and acceptance procedures, having essentially a vendor-designed contract, not controlling the operation and not demanding the right to say "Wait a minute, this is a low quality system and we're not going to accept it".
>
> Very few—if any—companies whose facilities are being managed by an outside vendor have renewed their contracts.

During the months following the Brandon interviews, *Computerworld* carried quite a few articles on FM by individuals with firms which utilize FM, by FM personnel, or by consultants. Generally these rebutted Brandon's claims, pointed out the merits of FM, and predicted a rosy future. Among the points made were the following[3]:

Two FM firms reported renewal rates of 80 to 90 percent. One article pointed out that even though a company may not review an FM contract it does not mean that it is fed up with FM. Rather, it may sign up with a different FM firm. (There has been at least one such switch within the hospital field.)

Market research firm predictions were reported as being extremely favorable for future FM growth. Input, Inc. predicted FM revenues would grow from $600 million in 1977 to approximately $1.3 billion in 1982. International Data Corporation predicted growth from $575 million in 1978 to $1 billion by 1982.

Almost all the articles pointed out the absolute necessity of having a well-reasoned contract with formal procedures for evaluation of vendor performance and for the acceptance of new systems.

Frederick A. Gross, president of Systems & Computers Technology Corpora-

> tion, stressed that "all the 'negatives' Brandon has stated . . . do not apply to an FM firm with subject matter expertise—a firm that specializes in its clients limited market and special needs."

Reinforcing the Gross comment is a market research study by International Data Corporation. It categorized the FM marketplace into three segments: government contracts, custom-commercial contracts, and industry specific contracts. It was only in the latter area wherein "the vendor can take advantage of 'learning curve' experiences which hold over from contract to contract" that continued growth of FM was predicted[4]. Fortunately most and possibly all firms now providing FM services to hospitals either specialize in the hospital marketplace or have a division of their firm which does so.

Four Types of FM Usage by Hospitals

Type 1

Classical FM: The FM firm provides the data processing manager and all personnel (data entry personnel excepted in some cases), does the system analysis and programming, and operates the computer room. There are some variations to classical FM. Under one variation, the computer is located at the FM firm site but is dedicated to a single hospital. Under a second variation, the FM firm provides FM for a data processing center which serves a few hospitals and undertakes in-hospital system development projects in the served hospitals. Under a third variation, the vendor may provide applications modules, offsite software maintenance, and onsite personnel only as needed to implement changes, to provide training, or to resolve problems. Generally it is small hospitals that may opt for this "as needed FM service". Twenty-five of Compucare's FM contracts are of this variation as are the two of Computer Sciences Corporation and two of Datacare.

All firms that have been in the classical FM business very long can bring to new clients a very attractive array of applications. All of them can provide proven business and financial applications software; most of them can provide on-line admission-discharge-transfer, outpatient registration and billing, and medical record index; and some of them can provide a hospital information system (HIS) or a hospital medical information system (HMIS).

Type 2

Turnkey HMIS FM: The firm is primarily a vendor of a comprehensive online hospital medical information system (HMIS). It offers to provide FM for any hospital which installs its HMIS. All of the vendors of turnkey HMIS do provide implementation assistance and software maintenance, so FM is not essential just for these purposes. The real need stems from the fact that turnkey

HMIS are not static systems. They are all growing as applications are added and others enhanced. Therefore, there is a continuing need to implement changes and train personnel. Thus, it can be attractive to some hospitals to opt for FM from the vendor of their HMIS system.

Type 3

Partial FM: The FM vendor may provide key personnel only or may only accomplish one function, such as programming.

Type 4

Short term FM: The firm provides FM for a limited period of time, perhaps 6 to 18 months. Short term FM is sometimes used by hospitals as they plan for and implement an HIS or HMIS.

From this point on this chapter will only be concerned with Type 1 and 2 FM usage for two reasons. First, Types 1 and 2 represent a total commitment to FM on the part of the hospital, whereas, Types 3 and 4 do not. Secondly, Types 1 and 2 are more commonly used.

Results of FM Survey, Spring, 1981

Vendors known to be providing FM services for hospitals were surveyed by questionnaire and/or telephone. The results are shown in Table 1.

The vendors contacted are listed in the first column. Sixteen are identified. Eight of these have provided FM to hospitals since 1973 or earlier.

The number of FM clients of each vendor is shown in column 2. A total of 178 hospitals were found to be receiving FM services. Of these 76% were serviced by the top four vendors. Also shown, when applicable, are the number of data centers and hospitals served therefrom.

The third column identifies the original orientation of the vendor toward hospital computing services. Did it enter the industry as a facilities management firm ("FM") or as a vendor of turnkey systems ("T")? The reason for this is that there is a growing amount of overlap in the services provided by these types of vendors. As will be seen, many vendors primarily considered to be classical FM firms, now include hospital information systems (HIS) and/or hospital medical information systems (HMIS) in their arsenals of available software. Conversely, many firms primarily considered to be vendors or turnkey HIS and/or HMIS offer FM to hospitals which install their turnkey systems. As it may be seen, 12 vendors were initially primarily FM oriented. They still are today. However, one, Medicus Affiliates, has acquired a company which markets a turnkey system, with or without FM by Medicus. Another FM vendor, Space Age Computer Systems, will now lease its software. In this respect, the classical FM firms which have developed HIS and/or HMIS are becoming quite like the turnkey system vendor who also provides FM.

Table 1. Providers of Facilities Management for Hospital Data Processing

Vendor	No. of[a] Hospitals Served	Vendor's Initial Orientation	FM & HIS	FM & HMIS	All HIS & HMIS
Compucare, Inc. McLean, VA	53 (4-14)	FM	9		9
Space Age Computer Systems, Inc. Washington, D.C.	34 (3-34)	FM	4	3	7
KeaMed. Hospital Systems Boston, MA	26 (1-3)	FM	1		1
Medicus Affiliates, Inc. Evanston, IL	22	FM		4[b]	16
Hospital Computer Systems, Inc. Colts Neck, NJ	12	FM	6		6
Technicon Med. Info. Systems, Inc. Santa Clara, CA	8 (2-7)	T	3	5	24
National Data Communications, Inc. Dallas, TX	7	T	1	6	14
Datacare, Inc. Roanoke, VA	4 (2-4)	T		3	10
Electronic Data Systems, Inc. Dallas, TX	2	FM			7
Pentamation Enterprises, Inc. Bethlehem, PA	2	FM		2	10
Computer Sciences Corporation El Segundo, CA	2	T			4
Atlas Computer Systems, Inc. Secaucus, NJ	2	FM	2		2
Others[c]	4	FM			
TOTAL	**178**		**26**	**24**	**110**

[a]Numbers in parentheses indicate number of data centers and number of hospitals served by data centers. For instance, 14 of Compucare's 53 hospitals are served through 4 data centers.
[b]Medicus provides FM for some installations of its subsidiary, Spectra Medical Systems.
[c]There are four additional firms which provide FM for one hospital each. These are: Gamma Systems Services, St. Petersburg, FL.; Jones Health Systems Management, Inc., New York, N.Y.; General Computer Corporation, Macedonia, OH.; Medical Systems International Corporation, Rockville, MD.

The number of hospitals in which each vendor has installed *and* provides FM for an HIS is shown in the fourth column. Note that four classical FM firms have HIS installations.

The number of hospitals in which each vendor has installed *and* provides FM for an HMIS is shown in the fifth column. Note that two classical FM firms have HMIS installations for which they provide FM.

The sixth column shows the total number of HIS and HMIS installations for each vendor. For the classical FM firms, the number in column (6) must equal the numbers in columns (4) and (5), at least until a classical FM vendor leases or sells HIS or HMIS software. However, for the turnkey vendors the number of column (6) will be greater than the total of the other two columns. That is, of course, because they sell many systems without FM.

Lest the reader be misled, it is pointed out that the total of the sixth column, 110 HIS or HMIS installed, is not the total of such systems installed by vendors. This is because there are other vendors that do not provide FM, such as HBO and Company, Shared Medical Systems, and MCAUTO. Actually, as of Spring, 1981, the total number of HIS installed, or in the process of installation, by FM firms and all turnkey HIS vendors is approximately 515. The corresponding number of HMIS installations is 85. (Note that neither of these numbers includes systems developed and implemented by hospitals using applications software available from hardware vendors, primarily IBM, NCR, and Burroughs.)

Future Growth in the Number of Hospitals Using FM

The 178 hospitals found to be using FM in Spring, 1981, certainly indicates that the growth rate was considerably slower than that estimated by Anderson *et al.* [1] as the basis for their prediction that 200 hospitals would be using FM by 1975. On the other hand, the 178 hospitals greatly exceeds the number of users recently reported by Dorenfest [5]. He found only 100 hospitals had acquired software from FM firms *and* software companies.

The future can best be predicted by examining growth over a period of time to establish a growth rate and then by modifying this growth rate before extrapolating into the future. The modification of the rate would be based upon other factors deemed to be important. The following analysis is in accordance with this procedure.

Earlier unpublished surveys conducted by the writer revealed that the number of Type 1 and 2 users (as defined in this article) had grown from:

31 hospitals in 1973 to
77 hospitals in 1978 to
178 hospitals in 1981.

Thus, the compound growth rate was 20% from 1973 to 1978, and 33% from 1978 to 1981.

The following factors will have some influence on the rate of growth for prediction purposes:

FM Usage by Small Hospitals

It is primarily one vendor, Compucare, that is capitalizing upon the small hospital need for FM support of the third variation of the classical FM kind—essentially this amounts to "as needed FM services." If this type of FM catches on and is pushed by FM vendors, it should have a very positive effect upon growth rate in terms of number of hospitals served and a lesser positive effect upon the total dollars spent by hospitals for FM.

Marketing of HIS and HMIS by FM Vendors

It is the writer's impression that the classical FM vendors have not marketed their services nearly as aggressively as have other vendors of financial systems, HIS and/or HMIS. Most classical FM vendors primarily provide services to hospitals in particular regions of the country rather than on a national basis. They probably lack expertise in and the sales force needed for marketing on a nation-wide scale. Until this deficiency is corrected, the classical FM vendors will be at a disadvantage in competing with turnkey HIS/HMIS vendors.

Hospital RFP's for HIS/HMIS with and without FM

Most hospitals have not seriously considered the advantage of FM. Accordingly, their requests for proposals (RFP) do not ask for bids with and without FM. Certainly such information should enable the hospital to make a more rational decision and one based upon input from all potential suppliers. It should be a prime objective of FM firms to spread the word and convince hospitals throughout the nation that the FM approach should be considered when planning to select and install HIS and HMIS. Until hospitals generally include an FM option in their RFP's, the FM firms will be at a disadvantage.

Larger FM Firms are Bullish

The writer found larger FM firms to be quite optimistic concerning their prospects for continued growth as did LaViolette.[6].

Satisfaction with FM by Hospital Users

The writer has not specifically surveyed hospital users of FM. However, over the years contacts with administrators of hospitals utilizing FM has lead the writer to conclude that almost all are more than satisfied with their FM services and in retrospect feel that they were quite correct in opting for FM.

Based upon the above discussion and the historical growth rate of FM, it is assumed that the growth rate for the next 4 years will be at the compound rate of 25% per year. At this rate the number of hospitals using FM by Spring, 1985, is predicted to be about 415 hospitals. It could be many more, if the small hospital "as needed FM services" concept catches on.

Some Lessons Learned

The suggestions contained in the 1974 article regarding the selection of FM vendor and contractual considerations continue to provide excellent guidance to the prospective FM user. The following suggestions are in addition to those in the 1974 article or are again stressed here in view of five years additional experience with FM:

(1) The contract should include an escape clause that permits cancellation of the contract by the hospital in the event the FM vendor fails to fulfill the contractual agreement in an acceptable manner.
(2) The contract should specify standards of quality, timeliness and documentation.
(3) Hospitals should avoid procuring FM services from firms which do not specialize in hospital FM or do not have substantial divisions which do so. Such firms will not have the health industry knowledge and expertise that should be expected of a firm that can truly assist a hospital. Neither will such firms be able to bring an arsenal of applications software designed specifically for hospitals—software that may be easily modified to meet each specific hospital's needs.
(4) One method many hospitals have used to select an FM vendor is the "request for proposal". By this means all potential vendors may be compared fairly and the best one for any particular hospital selected. Certainly as of this time all vendors should be considered, not just the one that approaches the hospital first. This will not only assist in the selection of the best one for the hospital, but will be useful in negotiating the contract.
(5) FM can be particularly attractive when a hospital wishes to interface hardware and/or turnkey systems from different manufacturers or vendors.
(6) In order to insure that hospital management does meet its data processing responsibilities, Compucare reports that a cornerstone to their FM contract is the establishment of a "hospital computerization committee" composed of department head users and top management. The committee emphasizes and assures that the hospital retains its responsibility to set data processing goals and to monitor the progress of the firm in meeting these goals. This, of course, is excellent doctrine whether the data processing department is FM or self-operated.

References

[1] Glenn Anderson, Gary L. Hammon, and Stanley Jacobs, "Facilities Management: Wave of the Future?", *Hospital Financial Management*, July, 1974.
[2] Articles by Howard Karten, *Computerworld*, August 14, 1978, p. 1, August 21, 1978 p. 7, and August 28, 1978, p. 12.
[3] Frederick A. Gross, "FM Works—If Supplier Acts as Specialist," *Computerworld*, May 14, 1979, p. 29.
[4] "Despite '75 Revenues, FM Losing Early Glamor," *Computerworld*, November 15, 1976, p. 53.
[5] Sheldon Dorenfest, "1980 Increase in In-house Finance Systems Signals Trend," *Modern Healthcare*, May, 1981, pp. 91–96.
[6] Suzanne LaViolette, "Facilities Managers' Rules Change As Hospitals Aim For More Control," *Modern Healthcare*, July, 1981, pp. 92–94.

III. AMBULATORY CARE SYSTEM

The use of computers to support ambulatory care is far more complex than the application of automation in a hospital setting. The units are smaller with little economy of scale, the data are required over longer periods of time, and the variability of individual practices is widest. Two approaches to deal with these difficulties are the use of standard, transportable products and the "do it yourself" approach. The latter is seldom cost-effective, but in this era of inexpensive equipment it is also quite seductive. The chapters in this section will address the basic issues and give some concrete examples.

The first thorough survey of systems which maintained an automated ambulatory medical record system (AAMRS) was conducted in 1974 and 1975 by Henley, Wiederhold *et al.* They identified sixteen operational systems which were sufficiently robust for a detailed examination. The significance of that report was that, while there were indications that the AAMRS would have an important impact on the delivery and management of health care, it was clear that much development and evaluation would be required before this potential would be reached. The first chapter of this section, by Kuhn, Wiederhold *et al.,* contains an update to that original report. It focuses on five followup sites. Four of these were included in the original study; three (COSTAR, TMR and RMIS) are described in this volume.

The next two chapters describe The Medical Record (TMR) and the Regenstrief Medical Record System (RMIS). The description of the TMR concept by Hammond, Stead, and their colleagues is included in this section because it provides an effective overview of the system. In the seven years since its initial publication, we have learned little about what we want the AAMRS to do; however, we have learned quite a bit about how to make an AAMRS work. Thus, the description is still timely.

The chapter by McDonald and his associates updates earlier descriptions of the Regenstrief Medical Record System. This system goes well beyond the typical AAMRS in that it includes user-oriented command languages whereby the user can customize reports to his needs and also define criteria for the printing of reminders. This ability to produce reminders from encounter data has created an environment for testing the effect of automated feedback on provider behavior. Most of our basic understanding of the impact of computer interventions on provider behavior comes from studies of this system.

Two other chapters describe systems used in larger settings. The one by Shapiro presents a system called SCAMP. It, too, contains tools for customizing outputs; it also supports a natural language query capability. The overview of the Core Record system, by McColligan, illustrates how a low cost AAMRS can be implemented in a large teaching hospital.

The next group of chapters describes COSTAR. This system is unique not only because of its ability to manage a computer stored medical record, but also because of the government commitment to produce and support a transportable product. The first chapter in the group, by Barnett and his coworkers, is descriptive. This is followed by two chapters which consider user satisfaction with (Kerlin) and proliferation of (Fiddleman) the system. The next two chapters (Studney and Kunstaetter, and Osburn and associates) are concerned with issues of local modification and transportability.

The final group of chapters focuses on the physicians' office. The chapter by Jelovsek and his colleagues considers the impact of the on-line TMR system in a medical office. The chapters by Forward and Schlager each describe a locally developed system. The final chapter of this section, by Brickman and his associates, provides a realistic appraisal of what the naive practitioner can expect from today's (1983) technology. They provide a useful set of questions to be asked by those considering the installation of an office practice system.

In conclusion, we have come very far since Henly and Wiederhold's first study. Systems are installed, in use, and paid for by patient care revenues. Unfortunately, we are not yet at the point where products oriented to patient care are routinely available to the office practitioner. To install a computer in an office practice which will do more than the financial and administrative functions entails a risk. Typically, practitioner investment in such a system will be motivated more by a personal interest than by a desire for economy or efficiency. Nevertheless, the availability of COSTAR, the lowering of equipment costs, and the relative ease of producing new products all suggest that effective tools will soon be available to help manage clinical data and support patient care. The challenge, of course, is in learning how to use and evaluate these tools.

Automated Ambulatory Medical Record Systems in the US*†

Ingeborg M. Kuhn, Ph.D., Gio Wiederhold, Ph.D., Jonathan E. Rodnick, M.D., Diane M. Ramsey-Klee, Ph.D., Sanford Benett, M.S., and Donald D. Beck

Introduction

From October 1974 through June 1975, a state-of-the-art review of Automated Ambulatory Medical Record Systems (AAMRS) was performed to determine how automated systems were being developed to improve the use of the patient's medical record in different types of health care settings. Automated Ambulatory Medical Record Systems are defined as systems that use computers to automate a substantial amount of the medical record for ambulatory care settings. The report of the 1975 study described in extensive detail the 16 sites visited and summarized the information collected for all other sites.[1] Some of the findings of the 1975 study were also reported by individual members of the study team in articles and at conferences.[2,3,4,5,6,7,8].

The 1975 study revealed that automated ambulatory medical record systems still required substantial development and evaluation. However, there were indications that effective systems were being designed and that, eventually, they would have an important impact on the delivery of health care and on the management of that delivery. The significance of the role played by the AAMRS management services indicated that the ideal automated ambulatory medical record systems of the future probably would have strong links with the

*This study was supported by a grant from the National Center for Health Services Research, HS04152-01 entitled "Automated Ambulatory Medical Record Systems Follow-up." A member of the site visit team was provided by The MITRE Corporation, McLean, Virginia. (Inquiries regarding the study and full report should be sent to Ingeborg M. Kuhn, Ph.D., 677 Beach Park Blvd., Foster City, CA 94404).

†This chapter was prepared expressly for this book. A more complete discussion of the topics is contained in Technical Report No. STAN-/CS-82-928, "Automated Ambulatory Medical Record Systems in the U.S.," by Kuhn, *et al.*, Department of Computer Science, Stanford University, Stanford, California, August 1982. A preliminary version was previously published in *Proceedings of the Fifth Annual Symposium on Computer Applications in Medical Care*, IEEE, 1981.

laboratory, the pharmacy, and the billing systems of each ambulatory care organization.

In recognition of a need for more current information on the nature of present AAMRS, a follow-up study was performed to determine whether the findings and conclusions of the 1975 study were still valid, and whether the field has progressed in ways that may have a significant impact on the design and implementation of future AAMRS. The follow-up study involved contacts to every site visited in 1975 for current information on the status of the medical record system, and visits were made during 1981 to six AAMRS sites throughout the country. Even though this study reviewed only a small sample of the AAMRS that are operational today, the systems and sites visited can be considered as representative of the systems developed for and implemented on minicomputers. The follow-up study team consisted of persons with expertise in medicine, medical information systems, health economics, and evaluation methodology. The sites and systems visited in the followup study are listed in Table 1. The sites visited in the 1975 study are listed in Table 7.

This chapter presents the findings on the current state of the AAMRS and comments on changes since the 1975 study. The following subjects are covered: Status of Systems Visited; Design Objectives, System Benefits, and Realization of Objectives and Benefits; Evaluation; Implementation, Acceptance, and Cost Justification; Transferability; System Characteristics; Technical Aspects; and the Future Potential of AAMRS. Tables are used to provide descriptive information on the systems visited in the follow-up study. Unless noted otherwise, the information presented in the tables describes the systems in general rather than the specific implementation of the system at the user sites visited.

Status of Systems Visited

Several of the promising systems visited in 1975 have matured to become operational and are undergoing transfer from a prototype demonstration in the research setting to the commercial market. Clearly, the leader in commercial availability of an indepth medical record system is COSTAR. Other systems in the process of becoming commercially available are TMR and RMIS.

There is still a need for research and system development to address the following issues:

The best match of the scope of automated services with the practice setting. For the smaller primary care practice, is automation of the medical record cost effective?

User Interface. Particular areas identified for further development include (1) encounter form design, (2) information presented to the user: surface behavior, explanations, CRT displays, graphic presentations, and (3) factors influencing effective human acceptance of the automated system.

User Accpetance and Motivation. In addition to user interface, acceptance

Table 1. Systems and User Sites Visited for the Followup Study

COSTAR	Computer-Stored Ambulatory Record System A modular system with the capability to totally computerize the medical record. This system was visited in the 1975 study. Developer: Laboratory of Computer Science, Massachusetts General Hospital, Boston, Massachusetts.
—NE	User: North End Community Health Center, Boston, Massachusetts.
—NC	User: North Counth Health Services, San Diego County, California.
TMR	Medical Record A modular system with the capability to totally computerize the medical record. This system was visited in the 1975 study. Developer: Division of Information Science, Department of Community Medicine, Duke University, Durham, North Carolina. Users: Renal Dialysis Unit and Nephrology Clinic, Veterans Administration Hospital; OB/GYN Clinic, Duke University Medical Center.
RMIS	Regenstrief Medical Information System A modular system that supplements and replaces some of the paper medical record. This system was visited in the 1975 study. Developer: Regenstrief Institute, Indianapolis, Indiana User: General Medicine Clinic, Regenstrief Institute, Indianapolis, Indiana.
ARION	Arthritis Research Information Office Network An administrative and medical data system developed by a private physician for practice management and data input to the ARAMIS project at Stanford University. The ARAMIS project was visited in the 1975 study. Developer/User: Arthritis Center, Wichita, Kansas.
FMIS	Family Practice Medical Information System A family-oriented information system providing administrative assistance to the family practice. This system was not visited in the 1975 study. Developer: Community Electrocardiographic Interpretative Service (CEIS), Denver, Colorado. User: Crow Hill Family Medicine Clinic, Bailey, Colorado.

problems may be linked to the need for better user training and documentation; or to the failure of the system services to adequately address the users needs.

System Transferability. What are the factors in system design and development that facilitate transfer from a prototype demonstration in a research setting to practical applications in a variety of settings?

Design Objectives, System Benefits, and Realization of Objectives and Benefits

The design objectives are more clearly defined than they were in 1975. The primary goals of current systems are defined in terms of improved quality of care and known administrative benefits. The major design objectives for the systems and sites visited during this study are summarized in Table 2.

Improved quality of care continues to be considered a major system objective. A well-organized display of historical, laboratory, and drug data in patient summaries and flow charts is available from the COSTAR, RMIS, and TMR systems. These systems also improve the ease of ordering lab tests or medicines and the ease of accessing the results. The pharmacy "subsystems" of TMR and RMIS are well thought out and provide additional information not usually available—such as telling the physician how much the prescription costs.

Increased medical record availability continues to be important in the large clinics where the traditional record is difficult to obtain for each visit. Record availability is a key objective at RMIS and access to the computer record has been taken one step further by having oncall physicians take a portable terminal home. These physicians can then have access to an emergency patient's record by a telephone communication link to the computer.

Management benefits are continuing to be an important objective. Financial and administrative services continue to be a major factor in facilitating system transfer, particularly in the commercial market. While reduction in administrative costs provides an incentive for installing an AAMRS, there is no indication that the potential for a reduction in the direct cost of providing health care is considered a justification for a system. There still is little evidence for the larger AAMRS that their benefits justify their costs. It is difficult to compare system benefits to costs, because of problems associated with the quantification of benefits and lack of detailed cost data.

Benefits that were seen as potentially important six years ago, but are considered less so now, include the use of medical diagnostic algorithms, notifying providers about drug interactions, and a heavy emphasis on meeting future PSRO requirements. At that time issues of privacy and confidentiality of systems were being debated; now these issues are less of a concern and are found primarily in the new systems.

It is still expected that the AAMRS will facilitate health services research. This is particularly true for those systems that provide the capability to search files and retrieve data on any combination of variables by means of query languages or report generators. The AAMRS have significant potential for assisting the acquisition of new knowledge in the following areas of research:

Health services research
Quality of care review methodology
Medical decision making
Epidemiology of disease
Medical education

Table 2. Objectives—Reasons for Implementing the Automated System

System Site	Costar NE	Costar NC	TMR	RMIS	ARION	FMIS
Improve quality of care through:						
Data Acquisition and Availability	XX	XX	XX	XX	XX	X
Record Accuracy and Legibility	XX	X	X	X		
Followup of Abnormal Tests	X	X	X	X		
Patient Surveillance	X	X	X	X	X	
Preventive Care	X	X	X	X		
Treatment Planning	X	X	X	X		X
Patient Compliance			X	X		
Feedback to Physician	X	X	X	XX		
Maintain continuity of care through:						
Record Availability	X	X	XX			
Improve patient services through:						
Patient Scheduling			X	X		
Pharmacy Management			X	X		
Laboratory Ordering and Reporting	X	X	X	X		
Facilitate research in the areas of:						
Quality of Care Review	X	X	X	X	X	X
Health Services		X	X	X		X
Medical Decision Making			X	XX	X	
Facilitate health provider training			X	X		X
Improve administrative services by:						
Providing Information for:						
Utilization Review	X	X	X	X		X
Budgeting		X	X		X	X
Long Range Planning		X	X			X
Funding Agencies	X	X				X
Faster Information Processing	X	X	X	X		X
Increased Accuracy	X	X	X	X	X	
Reduced Administrative Costs	X		X			X

XX—Major objective X—Objective

Clearly the realization of objectives is much greater than it was six years ago. The progress is most noticeable with the larger systems. Today, there is evidence that the medical record can be successfully stored on the computer, that the data can be used on a regular basis in the ambulatory health care setting, and that the systems can be transferred to other settings. Successful achievement of objectives is also evidenced by the availability of a variety of medical record systems as commercial products.

Evaluation

Formal evaluation studies of the impact of AAMRS were meager in 1975. This condition continues today, or the findings from such studies have not been reported. Of the few studies reported in the literature, most have addressed only a selected aspect of the effect of automation on the ambulatory care process. For example, the issue of quality assurance was the focus of a collaborative study between the Harvard Community Health Plan (HCHP) and the Laboratory of Computer Science, Massachusetts General Hospital. In this study, the impact of a computer-based ambulatory quality assurance program was assessed [10]. In a series of evaluation studies of the AAMRS at the Regenstrief Institute, the focus was on assessing the impact of computer reminders on physician behavior [11,12].

In considering ways to evaluate the impact of AAMRS on the delivery of health care, a distinction should be made between a site specific evaluation and an assessment of the general adaptability of a prototype AAMRS to other settings. Different factors should be taken into account. For a site-specific evaluation, the factors are resources required, installation time frame and costs, technical support needed, training of personnel, start-up, transition, and routine operation. For an assessment of the general adaptability of an AAMRS, the factors are system flexibility, modularity, programming language, computer and software support, and ease of modification.

For effective health planning and public policy making, there is a need for objective evaluation of both the impact and benefits of AAMRS. From the perspective of the marketability of an AAMRS, if a user or organization is willing to install a system and pay for it, that constitutes a practical evaluation of the system's worth. Additionally, continuing sales of AAMRS and evidence of profitable vendors may be considered as indicators of a product for which the benefits outweigh the cost.

Implementation, Acceptance, and Cost Justification

Major factors in the success of a system are provider acceptance and effective communication between the provider and the system designer. Provider acceptance appears to be best where the provider:

Was involved with system design,
Participated in the decision to install the system, and
Received training in the use of the system.

For RMIS, TMR, and ARION, some of the providers using the system were also involved in the design of the system. This gave them unique insights into the needs of the system and the concerns of their colleagues. For both COSTAR and FMIS, a large emphasis was placed on user training and ongoing interaction between the user/provider and the developer. This interaction has led to a gradual evolution of each of the AAMRS to better serve the providers and thereby gain their acceptance.

Another continuing important factor in successful implementation is strong leadership. Each of the systems visited possessed an individual or several individuals who were the driving force behind the system.

The importance of administrative services has been recognized for the systems that were originally designed to handle only medical data. Of the systems that were visited during the current survey, only one did not possess any financial subsystem.

There continues to be a lack of economic data to evaluate benefits derived from the AAMRS. Where cost savings are realized, they are most often realized from the administrative services provided by the system. While most system developers believe that the medical benefits of an AAMRS are the primary justification for the system, there is some evidence that the administrative services may be a primary justification to those who purchase and install the systems. For some of the modular systems, such as COSTAR, the administrative services are the first to be implemented, and in some cases, they are the only ones that are implemented [13].

New factors for successful implementation of an AAMRS were identified. It was observed that at sites that were actively installing an AAMRS in a new setting, user training and system documentation are critically important. There is a strong indication that the level of training has a direct correlation to the user's acceptance of the system. Each of the developers that were visited recognized this factor and have attempted to incorporate comprehensive training into the system installation process. Clearly, training and documentation go hand in hand in promoting acceptance. If a user does not know about the system's capabilities or how to use them and if there are no instructions for interacting with the system, it is unlikely that there will be strong incentives to effectively use the system. A few failures in attempting to use a system will lead to a high level of frustration and negative attitudes toward it.

Transferability

A conclusion of the 1975 study was that many systems or system components were suitable for transfer even though the development emphasis was to meet the unique local needs of the user institution. The current study has confirmed

Table 3. Transfer of Systems

System	COSTAR	TMR	RMIS	ARION	FMIS
Type of transfer					
Total System	X	X			X
Partial System	X		X		
Timesharing	X				X
Transfer agents					
Commercial Vendors	X	X	X		
Developer	X	X	X		X
Other	X				
Design characteristics					
High Degree of Standardization		X	X		X
Highly Flexible—allows modifications	X				
High Degree of General Applicability	X	X			X
Specialty Applications		X		X	
Site-Specific Design			X	X	
Modular Design	X	X	X		
Availability of technical support					
Software	X		X	X	
Documentation—Technical	X		X	X	
Documentation—User Manuals	X	X	X		X
User Training	X				X
Factors influencing transfer potential					
Strong Leadership/Key Individuals at:					
Development Site	X	X	X	X	X
Transfer Site	X				
Language—widely available and supported	X		X	X	
Low Cost				X	X
High Cost	X	X			
Minimal Installation Requirements				X	X
Hardware Restrictions				X	X
Existing Demand for the System	X	X	X		X
Software status					
Publicly Available	X			X	X
Proprietary	X	X	X		

that AAMRS are transferable, even those that were not developed with transfer in mind. Nearly every system visited in the earlier study that is still active has had some type of system transfer, either through the addition of new users or by transfer to a new site.

Table 3 shows the status of system transfer and the nature of factors relating to system transfer for each of the systems visited. In this table, three types of system transfer are identified. The first is total system transfer, which applies to the installation of a complete system at a new site. The second is partial system transfer, which applies to the installation of components or modules of a system at a new site. The third is timesharing, which applies to systems that offer services to several independent users on a timesharing basis.

For all systems visited, there is clear evidence of private sector interest to market the system or a component of the system. Today COSTAR is the most widely distributed system in the commercial market and has the greatest number of vendors. TMR is being marketed by its developers. Some components of the RMIS system are commercially available from Digital Equipment Corporation (DEC). Except for the ARION system, all of the system developers were also directly involved in transferring a part or all of their system to new sites.

Design Characteristics

Significantly different design concepts exist for systems undergoing transfer. To permit implementation at a variety of health care settings, the COSTAR system is designed to be highly flexible in terms of permitting modifications upon installation to meet the unique needs of the user. The design philosophy of the TMR system is that a particular health care setting can meet its unique needs by selecting an appropriate subset of reports from the large menu of standard reports which were designed to meet the needs of a variety of health care settings. Regardless of the design concept, all systems require a substantial amount of technical support from the vendor or from in-house capability for the initial implementation. Technical support is most important in the areas of software, documentation, and user training.

While the technical constraints on transfer are diminishing, there is no evidence of "turnkey" AAMRS systems being developed that also provide for the collection of a moderately high level of medical data. A "turnkey" system is a system that can be installed and implemented by the purchaser without any technical support from the vendor or other agent. One design feature leading to ease of transferability is the development of modular systems such as COSTAR. Such a system can be transferred in whole or in part depending on the needs of the user.

Design characteristics that influence the transferability of a system include

(1) the degree of general applicability;
(2) the ability to meet specialty needs; and

(3) the extent to which the system was designed to meet the specific needs of the site in which it was developed.

A system that has a high degree of general applicability may serve a wider market than one that is designed to meet specific specialty needs or the unique needs of a site. However, a system that has both general applicability and the ability to meet specialty needs may have an even broader potential market.

Factors Influencing Transfer Potential

In addition to design characteristics and availability of technical support, several other factors will have either a negative or positive influence on system transfer. These factors include:

Leadership characteristics at the development site and at the user site.
The nature of the computer language used and whether it is widely available and supported.
The cost of the system.
Ease of installation in terms of timing, training requirements, site specific system modifications, and impact on existing procedures.
The existence of any hardware restrictions.
The demand for the system, or the existence of a market for the system.

Software Status

The status of system software is characterized in terms of being publicly available or proprietary. Software that is publicly available includes software in the public domain and software that is available from the developer without restrictions on its use. While publicly available software may be obtained by anyone for free or for a small charge, there may not be adequate documentation or technical support available to permit duplicating the system. Proprietary software includes software that is owned by the developer or a commercial vendor and that is generally available only under licensing or other contractual agreements. The fees associated with proprietary software may be significant.

Market Characteristics

Marketing constraints on system transfer is an area of concern today. It is expected that the first lessons may be learned from COSTAR, because that system is on the market from a variety of vendors. It will be interesting to learn whether the cost of selling, installing, and supporting a highly flexible system can prove to be economically feasible. Also, to what extent does the use of an uncommon computer language represent a constraint?

Based on the follow-up study, the characteristics of a market for the larger AAMRS are:

A group practice with four or more providers. The benefits of an AAMRS increase as the number of providers and sites using the system increases. An exception is the FMIS, which is a timesharing system designed for the rural solo practitioner.

A need for automated financial and administrative services. The primary need and justification for the larger systems are to serve administrative needs of the practice. Practices have justified automated administrative services where there is:

A need for practice management data,

Evidence of lost charges, or

A problem with cash flow.

The potential to improve medical care by the acquisition of automated medical data services. The greatest potential for improved care from the use of medical services exists where there is:

A need for record availability, legibility, and common organization such as in groups with multiple sites, multiple providers providing care to the same patient, or a patient population with complex or chronic problems resulting in high volume medical data.

A need for data organization and report formatting. These needs may be found in certain subspecialties and for chronic diseases which require a great deal of lab work. These settings would derive significant benefits from the analytical and formatting powers of the AAMRS.

A desire to improve quality of care or to improve procedures for quality of care assurance review.

A setting where there is a desire to use the database for research.

Research needs resulting from either the institutional setting or individual physician motivation.

System Characteristics

The systems visited in 1975 and 1981 evolved from the characteristics of the setting in which they were designed. That is, system services first meet the needs of the local setting. Table 4 summarizes the system services available from the AAMRS visited in the current study. As the systems are prepared for transfer, modifications are made to make the systems suitable for different settings. New modules were added to the COSTAR system to facilitate transfer of the system. These modules provide financial services, and data retrieval and report formatting services. For TMR, the set of standard reports and services has been expanded to meet the anticipated needs of a variety of settings.

The most common vehicle for data capture is still the encounter form. Data entry methods are relatively consistent for all systems visited. Data entry is predominantly performed by clerical personnel. Experiments in physician interaction with direct data input and retrieval appear to have not been fruitful.

Table 4. System Services

System	COSTAR	TMR	RMIS	ARION	FMIS
Medical:					
Patient Summary	X	X	X		X
Encounter Form for Visit	X	X	X		
Medical Flow Sheets	X	X	X		
Encounter Reports	X	X		X	X
Hard Copy Output					
Integrated with Traditional Medical Record			X	X	
Fully Automated Medical Record	X	X			
Administrative: patient services and reports					
Appointment Scheduling	X	X	X		
Visit Reminders	X	X	X		
Visit Registration	X	X	X	X	X
Operating services and reports					
Medical Record Pull Lists	X	X	X		
Pharmacy Labels		X	X		
Drug Costs Reported to Physician		X	X		
Third Party Eligibility	X	X	X		X
Billing and accounting services					
Accounts Receivable and Billing	X	X		X	X
Patient—Bills and Statements	X	X		X	X
Third Party—Statements	X	X		X	X
Other Accounting Services	X	X		X	X
Accounting Data Input to Other Systems			X		
Utilization reports					
For Internal Use	X	X			X
For Outside Agencies	X	X			X
Other reports and data					
Quality of Care Review	X	X	X		
Health Care Planning	X	X	X		X
Data retrieval					
On-line Inquiry	X	X	X		X
Database Searches	X	X	X	X	X
Terminals in Off-Site Locations	X	X	X		X
User-Defined Reports	X	X	X		X
Other uses made of the system					
Medical Education		X	X		X
Research	X	X	X	X	X

While it was recognized in 1975 that encounter form design required careful attention and research, there is little evidence of improvements to the encounter forms in use today.

There continues to be a need for additional research in the design and use of the encounter form as well as a need for research in other areas of data entry. Other types of data entry that may prove suitable for AAMRS are on-line menu selection or voice data entry. On-line menu selection is currently used in some hospital information systems. Voice data entry is a more recent development without any applications in practical use today, but it has promise for the future. A potential advantage of these types of data entry is that the providers may find them easy to use and thus be more receptive to doing their own data input and retrieval.

Table 5 shows the capability of the systems for storing data in the computer. It can be seen that for the systems that can capture a large amount of medical data, there is little variation in the types of data that are stored in the computer. Even though a system has the capability to store large quantities of medical data, the amount actually stored varies among different installations.

Table 5. Content of the Automated Record

System	COSTAR	TMR	RMIS	ARION	FMIS
Medical data					
Medical History and Physical Exam	X	X	X	X	
Problem Lists	X	X	X		
Progress Notes					
Free Text	X	X			
Coded	X	X	X	X	
Patient Care—Plans/ Follow-up					
Lab Orders	X	X	X		X
Lab Results	X	X	X	X	
Drug Orders	X	X	X	X	
Disposition	X	X	X	X	X
Amount of medical data					
Comprehensive	X	X		X	
Medium	X		X		
Sparse	X				X
Administrative data					
Patient Identification	X	X	X	X	X
Patient Financial Information					
Account Status	X	X		X	X
Billing Information	X	X	X	X	X
Payment Information	X	X		X	X
Patient Visit Information	X	X	X	X	X
Appointments	X	X	X		

Technical Aspects

With respect to technological system design, it is the conclusion of the study team that there have been no significant changes or innovations since 1975. Systems were developed along lines intended. There was little evidence of new innovations in design or termination of significant features. It appears that the use of new technology has been definitely slower than the rate of development of new technology. Table 6 shows the designer, computer language, and computer used for each site visited in the current study along with associated costs.

The predominant hardware for AAMRS at the present time are minicomputers. Most of the developers planned to remain at the minicomputer level. The developer of TMR is in the process of implementing and testing a version of their system that will run on a microcomputer. The microversion is being developed in addition to the current minicomputer version, not as a replacement. If they are successful in this endeavor, it may open the possibility of placing an AAMRS into the small practice. The current minicomputer-based systems are too expensive for anything less than a group practice which has several providers. The development of a usable, well-documented system that runs on a microcomputer would do much to increase the number on systems in the field.

Hardware is not a determining factor in describing the differences of the observed AAMRS. The hardware is basically the same for most of the systems. The hardware being used today is generally reliable. While system reliability was discussed extensively during the 1975 visits, it is not an issue today.

As can be seen in Table 6, each of these systems has been written in a different programming language. This is insignificant as long as each of these languages gains widespread support and acceptance. The acceptance is important if personnel at the user site are expected to make site-specific modifications. Languages which earlier were considered uncommon, MUMPS and MIIS, have been promoted and are now more widely used. However, when compared to other high level languages in use today such as PASCAL, FORTRAN, or BASIC, the MUMPS and MIIS languages are still relatively uncommon.

The systems observed all run in an on-line mode. As far as could be determined, only RMIS ran a "batch mode" to enter data into the system. All of the systems use on-line CRTs to query the databases for patient medical and financial information. Two of the systems, COSTAR and RMIS, allow the user to define his or her own report format.

A major concern that was found in this area was a lack of software competence at the user sites. Except for RMIS and ARION, which are written in BASIC, the other systems visited in this study (Table 7) were written in languages that are not being taught in data-processing or computer science curriculums. At each of the sites visited, with the exception of those sites where the developer was in residence, there were no individuals that felt comfortable with making even minor software modifications. This puts the burden of tai-

Table 6. System Summary Description

System	COSTAR-NE	COSTAR-NC	TMR	RMIS	ARION	FMIS
Designer	Laboratory of Computer Science, Massachusetts General Hospital	Laboratory of Computer Science, Massachusetts General Hospital	Div. of Information Science, Dept. of Community Medicine, Duke University	Regenstrief Institute	Arthritis Center	Community Consortium
Language	MUMPS	MUMPS	GEMISCH	VAX BASIC	BASIC	MIIS
Computer system						
Type	DEC PDP 11	DEC PDP 11/70	Variety of DECs	DEC VAX 11/780	WANG	DEC PDP 11/50
Services	Shared	Dedicated	Shared	Shared	Dedicated	Dedicated
Size	Mini	Mini	Mini/Micro	Mini	Micro	Mini
Operating System	On-Line	On-Line	Timeshared	Timeshared	On-Line	On-Line
Terminals and Other						
User Interfaces	CRT, Printer	CRT, Printer	CRT, Printer	CRT, Printer	CRT, Printer	CRT, Printer
Num. of Terminals	20	30	20	45	2	25
Terminal Users	Medical, Clerical	Medical, Clerical	Medical, Clerical	Medical, Clerical	Medical, Clerical	Medical, Clerical
Costs						
Investment		$284,100			$25,000	
Operating Per Year	$33,420	$ 82,000			$22,000	$348,000
Source of Funds	Grants In House	Grants	In House	In House	Private Some Grant	Grant User Fees

Table 7. Sites Visited for the 1975 Study and Their Current Status

Site	Description
ARAMIS	Division of Immunology, Stanford University Medical Center, Time-Oriented Data Bank (TOD), American Rheumatism Association Medical Information System (ARAMIS). ARAMIS remains operational; the number of remote sites has increased; developmental work continues.
AUTOMED	Medical Data Systems Corporation, Olmsted Falls, Ohio. A private for-profit service bureau. Primarily financial services for solo practitioners and small groups. Some medical data in the record; amount depended upon individual user. This system is still in full operation.
BELLEVUE	Pediatric Outpatient Clinic, Bellevue Hospital, New York City. Hospital-based clinic and eMergency services. A supplement to the medical record, and medical record pulling service. The computer project serving the Pediatric Outpatient Clinic was terminated in 1976.
BUMED	The BUMED Computer Project, U.S. Naval Air Station Dispensary, Brunswick, Maine. Military dependents health services. A fully automated medical record. The project was cancelled at Brunswick in 1976.
CC	Cardiovascular Clinic. Oklahoma City, Oklahoma. Private group practice. A fully automated medical record system. Still in Operation.
COSTAR	Harvard Community Health Plan, Laboratory of Computer Science, Massachusetts General Hospital, Cambridge and Boston, Massachusetts. A fully automated medical record system. This system was visited for the current study.
DHD	Appalachia II District Health Department, Greenville, South Carolina, Public health clinics: program orientation. The system continues to be primarily administratively oriented, but is collecting medical data.
HIP	Casa de Amigos, Department of Community Medicine, Baylor College of Medicine, Houston, Texas. Community—neighborhood clinics. The Health Illness Profile (HIP) supplements the medical record. The system is still in existence, but at reduced level.
IHS	Indian Health Service, Tucson, Arizona. Clinics on the Papago Indian Reservation. The Patient Care Information System (PCIS) is a supplement to the medical record. The system is now in a microfiche rather than an on-line mode.
ITC	Insurance Technology Corporation, Berkeley, California. An information system used in processing Workman's Compensation Claims. The site visited in 1975 has been discontinued. Other sites remain operational.
LA	County of Los Angeles, Department of Health Services, Los Angeles, California. A Patient identification and record retrieval system serving hospital-based clinics in the County Medical Center and another system serving the East Los Angeles Child and Youth Clinic (ELACYC).

Table 7. (*Continued*)

Site	Description
	The County Medical Center system has increased the amount of medical information provided. The ELACYC system has not changed significantly.
MSIS	Multi-State Information System, Rockland State Hospital, Orangeburg, New York. Mental health services hospital and community-based clinics. The system collected primarily administrative data used to prepare required reports to supporting agencies.
	The system is now 100% user supported. Hardware and service improvements have been made.
MUSC	Department of Family Practice, College of Medicine, Medical University of South Carolina, Charleston, South Carolina. Residency family practice clinics. A fully automated medical record system, used in training.
	The orientation of the system is to provide a clinical, administrative, and research tool for practicing physicians. Several new features have been added.
RMIS	Regenstrief Institute, Indiana University Medical Center, Indianapolis, Indiana. Hospital-based clinics. Development of a medical record supplement.
	This system was visited for the current study.
TMR	Duke University Medical Center, Department of Community Health Services, Durham, North Carolina. Student health services and hospital-based clinics. A total medical record system was planned; only administrative services were operational in 1975.
	This system was visited for the current study.
YALE	Section of Medical Computer Sciences, Yale University School of Medicine, New Haven, Connecticit. A medical record information system for an HMO group practice with a goal to have the total medical record automated; minimally operational at time of visit.
	The system is no longer in existence

loring the system for the specific sites on the vendor or system developer. It is the opinion of the visit team that in order to implement one of these systems, it is essential to have some software competence at the site during the implementations.

Future Potential of AAMRS

It appears that the general development and growth of the type of AAMRS examined in this study is following that of Hospital Information Systems (HIS). The development, implementation, and acceptance of HIS have been slow, and most systems are primarily administrative/financial in character. It is predicted that the large AAMRS will take the same path.

In consideration of the current lack of evaluative information on AAMRS today, it is recommended that for anyone considering the implementation of any type of AAMRS, a visit should be made to sites where the system is operational. Particular attention should be paid at the site to how the system actually works; the requirements for implementation; costs associated with implementation, training, and routine operations; and any special or unique problems that need resolving.

The large AAMRS available today should be considered first generation systems that have evolved through modifications and additions to those systems examined in the earlier 1975 study. As lessons are learned from the transfer and commercialization of these systems, new generations should be developed that will be more responsive to the needs of the market and that will incorporate current and future technological advances. Important issues to be addressed in the development of new AAMRS include

(1) how to get the practitioner more involved with using the system;
(2) more efficient methods of data capture and entry; and
(3) improvement to user interaction with the system.

For the private physician or small group practice, there is strong evidence that the microcomputer will have a major influence on the development of ambulatory medical record systems. Individual physicians are acquiring the low cost microcomputers and slowly creating systems to meet individual needs. Because the use of a micro does require initiative on the part of the physician to acquire some level of technical computing competence, it is not clear how widespread the use of micros will be. Many of the initial systems will stress a few services out of the spectrum of the physician's personal needs. However, as soon as acceptable and effective medical record software packages are available for the micros, the use of computers as aids to the practice of medicine may change significantly.

In the short run, it is not expected that an AAMRS will have a measurable impact on patient outcome. That is because the AAMRS provides services to the process of providing health care. Even though the process may be improved, it is not clear that the patient will be healthier. However, the patient may be happier because the care was perceived to be better. The extent of AAMRS impact on the process of providing health care will depend on the providers' acclamation to the idea of an automated record, and whether they learn to use the system effectively.

In the long run, there is every indication that AAMRS can have a significant influence on patient outcome, particularly from the results of productive research using data from and the analytical capability of the larger AAMRS. Furthermore, future incorporation of new technologies such as artificial intelligence into AAMRS has great potential for improving medical decision making and medical education.

References

[1] Henley, R., Wiederhold, G., Dervin, J., Jenkin, M., Kuhn, I., Mesel, E., Ramsey-Klee, D., and Rodnick, J., "An Analysis of Automated Ambulatory Medical Record Systems," Technical Report #13, Office of Medical Information Systems, University of California, San Francisco Medical Center, June 1975, NTIS #PB 254 234.

[2] Kuhn, I. M., "Looking at Outpatient Medical Record Systems," *Hospitals, J.A.H.A.*, October 16, 1977, pp. 145–152.

[3] Rodnick, J. E., "The Use of Automated Ambulatory Medical Records," *The Journal of Family Practice*, Vol. 5, No. 2, 1977, pp. 253–264.

[4] Rodnick, J. E., and Wiederhold, G., "A Review of Ambulatory Medical Record Systems in the United States: Charting Services That Are of Benefit to the Physician," *MEDINFO 77*, D. B. Shires and H. Wolf, Eds., New York: North-Holland Publishing Co., 1977, pp. 957–961.

[5] Wiederhold, G., and Kuhn, I., "Effective Services of Automated Ambulatory Medical Record Systems," *Policy Analysis and Information Systems*, in Vol. 1, No. 2, 1978, pp. 21–32.

[6] Wiederhold, G., and Mesel, E., "Automated Ambulatory Medical Record Systems," *Proceedings of the 1975 MUMPS Users' Group Meeting*, J. Zimmerman, Ed., Washington University, Biomedical Computer Laboratory, St. Louis, MO., 1975, pp. 182–191.

[7] Wiederhold, G., "Alternatives in Database Design for Ambulatory Records," *Proceedings of the Third Illinois Conference on Medical Information Systems*, 1976, pp. 78–87.

[8] Wiederhold, G., "Databases for Health Care," in *Lecture Notes in Medical Informatics*, Lindberg, D. A. B. and Reicherts, P. L., Eds., New York: Springer-Verlag, 1981.

[9] Kuhn, I. M., and Wiederhold, G., "The Evolution of Ambulatory Medical Record Systems in the U.S.," *Proceedings, The Fifth Annual Symposium on Computer Applications in Medical Care*, H. G. Heffernan, Ed., IEEE Computer Society Press, 1981, pp. 80–85, IEEE Catalog Number 81CH 1696-4.

[10] Winickoff, R. N., Barnett, G. O., Coltin, K. L., and Morgan, M. M., "A Computer-Based Ambulatory Quality Assurance Program," Final Report, Harvard Community Health Plan and the Laboratory of Computer Science, Massachusetts General Hospital, October 1979.

[11] McDonald, C. J., "Protocol-Based Computer Reminders, the Quality of Care, and the Non-Perfectability of Man," *New England J. Med.* **295,** 1976, pp. 1351–1355.

[12] McDonald, C. J., Wilson, G. A., and McCabe, G. P., Jr., "Physician Response to Computer Reminders," *J. of Am. Medical Assn.* **244,** 1980, pp. 1579–1581.

[13] Fiddleman, R. H., "Survey of COSTAR Installations and Vendors," MITRE Technical Report MTR-81W122, The MITRE Corporation, June 1981.

Data Base Management System for Ambulatory Care*†

W. E. Hammond, Ph.D., W. W. Stead, M.D., S. J. Feagin, B. A. Brantley, M.D., Ph.D., and M. J. Straube

Introduction

Despite the fact that over 90% of patient/physician encounters will occur in the ambulatory care setting, considerably less data is available about this group than the smaller inpatient group. The rising cost of health care delivery has stimulated an increased dependency on ambulatory care with a concomitant need for better resource utilization. Concurrently, more information about the ambulatory encounter is being requested by a variety of private, state and federal agencies for planning, utilization review, standardization, and research. The consequence is that any physician, whether in solo or group practice, has an increasing need for timely and efficient organization of administrative, financial and medical data. In order to meet these additional information demands, most physicians have had to increase their manpower commitment for data collection, processing and dissemination. These requirements enhance the desirability and perhaps create the necessity of using the computer in certain clinic environments.

Most attempts to integrate the computer into the ambulatory care setting have had, at best, only limited success. The primary limiting factor has been that computers could not be cost justified. Other important factors include inflexibility, unreliability, nonsuitability, lack of physician acceptance, and awkwardness of use. The computer will become an important component in the health care delivery system process only if it can reduce manpower requirements while meeting the increasing information demands.

A careful examination of staffing, provider objectives and behavior, patient expectations, and time-flow considerations reveal a very narrow window in which the computer can influence these factors. Operational cost reduction may be achieved with increased health care availability by a reduction of cler-

*This work was supported in part by HEW Grant No. 5-R01-MB-00123 (Bureau of Health Resources Development).

ical and financial personnel, by immediate billing and rapid claims processing, by increasing physician throughput, by minimizing the cost of health care mistakes, and through proper management to provide complete resource utilization. Health care may be improved through more accurate and more available medical records, prompting the provider on follow-up, monitoring time lines for preventive care, and providing consultative services. Furthermore, the organization required for computerization of a medical record makes it easier to follow chronic problems, patient functional status, treatment effectiveness, and interactions among problems.

This chapter describes a data base management system (DBMS) which attempts to meet the above objectives and discusses our experiences in introducing the computer into several ambulatory care settings. The chapter also discusses briefly the software design and details some functions of the applications programs.

Variations of the DBMS have been used in several clinics associated with Duke University Medical Center for the past several years. Initial system development began with the Division of Obstetrics in 1970 [1]. This system supports an on-line obstetrical record and has over 14,000 patient records in its data bank. The obstetrical system provided experience in dealing with a detailed medical problem of limited time span. The University Health Services Clinic (UHSC) served as the pilot project for a primary care system which was initially implemented in 1974. This clinic provided experience in areas of administrative and financial management. The UHSC serves approximately 150 patients daily and has over 26,000 active patients in its data bank. The Family Medicine Center (FMC), whose computer system was implemented in 1975, provided an opportunity to develop programs useful in primary care education and evaluation. This clinic has approximately 90 encounters daily and has an active file of 8500 individuals. The Renal Outpatient Clinic began a limited use of the system in 1977. This activity has subsequently been shifted to the renal dialysis clinic at the Veterans Hospital in Durham. The renal clinics provided an opportunity for working with patients with detailed medical data and highly interactive and complicated problems and also gave us experience in the direct input of data by the physician.

The data base management system is written in GEMISCH [2] and runs under Digital Equipment Corporation's RSX 11D operating system. The system supports 16 simultaneous, interactive users as well as several high speed and medium speed printers on a PDP 11/45 with 120K words of memory. Most video terminals are driven full duplex at 9600 baud, and the average response time is less than 0.5 seconds.

System Philosophy

The data base structure is focused on patient care data, and from this data base, all other functions, including financial and management, are derived. Consequently, the definition of the data base is extremely important. The detail of the data base is limited by cost in time and money of both collection and storage, and therefore each data item must be specifically considered and defended for inclusion into the data base. Data should not be collected simply because it is available and might be useful. Data should be collected only if its use in a particular decision-making process can be identified. Our experiences with data definition have made us aware of the difficulties in following these guidelines. Providers have misconceptions about what data they record and what data they actually use. When asked to identify what data is necessary for patient care, physicians seem to remember their medical school training, and required data seems to grow without bounds. It is very easy to fall into the trap of identifying exceptions and trying to define a data base which will handle all problems. Instead, the most frequent circumstances should be defined and free text narrative should be used to handle the exceptional cases. Data should also have a specific time limit for remaining in the active data base.

Although the record is defined primarily for patient care, this feature is of lesser importance than administrative functions when trying to interest a clinic or a physician to use an automated system. In fact, the computer must perform the administrative and financial functions of the clinic to displace, at least in part, its added cost. Furthermore, this area provides one of the few measurable parameters for cost-justifying the system.

The input of data into the computer is the next critical area of design. Logon procedures must be uncomplicated but still provide adequate security. Data entry must be simple, fast, unambiguous, and both flexible and yet structured enough to insure correctness. Mistakes must be easily corrected. We have designed two modes for the input of data. One is the direct, real time, interactive entry of data by the person collecting it, and the other is the indirect, delayed entry of data through the use of encounter forms. In the first case, the data originator interactively enters parameters of his choosing and holds the main responsibility for directing program flow. The indirect mode is a more rigidly structured conversational flow. Both modes require a certain minimum of data to organize correctly the resulting medical record. Multiple and unique levels of password protection aid in record confidentiality and security.

Our experiences have suggested that a successful DBMS must be easy to change. Different clinics have different needs and different ways of doing things. Our design has been influenced by this fact and permits modification through dictionary files independent of programming.

Software Design

The DBMS consists of several program modules which perform all tasks related to the function of that module. A monitor section logically connects these units to provide for the patient and information flow through the clinic. The modules provide for the following functions: appointments, collection of demographic and financial data, problem orientation, laboratory studies, therapies, time-oriented data, and accounting. The programs also support such medical functions as protocols, drug interactions and MD-assist algorithms and such administrative functions as patient billing, third party claim filling, and numerous financial and management reports.

A major innovation of the system is the flexibility provided through the use of dictionary files. The dictionary is a self-contained direct access file which defines various clinic specific parameters such as a definition of the clinic environment (clinic name, types of visits, etc.), content of minimal data base, provider identification, problem names and codes, studies, therapies, and supplies. Cost data is included with each item as appropriate. The dictionary also provides for protocol definition of data to be collected and studies to be performed as a result of either problem selection or visit type. Modifying a parameter in the system becomes a simple task of altering its entry in the dictionary. The dictionary is also used to provide flexibility in displaying menu choices on displays which may vary from clinic to clinic.

Codes are used extensively within the system to represent most parameters, providing standardization and reduced record size. Two important features of the software make this an acceptable choice. The first is the opportunity to use a narrative entry for any variable when the coding scheme fails. The second is a data entry algorithm which provides a very efficient translation between text entry and codes.

Speed of data entry is further enhanced through the use of read-ahead programming. That is, if the appropriate data has already been entered, the prompting frame for that data entry is skipped. Finally, the computer prompts with as much anticipatory data as possible. For example, since most of our patients are from Durham, NC, this data is preloaded from the dictionary into the collection frame for city, state. Only if another city is correct does further entry have to be made.

Application Program

One requirement which all practices have in common but accomplish with varying degrees of success is that of an appointment system. Because of the value of a provider's time, he wishes to schedule his practice to the fullest extent; yet he realizes the necessity for leaving time for emergencies and walk-ins. Problems are further compounded by the changing schedules of both providers and patients, by the number of no-shows, and frequently by multiple scheduling of patients, particularly in a group multispeciality practice when more than one provider may see the same patient.

The appointment system is integrated with the medical record, eliminating redundancy of data collection and entry. Data are more timely and accurate, and for the price of a single data entry, many functions manually achieved through multiple index files are economically and efficiently achieved. Appointments can be made by provider, by clinic, and/or by time. Appointment lists can be printed and workload assessments can be made. Postcard reminders to patients with appointments are generated automatically and decrease significantly the number of no-shows. In one clinic, the use of the postcard reminder reduced the number of no-shows by 30%. Automatic rescheduling is performed when required. All of the patient's appointments can be easily reviewed. This one feature can represent considerable savings to patients in reducing the number of separate visits to a clinic.

A computerized appointment system also permits analysis of no-shows and cancellations. By adding checkin and checkout times for various encounter points within the clinic, patient flow can be evaluated as a function of various parameters such as provider, problems, pathway, or time of day. Comparison of provider encounter times with procedure codes permits evaluation of charges for office visits. Timing data can also be used to evaluate the increase of throughput as a function of the use of paramedical personnel.

Initial registration data are collected on-line which provides improved control of data collection. Registration requires approximately 3 minutes. Forced input and computer checking of data to the fullest extent possible insures a nearly complete data base. Delayed redisplay of data just entered for verification by both patient and data terminal operator also increases accuracy of data. For a subsequent visit, the data base is easily updated in approximately 15 seconds, again by both patient and data terminal operator. This updating is particularly important and cost-effective for maintaining correct billing addresses and insurance data.

Direct Entry

Problems or diagnosis codes are entered interactively. Selection of a given problem leads the data enterer through a series of displays which prompt for additional problem entries as either causes or manifestations of the original problem. These linkages are clinic specific and are defined in the dictionary. Problems are classified as primary, secondary or minor. Minor problems are automatically assigned to the bottom of the problem list. A recommended workup, with results if available, is optionally displayed with each problem.

The studies summary provides access to the last result of each test. The dictionary controls display of abnormal results and permits clinic-specific grouping of tests to form individualized packages.

In addition to providing a dictionary-defined, clinic specific formulary of the most commonly used drugs, the therapy section checks for reactions to medications taken in the past and monitors for drug/drug interactions. Prescriptions are printed and therapies which will expire prior to the next scheduled visit are flagged for the provider.

The primary purpose of the time-oriented flow sheet is to permit the provider to follow critical parameters reflecting the current functional status of the patient. Parameters may be added to the flow sheet through manual and automatic methods: the provider may select either a coded or free text parameter; parameters may be added through a protocol linked to a given problem (hypertension adds weight and blood pressure to the flow sheet); all abnormal studies post to the flow sheet; and all therapies are added to the flow sheet.

Indirect Entry

In some clinics, it may not be possible or desirable for the provider to interact directly with the data base through a video terminal. In these cases, a preencounter summary is generated for the provider. This summary, illustrated in Figure 1, contains the permanent problem list and a visit profile which graphically presents all problems, medications, and providers dealt with over the past 10 encounters. This visit profile permits a rapid evaluation of what has happened to the patient, how frequently the patient is seen and for what, continuity of care through various providers, and of the types and, to some extent, the effectiveness of the medications.

```
FAMILY MEDICAL CLINIC - 08/01/77                             * CRANE, ICABOD
SSN: 999-99-9999       ID:  01-999-99                        *   999 DUKE CHAPEL ROAD
Resp. MD: WELBY                                              *   DURHAM, NC 27704
Employer: PARTS MAN                                          *    PH: 383-9999
      PH: 684-6421                                           *   DOB: 08/19/39
NK: WIFE, MARY SUE CRANE                                     *
    same address                                             *   37 yo wh mar male
                                                             ******************************
VISITS: 76(4),  77(8)                              LAST PE: 05/03/76
-------------------------------------------------------------------------------------------
RECORDED                   PROBLEM LIST                                LATEST ENCS
-------------------------------------------------------------------------------------------
03/27/77         PEPTIC ULCER                                        04/06/77   2

01/27/77         DEPRESSIVE NEUROSIS                                 03/09/77   5

04/02/76         ANXIETY NEUROSIS                                    02/23/77   5

04/02/76         ALCOHOL ABUSE                                       03/09/77   7

-------------------------------------------------------------------------------------------
                                     VISIT PROFILE
                                        YEAR   <----------77----------------X-76--->
             PROBLEM                   MONTH:   6   4   3   3   2   2   2   1   9   6
-------------------------------------------------------------------------------------------
SPRAIN/STRAIN ANKLE                             *
PEPTIC ULCER                                        *   *
ALCOHOL ABUSE                                               *   *   *   *   *       *
DEPRESSIVE NEUROSIS                                         *   *   *   *   *
ANXIETY NEUROSIS                                                *   *           *
ANOREXIA                                                                        *
...........................................................................................
WELBY                                           *   *   *   *           *   *
KILDARE                                                         *   *
SMITH                                                                           *   *
...........................................................................................
MAALOX                                          *   *   *
LIBRIUM                                         *   *   *   *   *   *   *   *
VALIUM                                                                          *
```

Figure 1. Sample preencounter printout.

The pre-encounter summary also provides information to the provider concerning outstanding laboratory tests, suggested tests or therapeutic interventions as a result of previous studies or treatments, and preventive care prompters. The last values of subjective, physical, laboratory and therapy parameters are included from the flow sheet with blanks provided for new values.

The data collection form can be either the patient specific pre-encounter form or a preprinted form. Data entry is provided through a single display frame which is optimized for rapid input of data. For a trained operator, all encounter data can be entered as a string of labels and codes without prompting. Entry times vary from 15 to 30 seconds.

After the patient is checked out, an itemized bill is displayed along with other accounting and insurance data. Payments may be made and recorded at this time.

The data base is independent of the mode of entry, and data entered via one mode may be viewed in the alternate mode. A visit may be recalled and viewed in complete detail at any time.

Various hardcopy outputs are available. Figure 2 includes a total record and an encounter summary. A number of special retrieval reports are also available (Figures 2(a)–(e)).

An important subset of the clinic operation is that of financial management. Computers have been involved in ambulatory care primarily for purposes of billing the patient. In most cases, little, if any, data is retained, and the computer functions as a sophisticated adding machine and typewriter. Few clinic or practice managers know how much it costs them to collect a patient fee or, in fact, how long it takes them to collect. With rising costs, physicians must be more concerned with timely collections from a greater percentage of patients. In many delinquent cases, the patient obviously has the ability to pay. The immediate collection of the fee is becoming more important and has begun to be practiced in a number of clinics. In order to do this, patients must either be responsible for collecting from the third party payor or the clinic must be able to determine what the third party payor is likely to pay and bill the patient for the remaining amount. In addition to getting the bill to the patient as rapidly as possible, completeness and clarity of the bill are important factors to enhance early payment. Figure 3 shows a sample bill which is visit oriented and clearly identifies all transactions associated with a given visit. Aging is automatic since all dates are clearly identified. The accounting module also permits rapid review of the full accounting details of any patient visit at any time.

The automatic filing of third party claims is becoming increasingly desirable. Two major reasons are the amount of detailed data that is required about the encounter, particularly from Medicare and Medicaid, and the need to file the claim as quickly as possible. In our system, claims are printed nightly. A systematic review of rejected claims can suggest dictionary modifications to minimize these costly occurrences. It appears reasonable to be able to process any claim within 30 days of the visit.

```
DUKE UNIVERSITY MEDICAL CENTER         PAGE 1                  * BROWN, CHARLIE
ENCOUNTER SUMMARY: 08/01/77  FORM #: 33322                      *  GEN DEL
DEPARTMENT OF MEDICINE   W. STEAD                               *  HALIFAX, NC 20801
  PLACE OF VISIT: OFFICE                                        *  919-456-1231
 MD: MARCUS WELBY                                               *
 PO BOX 100                                                     * 999-99-9999
 ROANOKE RAPIDS, NC 27870                                       * 11/22/40 BLACK MALE
                                                                ******************************

A 36 yo BM with GLOMERULONEPHRITIS-FOCAL SCLEROSING, NEPHROTIC SYNDROME and
HYPERTENSION-DIASTOLIC presents SCHD F/U.

                              SUBJECTIVE
Decreased dietary compliance
PRESENTING THERAPY:  Na 2 per day
                     Furosemide 2 per day
                     Prednisone 4 per day

                              OBJECTIVE
PE - Weight 76 kg; Bp-sitting 135/95; Pedal edema/4 2; reinforce diet
LAB - UT-PROT 2.5, GLU R 97, BUN 20, NA 137, K 4.1, CL 95, CO2T 26, T PROT 6,
ALB 3.4, CA 9.1, PHOS 3.1, CHOL 250, URIC A 6.7, CREAT 1, T BILI 1, ALK
P 90, LDH 160, SGOT 40

                              PLAN
STUDIES ORDERED: Urinalysis, Ut-prot, Chem 6 and Chem 12

THERAPY SUMMARY
 1. NA GMS                      2           06/13/77 W. STEAD
 2. FUROSEMIDE 40 MG T          4 PO QAM    08/01/77 W. STEAD
 3. PREDNISONE 20 MG T          1 PO QID    08/01/77 W. STEAD

                                            --------------------------------
```

Figure 2. Sample Total Medical Record. (a)–(f) show all possible screen displays.

```
PRIMARY CARE SUMMARY: 08/01/77         PAGE 1                  * BROWN, CHARLIE
DEPARTMENT OF MEDICINE                                          *  GEN DEL
W. STEAD                                                        *  HALIFAX, NC 20801
                                                                *  919-456-1231
RPV: MARCUS WELBY                                               *
     PO BOX 100                                                 * 999-99-9999
     ROANOKE RAPIDS, NC 27870                                   * 11/22/40 BLACK MALE
                                                                ******************************

------------------------------------PROBLEMS-------------------------------------------------
NO.  ONSET RESOLVED                              (*indicates active problem)
 1   06/09/77              *Glomerulonephritis-focal sclerosing
 2   06/09/77              *  Nephrotic syndrome
 3   06/23/77              *    Steroid administration
 4   08/01/77              *  Hypertension-diastolic
 5   02/13/50-02/14/50      Appendectomy
 6   01/00/57              *Cigarette abuse
```

Figure 2(a).

```
PRIMARY CARE SUMMARY: 08/01/77          PAGE 2              * BROWN, CHARLIE
DEPARTMENT OF MEDICINE                                      *  GEN DEL
W. STEAD                                                    *  HALIFAX, NC 20801
                                                            *  919-456-1231
RPV: MARCUS WELBY                                           *
     PO BOX 100                                             * 999-99-9999
     ROANOKE RAPIDS, NC 27870                               * 11/22/40 BLACK MALE
                                                            ******************************

-----------------------------------STUDIES SUMMARY-----------------------------------------------

06/13/77  HGB              15.5
06/13/77  HCT              49
06/13/77  RBC              4.9
06/13/77  WBC              7.3
06/13/77  DIFF
          POLY             69
          LYMPH            25
          MONO             5
          BASO             1
06/13/77  PLATLET          NL
06/13/77  MCV              90
06/13/77  MCHC             35
08/01/77  SED WEST         40
06/13/77  VDRL             NL
07/01/77  U-PH             6
07/01/77  U-SPGR           1.015
07/01/77  U-PROT       +   2+
07/01/77  U-GLU            0
07/01/77  U-KET            0
07/01/77  U-BENZ           0
07/01/77  U-SEDIMENT
          U-WBC            1-3
          U-RBC            0
          U-CAST           NL
          U-S-TXT          NONE
06/13/77  UT-VOL           1400
06/13/77  UT-CREAT         1713
08/01/77  UT-PROT          2.5
06/13/77  ST GUAIAC        0
06/13/77  CHEST            NL
06/13/77  IVP              NL
08/01/77  GLU R            97
08/01/77  BUN              20
08/01/77  NA               137
08/01/77  K                4.1
08/01/77  CL           -   95
08/01/77  CO2T             26
08/01/77  T PROT       -   6
08/01/77  ALB          -   3.4
08/01/77  CA               9.1
08/01/77  PHOS             3.1
08/01/77  CHOL             250
08/01/77  URIC A       +   6.7
08/01/77  CREAT            1
08/01/77  T BILI           1
08/01/77  ALK P            90
08/01/77  LDH              160
08/01/77  SGOT             40
```

Figure 2(b).

```
PRIMARY CARE SUMMARY: 08/01/77          PAGE 3                  * BROWN, CHARLIE
DEPARTMENT OF MEDICINE                                          *  GEN DEL
W. STEAD                                                        *  HALIFAX, NC 20801
                                                                *  919-456-1231
RPV: MARCUS WELBY                                               *
     PO BOX 100                                                 * 999-99-9999
     ROANOKE RAPIDS, NC 27870                                   * 11/22/40 BLACK MALE
                                                                ******************************

06/13/77  EKG          NL
06/13/77  MORPH        NL
06/13/77  SICKLE PRE   NEG
          GLU F        94
          GLU 2H       105
06/13/77  GTT
          GLU F        94
          GLU 1/2H     190
          GLU 1H       160
          GLU 2H       105
          GLU 3H       100
          GLU 4H       90
          GLU 5H       100
06/13/77  ASO          NL
06/13/77  SEP          NL
06/13/77  UEP          NL
06/13/77  LE PREP      0
06/13/77  FANA         NL
06/13/77  ANTI DNA     NL
06/13/77  CH50         48
06/13/77  UR-CC-C      NL
06/13/77  KID BX       INC GLOM STALK, SEG IGG & IGM
```

Figure 2(c).

```
PRIMARY CARE SUMMARY: 08/01/77          PAGE 4                  * BROWN, CHARLIE
DEPARTMENT OF MEDICINE                                          *  GEN DEL
W. STEAD                                                        *  HALIFAX, NC 20801
                                                                *  919-456-1231
RPV: MARCUS WELBY                                               *
     PO BOX 100                                                 * 999-99-9999
     ROANOKE RAPIDS, NC 27870                                   * 11/22/40 BLACK MALE
                                                                ******************************

                              THERAPY SUMMARY

-------------------------------------------------------------------------------------------
                              CURRENT THERAPIES
-------------------------------------------------------------------------------------------
         THERAPY                      SIG                      STARTED      MD

     1.NA GMS                         2                        06/13/77  W. STEAD
     2.FUROSEMIDE 40 MG T             4 PO QAM                 08/01/77  W. STEAD
     3.PREDNISONE 20 MG T             1 PO QID                 08/01/77  W. STEAD

-------------------------------------------------------------------------------------------
                              PAST THERAPIES
-------------------------------------------------------------------------------------------
    THERAPY                                                    REACTION

   NAFCILLIN                                                   NONE

-------------------------------------------------------------------------------------------
                              IMMUNIZATIONS
-------------------------------------------------------------------------------------------
     IMMUNIZATIONS          DATE

     DPT BOOSTER             03/15/69
     SMALL POX               03/15/69
```

Figure 2(d).

PRIMARY CARE SUMMARY: 08/01/77 PAGE 5
DEPARTMENT OF MEDICINE
W. STEAD

RPV: MARCUS WELBY
PO BOX 100
ROANOKE RAPIDS, NC 27870

* BROWN, CHARLIE
* GEN DEL
* HALIFAX, NC 20801
* 919-456-1231
*
* 999-99-9999
* 11/22/40 BLACK MALE
****

---------------------------------FLOW SHEET---------------------------------

	06/13/77	06/17/77	06/23/77	07/01/77	07/15/77	08/01/77
1 WEIGHT KG	85	81	78	74	73	76
2 BP-SITTING	130/75	120/80	128/80	125/70	128/85	135/95
3 PEDAL EDEMA/4	4	3	2	1	1	2
4 SED WEST	60			45		40
5 CL	110	108	107	101	100	95
6 T PROT	3.0	3.2	4.2	5.4	5.6	6
7 ALB	1.6	1.8	2.2	3.1	3.2	3.4
8 CA	8.4	8.5	8.8	9.1	9.2	9.1
9 CHOL	400	390	350	300	290	250
10 URIC A	5	5.5	7	6.7	6.5	6.7
11 U-PROT	3+	3+		2+		
12 UT-PROT	8.3			4.6	2.9	2.5
13 KID BX	INC GLOM S					
14 NA	0>2	2	2	2	2	2
15 FLUID	0>1000	1000	1000	1000>0		
16 FUROSEMIDE	0>2	2	2	2	2	2>4
17 PREDNISONE			0>4	4	4	4

Figure 2(e).

FAMILY MEDICAL CLINIC
999 S MAIN STREET
DURHAM, NORTH CAROLINA 27710

DATE: 08/01/77

PAPPY YOKUM
10 HOG LANE
DOGPATCH, USA 99999

==

Patient:	YOKUM, PAPPY					
6/02/77	OFFICE VISIT (Dr. Welby), lab, med				$ 32.50	
	PAYMENT	06/02/77	CHECK	$ 10.00		
	PAYMENT	07/01/77	BCBS	$ 26.00		
	ADJUSTMENT	06/02/77	ADD CHRG	$ 3.50		
				Balance this visit:		$ 0.00
7/15/77	OFFICE VISIT (Dr. Welby), lab				$ 47.50	
	PAYMENT	07/15/77	CASH	$ 20.00		
			BCBS INS FILED			
				Balance this visit:		$ 27.50
			Balance for PAPPY			$ 27.50
Patient:	YOKUM, MAMMY					
7/18/77	EMERGENCY ROOM (Dr. Kildare), lab, med, xray, sup				$ 44.43	
	PAYMENT	07/18/77	CHECK	$ 15.00		
			BCBS INS FILED			
				Balance this visit:		$ 29.43
			Balance for MAMMY			$ 29.43
				TOTAL amount owed:		$ 56.93

PLEASE RETURN BELOW PORTION WITH PAYMENT

NUMBER	NAME	AMT. OWED	AMT. PAID
1-999-99	YOKUM, PAPPY	27.50	________
2-999-99	YOKUM, MAMMY	29.43	________

Figure 3. Sample bill.

The availability of more sophisticated financial reports and aids is often unappreciated because clinic managers lack training and education in these techniques. The introduction and availability of such programs must be accompanied by appropriate training and education.

Another area of use of the medical record that can be considerably expanded is the use of the computer in the management of the practice. Simple demographic statistics provide valuable management data. For example, how evenly distributed are the patient population groups? Is the practice expanding? Does the practice meet the needs of the community? Are the patients primarily an old group who will have more problems, be more tiring, often be more depressing, and not provide variety to the provider? How many consultants are used? Responsibilities of management are to assess use of facilities and staff personnel, to determine the appropriate mix of patients and their scheduling, and to know what ratio of scheduled versus drop-in patients can be handled.

The primary function of the medical record is to assist the physician in the care of his patients. Unfortunately most physicians either do not perceive or are not willing to acknowledge a need in this area. Many providers contend that their organization of patients' medical records is sufficient to meet not only their needs but those of other interested parties. However further conversation, as well as a review of their records, often reveals significant discrepancies in this area. It is with this intrusion that perhaps the physician begins to see the computer as a threat. Careful study reveals that the physician is not always aware of the complete status of a patient. The most glaring example is that of medications. Physicians are often surprised at the length of time that a patient has been on a given drug. In one informal study, physicians, at least in terms on recording in the medical chart, were aware of less than 30% of abnormal laboratory findings. Continuity of care, between providers, between clinics, between primary, secondary and tertiary settings, and between consultant and refering physician is enhanced by a shared computerized data base.

The contribution of the computer to preventive care is significant. During the past year, we have frequently done patient recalls for a particular function. Before flu season started last year, the computer generated addressed postcards for all patients registered to the clinic who were over 60 years old or who had a diagnosis of some type of respiratory shots. Women patients are routinely requested to come for a pap smear, and all patients are monitored for vaccination updates.

Another function of the computerized system is to aid the physician in evaluating the quality of his own patient care as well as formal review. For example, the computer can print out all patients with a given problem and display selected data. By comparing the records and the course of the disease among the patients, the physician is able to evaluate what patients need additional attention. Such longitudinal review is also useful in patient care. The computer can automatically select records for further manual review if certain preselected conditions are not met. Abnormal laboratory data can be specifically monitored and called to the provider's attention.

In another area, the computer can be used as a consultant in problems which the primary care provider does not see frequently enough to always know the most recent or best courses of action. One example of this assisting type program is an acid-base evaluation program. These types of programs can be developed and maintained in a university and made available to community practices, or practices can develop and share their own programs.

The computerized data base is also of value in providing a better understanding of the natural history of health problems, complaints, and natural diseases. Treatment of acute problems is often more effective than treatment of chronic problems because of the instant feedback. As different medicines are used, the provider is able to evaluate their effectiveness because of the shorter time frame. With chronic problems extending over months to years, changes may be so subtle as not to be recognized, and data is usually not available in such a fashion as to be able to understand the course of the disease and to evaluate the treatment.

The automated system provides an excellent tool for provider education. It is extremely important that medical schools be aware of what problems primary care physicians are seeing in order to properly prepare students for that type of practice. For example, anxiety, depression and obesity are within the ten most frequently seen problems in most practices. These problems are ones for which most students receive little preparation in their education. The computer makes it easy to provide self-education packages to the physicians, frequently using his own patients as examples.

The patient's role in primary care has recently been emphasized in a number of publications. For obvious reasons, patients do not alwas comprehend the provider's explanation of their problems, how the problems relate to their everyday health, the medications and what their effects might be, and other general information. Such data, enhanced by a strong emphasis on the patients' responsibility for their own health, could significantly reduce the demands for and the cost of primary care. Patient specific information can easily be printed from the computerized medical record.

Conclusion

The success of the integration of computers into the ambulatory care setting involves adequate hardware and software capability, appropriate applications, acceptable human/computer interfaces, physician acceptibility, and cost effectiveness. As advances are made in each area separately, the platform to success becomes one notch higher. The final step to success and acceptance now seems obtainable within the next few years [3].

References

[1] "Five-Year Experience in the Development of a Computerized Obstetrical Medical Record," W. W. Stead, R. G. Brame, W. E. Hammond, F. R. Jelovsek, E. H. Estes and R. T. Parker. Journal of Obstetrics and Gynecology **49,** No. 4, April 1977, pp. 502–509.

[2] "GEMISCH—A Minicomputer Information Support System," W. E. Hammond, B. A. Brantley, S. J. Feagin, S. C. Lloyd, W. W. Stead, and E. L. Walter. *Proceedings of the IEEE* **61,** No. 11, November 1973, pp. 1575–1583.

[3] This paper was originally published in 1977. Since that time TMR has been available commercially. For further information about the system and its utility, readers are referred to the five papers on TMR which appear in SCAMC-82 (Editor's Note).

Regenstrief Medical Record System*†

Clement McDonald, M.D., Lonnie Blevins, Tull Glazener, Jeff Haas, Larry Lemmon, and Jim Meeks-Johnson

Introduction

The information needs of a practice vary with the size of the practice, the subspecialty, the patient mix, the expectations of the patients, the goals of the practitioner, and the mechanism of reimbursement. Moreover, even within a given practice these needs change because of shifts in the demographic distribution and the size of the patient population and changes in the diagnostic and therapeutic armamentarium that occur with time. Thus, it is extremely important that computer systems that serve medical practices be flexible enough to conform to the needs of different kinds of practices over time.

In this chapter we will describe two techniques—the use of stored data definition and user-oriented command languages—for achieving the above goals and their application to the Regenstrief Medical Record System. We don't pretend we've developed the ultimate in universal medical record systems, but we will try to show (while describing some of the features of the Regenstrief Medical Record System) how these techniques make it easier to shape a particular implementation to specific practice needs.

Advantages of Data Base Management Systems

The most important idea advanced by the inventors of data base management systems is that of extracting the information about files and their contents from

*The system described herein is installed at Wishard Memorial Hospital, Indianapolis, Indiana. We wish to thank Wishard's Administration, without whose foresight and support none of this would have been possible and the physicians and nurses of the General Medicine Clinic, especially Ms. Doris Paradise, for their close cooperation, their many good suggestions and their forebearance during the early stages of this development. The studies mentioned in this paper were supported by the National Center for Health Services Research, grants HS 02485 and HS 04080.
†Reprinted with permission from *Journal of Medical Systems,* 7, Plenum Publishing Corp., © 1983; modified from a version previously published in *Proceedings of the Sixth Annual Symposium on Computer Applications in Medical Care,* IEEE, 1982.

programs and storing this information in an independent data file called a stored data definition (SDD). In programs that use data base management systems (DBMS), statements containing information about the location of a file, the offset and length of the data items (fields) within its records—the kinds of information commonly included in format statements of pre-DBMS programs—do not appear. Instead one finds logical names of files and calls to data base access routines. The access routines obtain the file and field specific information needed to transfer data to or from the file from the stored data definition.

The stored data definition provides independence between programs and the data they access and a number of consequent advantages. It reduces programming work and errors. Rather than writing specific code for accessing each of the different files and their contained fields, as was required previously, programmers use a set of preformed access routines.

It eases the inevitable process of change. In the pre-data base era, when a new field had to be added to a file or an existing field had to be changed, all the file access statements across all the many programs that use the data items would have to be found and changed, and a difficult period of program and file conversion would ensue. With data base systems, most such changes become a matter of editing the description for a single field within the stored data definition. (In some systems this can be done while applications are still running; in others it requires stopping the application, sometimes rebuilding files and then restarting the application.)

There are other advantages. If the stored data definition includes information about edit checks required on data entered into the files and prompts for requesting information about these items, the data base routines can also control entry of data from CRT terminals. This assures a uniform perspective of the data to the user and helps to achieve the goal of nonredundancy, i.e., a given data item should be stored in one, and only one, place in the system of data files. In addition, since the sorted data description is itself a file, it can be displayed. Thus, it provides self-documentation of its structure.

Finally, the stored data definition permits ready development of general purpose programs, such as report generators and file packing programs. With a utility program, one program performs a given function on all files within the data base system. This may be the greatest advantage of all. In the pre-data base era, separate programs would usually have to be written to report the contents of each file; another set of programs would have to be written to load the initial contents of those files; and another to recover space from files when records were deleted from them. In a data base management system, it is relatively easy to develop general purpose programs for these functions since the file specific information which distinguished the various storing, reporting and packing programs in pre-data base days is no longer part of the programs. Such utility programs have their own cascade of benefits. The most important of these is that they permit the nonprogrammer to store, report and manipulate the contents of his data files without having to learn programming or to translate his request into a programming project.

The Regenstrief Medical Record System is built on a true data base management system which confers many of the advantages listed above. The system is not a pure data base system since it employs one major file, the medical record file (MRF), which is not a true data base file.

In the following, we describe the Regenstrief Data Base System (RDB), its utility programs and the advantages they provide, the medical record file and its associated language (CARE), and finally, the operation of the medical record system in the context of a typical patient visit cycle.

Regenstrief Data Base Management System (RDB)

Technical Features of RDB

Here we provide a technical description of RDB for those interested in such issues. (Others may skip to the next section.)

Records within RDB files are of fixed length and fixed format. That is, within a given file, all records have the same length and fields appear in the same location in every record. A record within a given data base file can be accessed directly by its location (its logical position number within the file) or indirectly by content (via hash or index sequential access lookup tables). In addition, a RDB record can be accessed via link pointers from records in other files.

RDB data base files may be aggregated into hierarchical and/or network structures. Relationships between records within such structures may be one-to-one, many-to-one, one-to-many or many-to-many. A field may be defined as one of 12 different data types including integer, floating point, string, radix, date, time, pointer and other special types.

The stored data definition includes a field prompt (for requesting entry of data into a field) and data verification checks for each field. Entry checks can include checkdigit calculations and COBOL-like pictures for specifying the format of items such as telephone and social security numbers. The data type is always verified (e.g., 2/32/81 would be rejected as an illegal date).

There are two means of accessing RDB files. The first is via a set of general purpose utility programs. The second is via application programs that call standard RDB access routines. Users of our system employ both kinds of programs on a daily basis.

RDB Utility Programs and Command Languages

The RDB utility programs have two special advantages. The first is that they permit nonprogrammers to access and manipulate data base files. In fact, in our institution nonprogrammers have built simple applications such as mailing lists themselves. The second is that the format and content of most batched reports, and the specification of most maintenance procedures is defined in terms of RDB commands (as shown below). Thus new reports can be created, and old report formats can be modified, without the need for programming.

The RDB utility system includes commands for storing (STORE), editing (EDIT), sorting (SORT), reporting (REPORT), and cross-tabulating (XTAB) the contents of records contained within single data base files or hierarchical structures of data base files. There are also commands for creating new data base files from old ones. EXTRACT creates "flat" records that are amenable to sorting and statistical analyses from hierarchical record structures. JOIN adds records from two "flat" files side to side according to specified match fields. PROJECT squeezes out records that are duplicate with respect to specified match keys. APPEND lengthens a file by adding another file to its end. Finally, there are maintenance commands such as PACK, which recovers freespace by squeezing out records that meet deletion criteria.

With most of these commands the user can select subsets of fields and/or records for processing. For example, the following STORE command selects three fields for processing:

STORE PATIENT'S NAME, HOSP# AND TELE#.

The above command tells the computer to request input from the terminal operator about a new patient's name, hospital number and telephone number. The computer asks the same set of questions on the next patient and so on, until the operator terminates the entry session.

To select records for processing, the user adds an IF and selection criteria to the command, as illustrated by the following.

REPORT PATIENT'S NAME AND ADDRESS IF SEX = "M"
AND (TODAY—BIRTHDATE/365) GT 40.

The above command stimulates a number of questions to the user, such as the name of the format for the report and the device to which the report is to be transmitted. It then generates a report containing the name and address of all patients who are male and over age 40. Notice that the birthdate, not the age, is stored in the patient's record. Thus, the birthdate is arithmetically converted to an age in the above.

The same command language can process hierarchical structures of records. Within the medical record system, the patient's registry records, the records that represent the patient's registry records, the records that represent the patient's appointments and the records that represent the physicians to whom the patient has an appointment, constitute such a hierarchy. The following command lists information contained in that hierarchy, i.e., the patient's name and address and for each patient, the dates and physician's name of each scheduled appointment:

REPORT PATIENT'S NAME AND ADDRESS AND
APPOINT'S (DATE AND DOCTOR'S NAME).

Note that the use of the possessive form is carried through from the naming of the root level file to other files in the hierarchy. Users are limited in the files they can traverse through the RDB Utility System by their RDB passwords and associated keys to the various protection locks within the system.

A number of RDB commands can be processed in sequence. A sequence of commands could include one command to select records for processing, another to sort them, another to generate a report, another to re-sort the initial file and so on. The user can save such command sequences in a named file (for example, NIGHTLY-PROCESS) and then evoke the entire sequence of commands by name as a batch process by using the following command:

BATCH NIGHTLY-PROCESS.

The RDB command language has a number of other special features such as the ability to react to system errors, to call programs other than the standard utility programs, to expand macro statements and to ask the user for specific input at the time a command is executed.

RDB utility programs accomplish most of the batch oriented data base processing for the medical practice. For example, they produce most reports about information contained in RDB files. They are used to maintain the various tables employed by the application programs, such as the term dictionary and the physicians' schedule, the physicians' registry file, locations file, etc. They are used for generating workload statistics, for example, the number of patients seen by physician, by day of week, by month; for writing tapes containing billing transactions; and for file maintenance, i.e., recovering space occupied by deleted records. Finally, they are used for ad hoc reports and retrievals of RDB files.

[*Editor's Note*: Digital Equipment Corporation plans to market the Regenstrief Data Base System under its EAS Library Project. Full details about how to define RDB files, link them into structures, and the capabilities of the RDB command options and syntax are given in *RDB Utility User's Guide No. QA030* (from Digital Equipment Corporation, Maynard, Massachusetts).]

Application Programs, Field Lists and the Program Control File

Application programs communicate with data base files through a standard set of interface routines. Application programs accomplish almost all of the interactive processing in our system including the storage of medical data, the registering of patients, scheduling of appointments, inquiry into the medical record file, etc.

In application programs, the unit of processing is a named subset of data base fields, which we call a field list. These are similar in purpose and content to the lists of fields that are shown in the RDB commands illustrated above. Field lists and their associated format statements define much of the interface between application programs and their users. They define all display screen headers, the body of many display screens and the set of elements requested for many input screens, e.g., the patient registration screen.

With a few exceptions, the program does not care exactly what subset of fields is contained within a given field list. The name of a given field list appears within an application program; its exact contents do not. Each program has an associated control file which contains the list of fields that belongs to each of the field lists named within the program, the formats associated with these field lists, the program's help messages and error messages and other items that control how the program operates in a given environment. Since this control file is a text file, the system manager can change the format and contents of many of his display and entry screens by appropriately editing the control file.

Most of the files used by the Regenstrief Medical Record System as well as our other two systems, the Regenstrief Clinical Laboratory System and the Regenstrief Pharmacy System, are RDB files. Examples of RDB files are the patient registry file, the physician registry file, the term dictionary which contains the names of drugs and laboratory tests, the patient appointment file, physician schedule file, and files used for temporary storage of clinical information on the way to and from the medical record file.

Medical Record File (MRF) and Its Associated Functions

MRF Structure and Contents

The medical record file is the only major file within the medical record application that is not a data base file. The medical record file is large (currently 80 million bytes). It contains records for 60,000 patients. Because of its size we took particular pains to assure that it was compact. The individual results within a patient's storage area are variable length and stored sequentially. In addition, information within the file is highly encoded. Dates are compressed into the number of days from a fixed reference point, permitting storage of dates from a 180 year span within 2 bytes. The names of drugs, tests and clinical findings are represented within the file as integer codes that reference the system's term dictionary. While the system stores numeric codes within its record, it uses names when communicating with users. For example, when the user asks for results of a potassium determination, he does not type in the code number for potassium. He types in "K^+". Similarly, when the computer reports results of a potassium determination to the user, it identifies them by name.

Only three different record types are allowed within the medical record file. These correspond to observations with numeric values (for example, serum potassium of 4.5), those with multiple-choice, single answer values (such as VDRL) and those with multiple-choice, multiple-answer values (such as chest x-ray). Each of the record types may have free text stored instead of, or in addition to, the standard answers. Consequently, the practice is free to choose the degree to which it stores narrative reports as codes rather than free text.

Because the logic for processing these three data types is hardcoded into the programs that access the medical record file, the medical record file and associated processing routines do not constitute a true DBMS. Nonetheless, the medical record file system has a number of parallels with data base systems. Firstly, the kinds of information that can be stored in the medical record file (within the constraints described above), are defined by an RDB file (the term dictionary), that can be construed as a stored data definition. The practice is free to create its own unique term dictionary and to add terms to that dictionary as time goes on.

Also, as is true of the data base system, the user may define named sets of terms. LYTES, consisting of sodium, potassium, chloride and CO2 is a typical set in our current system. Set names facilitate storage and retrieval of medical information. On input of results for serum electrolytes, the user would type in LYTES. The computer would return a screen with space for entry of each of the four electrolyte results. On the output side, by typing in LYTES, the user can obtain an electrolyte flowsheet.

Finally, the medical record system has its own special retrieval system called CARE. This system is an analogue to the RDB command language for RDB files.

The CARE system permits nonprogrammers to perform complex queries of the medical record file or to provide quality assurance of medical care (as will be described below). The underlying process is the same in both cases. The CARE system requires a file defining the patient records that are to be searched and a set of CARE statements defining the search criteria, as inputs. Given these inputs, it searches each of the selected patient's medical records for the criteria of interest. When it finds a patient who meets these criteria, it saves one or more records in an output file.

In the case of the medical record retrieval, the patients of interest are usually all registered patients. Consequently, the patient registry is the file defining the patient records to be searched. In the case of quality assurance, the scope of the search is the set of patients expected in clinic the next day. Consequently, its input file is a subset of the records within the appointment file.

The output of a retrieval and a quality assurance run are different. In the case of a retrieval, the output contains patient observations selected from the medical records by the CARE statements. These then can be further analyzed. In the case of quality assurance, the output is a set of reminder messages to the responsible physicians.

The building block of a CARE statement is a clause. As in English, a clause consists of subject, verb and predicate. The subject may be any of a number of constructs. It may be any parameter name, such as "K+" defined in the term dictionary. It may be a set of parameters, such as "LYTES". It may be a variable defined by a CARE statement, such as "LAST THYROXIN CHANGE". Such variables are defined in a backward chaining mechanism analogous to that employed by MYCIN [1]. The subject can also be an arithmetic statement involving previously defined variables; it can be the change, the area, the slope or the interval between the observations of a named parameter (for example, CHANGE OF "K+") or a selection of one of a sequence of observations, MAX OF "K+". Finally, the subject can be one of the above processes applied to a subset of observations within the patient record. The subset construct is analogous to the relative clause in English.

The verb in the clause is simply a separator. It may be any of "to be" verbs, WAS, WERE, ARE, IS, etc. A clause may or may not have a predicate. A clause that does not have a predicate is true if the defined subjects exist within the patient's record. A clause with a predicate is true when the subject satisfies the conditions of the predicate. The predicate may be a single criteria, for example, > 5 or BEFORE 1/1/71; a combination of criteria, for example, > 5 & BEFORE 1/1/71; or it may be the date or value of a previous observation, for example BEFORE "LAST GLUCOSE". The syntactic constructs that are valid predicates are also valid subset definitions.

To find patients who had congestive heart failure and a low potassium at any time in their course, one would write the following query:

```
IF ("CXR" WAS = "CHF" OR "DX'S" WAS = "CHF") AND
"K+" WAS LT3.5
  THEN SAVE
```

The above query saves a record for each patient who meets the criteria of the conditional part of the clause (that which precedes the THEN).

The next two statements accomplish quality assurance. They identify patients who have not had a follow up T4 index after their last thyroxin dosage change.

```
DEFINE "THYROXIN CHANGE" AS LAST CHANGE [NE 0] OF
  "LEVOTHYROXIN" EXISTS
IF "THYROXIN LAST" WAS GT 0 AND NO "T4 INDEX" WAS
  AFTER "THYROXIN CHANGE"
  THEN SAVE
```

(There is no documentation within the record of a measure of T4 index following the last change in levothyroxin dosage.)

To find the average serum SGOT level in the first year following the start of therapy, as might be of interest in a drug toxicity study, one could define a query similar to the following:

```
DEFINE "INH START" AS FIRST "INH" EXISTS
IF AVERAGE OF "SGOT" [AFTER "INH START" & BEFORE "INH
  START" + 1 YEAR] EXISTS
  THEN SAVE AS "INH.SGOT"
```

In this case, a record will be saved for each patient who had a serum SGOT level within the first year of INH treatment. The average value of the SGOT during that year will be stored in a field called "INH.SGOT". (The full syntax for the CARE language is given in [2].)

This last example brings us to the linkage between the data base system and the medical record file. As we mentioned above, the CARE system takes two files as the input files and produces one file as output. The input file that defines the patient records and the output file are data base files. Thus, the user can select patients for retrieval on the basis of information in any of the data base files using the RDB command language. Since the output file is also a data base file, it can be cross-tabulated, reported and/or searched by means of RDB utility programs. The output file can also serve as input to a succeeding CARE retrieval when required.

Use of Medical Record System in Practice

The Regenstrief Medical Record System has been in operation at Wishard Memorial Hospital since 1974. Currently it carries medical records for over 60,000 patients. The system obtains data regarding inpatient, outpatient and emergency room care. Records of laboratory tests and drug usage are obtained directly from the Regenstrief Clinical Laboratory System and Regenstrief Pharmacy System, respectively. Results from most other diagnostic studies including EKGs, nuclear medicine scans, x-rays, CAT scans, sonars, endoscopies and bone marrows are manually coded and entered into the computer by coding technicians. Diagnostic impressions from hospital stays, emergency room visits and clinic visits are similarly encoded and entered.

The system runs on a Digital Equipment Corporation VAX 11/780 that it shares with the automated clinical laboratory system and automated pharmacy system.

In the usual patient cycle, activities are performed prior to each visit, at the visit and following the visit. The ensuing discusses the usage of the Regenstrief Medical Record System through a single patient visit cycle.

Pre-Visit Activities

Appointments. The scheduling of the patient appointment begins the patient's visit cycle. Typically these are made by telephone. All appointments are scheduled by the Regenstrief Medical Record System. The system permits each physician to have his unique schedule and to practice in one or more clinics. The unit to which appointments are made is the schedule slot. Slots can be of varying size and have associated weights. Different appointment types also carry associated weights. The number of visits that may be allocated to a given slot depends upon the weight of the slot and of the scheduled visit. By adjusting the weights and the size of the slots, one can accommodate schedules that range from a single patient per slot through block scheduling within a slot.

In operation, the computer searches for an available slot that satisfies the criteria stated by the terminal operator. Criteria may include a specific day of the week (for example, Tuesday), a time interval (for example, before 10 a.m.), a relative date range (for example, within three weeks) and/or a physician's identity. The system manager can modify the contents and format of screen header by editing the associated field list in the program's control file.

On the day before the visit, a number of standard patient reports are generated by the data base utility program. These include alphabetic listings of patients expected on the visit date, the total future appointments scheduled for each physician, etc.

Three patient specific reports are also produced for each patient by application programs the night before the scheduled visit. (These can also be produced on demand.)

Patient Encounter Form. The first of these is the patient's encounter form (Figure 1, the encounter form). This report is a two-part carbonized document produced by an application program. As you can see from the figure, it is divided into four different areas. The upper right-hand area contains the problem list as recorded at the previous visit and provides room for the physician's visit note. (In the figure we have shown the report as it appears after the physician has used it.)

In the left-hand upper section is a list of flowsheet observations. In the Medicine Clinic these include weight, systolic and diastolic blood pressure and pulse. The CARE system can be used to tailor the contents of parts of the encounter form. In particular, it can be used to add items to the parameter list according to the contents of the patient's record. In the example, dyspnea on exertion, PND, edema, prematures per beat were added because of the patient's use of digoxin. Such patient specific parameters facilitate the maintenance of problem-oriented flowsheets.

In the middle section on the right side is the orders area. In this area the computer provides a "menu" of diagnostic studies. The menu includes a fixed list of frequently ordered tests for each clinic. CARE statements may also be used to add patient-specific test options to this menu. Physicians order these tests by circling them. To order a test not included in the list, the physician writes its name.

The bottom section of the form presents the prescriptions which were active as of the last clinic visit. Physicians may continue a maintenance medication by simply writing the amount of medication to be dispensed at this visit in the space provided at the right of each prescription. To discontinue a prescription, the physician writes "D/C" in this space. To modify a previously written prescription, the physician writes a different sig in the space below the old one. New prescriptions are written in the available space at the bottom of the prescription area. The only prescription writing required is that which you see in Figure 1.

Flowsheet Summary. The second report for each visit is the patient's flowsheet. This is a paper report which displays all information contained within the patient's medical record in flowsheet format. The contents of the flowsheet for our sample patient (see Figure 2) is typical of the records stored in our system.

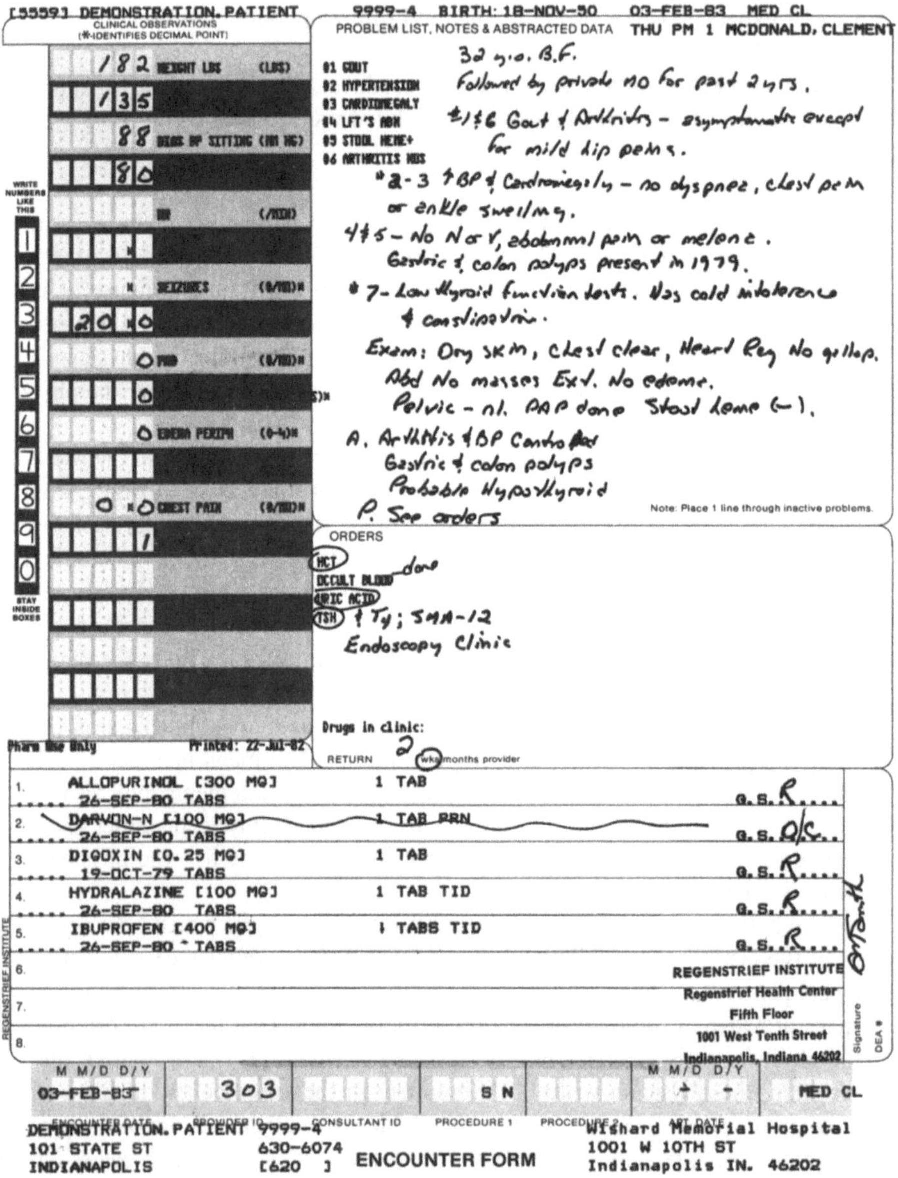

[5559] DEMONSTRATION, PATIENT 9999-4 BIRTH: 18-NOV-50 03-FEB-83 MED CL

CLINICAL OBSERVATIONS (✱-IDENTIFIES DECIMAL POINT)

182 WEIGHT LBS (LBS)
135
88 DIAS BP SITTING (MM HG)
80
HR (/MIN)
SEIZURES (0/MO)✱
20 0
0 PND (0/MO)✱
0
0 EDEMA PERIPH (0-4)✱
0 ✱ 0 CHEST PAIN (0/MO)✱
1

WRITE NUMBERS LIKE THIS: 1 2 3 4 5 6 7 8 9 0 — STAY INSIDE BOXES

PROBLEM LIST, NOTES & ABSTRACTED DATA — THU PM 1 MCDONALD, CLEMENT

01 GOUT
02 HYPERTENSION
03 CARDIOMEGALY
04 LFT'S ABN
05 STOOL HEME+
06 ARTHRITIS NOS

32 y.o. B.F.
Followed by private MD for past 2 yrs.
#1 & 6 Gout & Arthritis – asymptomatic except for mild hip pains.
#2-3 ↑BP & Cardiomegaly – no dyspnea, chest pain or ankle swelling.
4 & 5 – No N or V, abdominal pain or melena. Gastric & colon polyps present in 1979.
#7 – Low thyroid function tests. Has cold intolerance & constipation.
Exam: Dry skin, Chest clear, Heart Reg No gallop. Abd No masses Ext. No edema. Pelvic – nl. PAP done Stool heme (–).
A. Arthritis & BP Controlled
Gastric & colon polyps
Probable Hypothyroid
P. See orders

Note: Place 1 line through inactive problems.

ORDERS

HCT
OCCULT BLOOD — done
URIC ACID
TSH & T4; SMA-12
Endoscopy Clinic

Drugs in clinic:

RETURN 2 wks/months provider

Pharm Use Only — Printed: 22-JUL-82

1. ALLOPURINOL [300 MG] 1 TAB — 26-SEP-80 TABS — Q.S. R
2. ~~DARVON-N [100 MG] 1 TAB PRN~~ — 26-SEP-80 TABS — Q.S. D/C
3. DIGOXIN [0.25 MG] 1 TAB — 19-OCT-79 TABS — Q.S. R
4. HYDRALAZINE [100 MG] 1 TAB TID — 26-SEP-80 TABS — Q.S. R
5. IBUPROFEN [400 MG] 1 TABS TID — 26-SEP-80 TABS — Q.S. R
6.
7.
8.

REGENSTRIEF INSTITUTE
Regenstrief Health Center
Fifth Floor
1001 West Tenth Street
Indianapolis, Indiana 46202

Signature — DEA #

M M/D D/Y 03-FEB-83 — ENCOUNTER DATE | 303 — PROVIDER ID | CONSULTANT ID | S N — PROCEDURE 1 | PROCEDURE 2 | M M/D D/Y — APT DATE | MED CL

DEMONSTRATION, PATIENT 9999-4
101 STATE ST 630-6074
INDIANAPOLIS [620]

ENCOUNTER FORM

Wishard Memorial Hospital
1001 W 10TH ST
Indianapolis IN. 46202

Figure 1. Encounter form.

The RMRS will accept history and physical and progress note information in free text or coded (multiple choice, multiple answer) format. We currently enter very little such information because the costs of such entry would be added costs since physicians now handwrite this information in the paper record. If these items were currently transcribed, the cost of their entry could be more easily justified.

The practice can shape the flowsheet to their own requirements within the constraints that the columns represent dates and the rows represent parameters. The header and footer of the summary report contain fixed textual information plus information from the patient's registry file. Since both of these are defined by field lists as described above, the practice can tailor them to suit their own needs.

The organization of data within the flowsheet is in a three level hierarchy. The first level is the block. Parameters included within a single block are displayed as a single large matrix with dates at the top of each column. The second level is the group. A group is a labeled set of parameters separated from other groups within a block by a blank line. The practice may group parameters within a block into an arbitrary number of groups. Clinic data and SMA12 and treatments shown in Figure 2 are examples of blocks.

Vitals and treatments are examples of groups. The final level is the individual parameter. The practice can specify that parameters appear in any chosen order.

The practice can tailor other aspects of the summary report as well. It can choose to list the dates within the blocks in chronologic or reverse chronologic order (so the most recent information appears first). It can "compress" many dates (e.g., 7 or 60 days) into a single column to obtain more compact time frames. It can choose alternate sized print fonts if the printer permits. Using compressed print, up to 15 columns of data can be displayed across an 8½ × 11-in. page.

Quality Assurance Report. The third report produced prior to each visit is the surveillance report. This is the result of a search through the patient's record for conditions requiring physician attention. The search is directed by rules written by a physician in the CARE language described above. We currently use over 1400 rules to review each patient's record. These rules and their medical rationale are described in full in [2].

The output of this process is an RDB file containing the patient identifier, the message to be reported and other such information. The printed report is produced by the RDB report generator described above. An example is shown in Figure 3. We have shown in a number of studies that the provision of such reports to physicians strongly influences their behavior [3,4].

Visit Date Activities

Using an on-line terminal the clinic secretary identifies the patient to the computer by hospital number or name, verifies that he/she has an appointment and updates the registration record if any data is either missing or out of date.

BIRTHDAY 06-DEC-09 [5559]

WISHARD MEMORIAL HOSPITAL ¦DEMONSTRATION.PATIENT #9999-4

DX & COMPLAINTS

DX & COMPLAINT 10-AUG-79 HYPERTENSION, CARDIOMEGALY, LFT'S ABN, STOOL HEME+, ARTHRITIS NOS
13-OCT-78 GOUT, HYPERTENSION, CARDIOMEGALY

CLINIC DATA	26-SEP -80	16-MAY -80	19-OCT -79	07-SEP -79	10-AUG -79	22-JUN -79	16-MAR -79	13-OCT -78	18-AUG -78	10-MAR -78		
VITALS												
PULSE	72	82	80	68	72	.	.	80	.	85	/MIN	(60-100)
SYS BP SITTING	136	160	130	138	134	180*H	150	140	180*H	140	MM HG	(100-160)
DIAS BP SITTIN	90	84	72	84	80	110*H	84	90	100*H	80	MM HG	(60-90)
WEIGHT LBS	196	196	195	200	200	200	201*H	198	206*H	210*H	LBS	(50-200)
TREATMENTS												
ALLOPURINOL	300	300	300	300	300	300	300	300	300	300	MG	
COLCHICINE	D/C	1.3	0.6	0.6	0.6	1.3	0.6	1.3	1.3	1.3	MG	
DARVON-N	300	300	30	30	30	30	30	.	.	.	MG	
DIGOXIN	.	.	0.250	.	.	0.500	.	.	.	0.500	MG	
HYDRALAZINE	100	100	100	100	100	100	75	75	75	75	MG	
IBUPROFEN	1600	1600	1600	1600	1600	1600	1600	1600	1600	1600	MG	

SMA12	16-MAR -79	18-AUG -78	16-SEP -77	10-SEP -76	16-JUL -76	23-MAY -75		
SMA12								
CALCIUM (SMA)	9.4	9.8	9.8	9.8	9.4	9.7	MG/DL	(8.4-10.6)
PHOS (SMA)	3.2	3.1	3.7	2.7	2.9	5.2*H	MG/DL	(2.5-4.8)
GLU (SMA)	60*L	90	85	113*H	100	138*H	MG/DL	(70-110)
BUN (SMA)	24*H	27*H	21	29*H	22	26*H	MG/DL	(7-22)
URIC ACID (SMA	5.1	5.4	5.5	9.4*H	8.2*H	>12.0*H	MG/DL	(2.6-7.1)
CHOL (SMA)	173	190	183	187	200	175	MG/DL	(120-280)
TOT PROT (SMA)	7.8	8.1	7.9	7.9	7.8	8.0	G/DL	(6.4-8.2)
ALBUMIN (SMA)	4.2	4.1	4.2	4.2	4.0	3.7*L	G/DL	(3.8-4.8)
BILI (SMA)	0.4	0.5	0.5	0.6	0.5	1.3*H	MG/DL	(0.0-1.0)
ALK PHOS (SMA)	90	75	82	92	77	86	UNITS	(25-115)
LDH (SMA)	286*H	215*H	300*H	250*H	220*H	178	UNITS	(100-190)
SGOT (SMA)	47*H	46*H	54*H	39	26	38	UNITS	(8-40)

ELEC PROF	30-JAN -81	26-SEP -80	16-MAY -80	16-MAR -79	18-AUG -78	16-SEP -77	10-JUN -77	17-DEC -76	10-SEP -76	16-JUL -76	04-JUN -76	11-JUL -75		
ELEC PROF														
SODIUM	.	142	143	141	144	145	142	142	143	144	.	145	MEQ/L	(135-145)
POTASSIUM	3.5	3.9	4.7	3.9	3.6	4.3	3.7	3.5	4.0	3.4*L	2.9*L	4.4	MEQ/L	(3.5-5.5)
CHLORIDE	.	108*H	110*H	107*H	109*H	110*H	108*H	107*H	108*H	110*H	.	107*H	MEQ/L	(95-105)
CO2 (TOTAL)	.	25	22	24	26	25	26	25	27	26	.	25	MEQ/L	(22-27)

ENDOCRINE	16-MAR -79		
THYROID TESTS			
T4 INDEX	2.5*L		(4.5-11.5)
TSH'	100.0*H	NG/ML	(0.2-5.6)

```
CYTOLOGY
 CERVICAL PAP   16-MAY-80 NO TUMOR CELLS

RADIOLOGY'
 CHEST X-RAY    10-AUG-79 LUNGS NEGATIVE, CARDIOMEGALY, SAME
                18-AUG-78 DEG SPINAL CHG, CARDIOMEGALY, WORSE, LUNGS NEGATIVE
 COLON X-RAY    24-AUG-79 FILLING DEFECT, RECTAL POLYP
 KNEE X-RAY     18-AUG-78 BILATERAL, JOINT NARROWING, OSTEOPHYTE, OSTEOARTHRITIS, LEFT, JOINT EROSION, JOINT EFFUSION
                            EDEMA PERIPHL, OSTEOSCLEROSIS, GOUT ?
 UPPER GI X-RAY 09-NOV-79 AIR CONTRAST, FLUOROSCOPY, GASTRIC POLYP, ESOPH STRICT, OTHER ABN, ENDOSCOPY RECND

    * ABNORMAL RESULT
    ----------------------   RHC  SUMMARY  22-Jul-82  ----------------------

DEMONSTRATION.PATIENT          PAGE 1
#9999-4                                  PLACE IN FRONT OF CHART
```

Figure 2. Flowsheet summary.

```
SURVEILLANCE REPORT   DEMONSTRATION PATIENT   9999-4   BORN 18-NOV-50


Please record weight, PND, exercise tolerance, etc. in space allocated on
Encounter Form to document effect of change (-0.25) on 19-OCT-79) of
digitalis dose.

All women should have at least 2 annual CERVICAL PAP tests on record.
It has been more than a year since this patient's first test which was
done on 16-MAY-80.

Consider use of TSH to distinguish primary from secondary hypothyroidism
and/or to interpret borderline low T4 INDEX of 2.5 on 16-MAR-79.
```

Figure 3. Quality assurance report.

The clinic chart has four surfaces on which to place records. One of the surfaces is dedicated to the flowsheet report. At each visit the secretary replaces the old computer generated flowsheet with a new cumulative one produced the night before.

Before the physician sees the patient, he/she usually reviews the last visit record and the most recent portions of the flowsheet to review the patient's course and identify results that may have been obtained since the last visit.

The surveillance report calls attention to other issues that might require specific attention. If the patient is a walkin (in which case the summary report would not have been printed the night before), it is possible to obtain a flowsheet on demand. However, more commonly physicians use the on-line inquiry terminal to review the patient's record.

A typical dialogue with the Inquiry program is as follows:

PATIENT'S NAME?	999999-6
DATE RANGE?	1/1/81 TO 1/1/82
PARAMETERS OF INTEREST?	LYTES, URIC ACID, BP MEDS, PB'S

The above dialogue exemplifies the use of Inquiry to review a patient's antihypertension regimen and its effect on relevant outcome variables. Notice that three parameter sets are requested in this dialogue. With this program the user can create arbitrary flowsheets unique to the specific clinical issue of interest. As was the case for previous reports, the header content and format can be modified by the system manager.

After seeing the patient, the physician writes his notes, updates the problem list (to the degree he is so inclined), orders tests and/or procedures, records the time interval until the next scheduled appointmnt, and writes his prescriptions—all on the encounter form, which he signs.

At checkout, the secretary identifies the physician who saw the patient, makes the return appointment according to the physician's request, reviews the outstanding appointments and cancels any that are no longer needed. The bottom part of the encounter form serves as the prescription, and the first part is put in the chart as the record of the visit.

Post-visit Activities

Following the visit, information gathered at the clinic visit and diagnostic studies that return following the visit are entered into the computer. The Regenstrief Medical Record System has no biases as to the optimal degree to which the medical record should be stored in the computer. For the foreseeable future, all practices will require a hardcopy record at least for the storage of graphical, (e.g., EKGs,) and signatory data, (e.g., release forms and other such items.) Regardless of the percentage of the medical record contained in the computer, a computerized record can be useful and complementary to the paper record.

For example, by entering all diagnostic studies into the computer, we get the results into the manual chart without the cost of pulling the chart and entering all the individual lab slips. This is done by putting the cumulative flowsheet into the medical record chart at the time of the next patient's visit when it would have to be retrieved anyway. At the same time, this approach provides on-line availability of the laboratory and diagnostic studies at all times.

The decision about how much of the medical record should be stored in the computer is strictly an economic one. It depends upon the availability of machine-readable patient information from other automated systems within the practice environment, the frequency of use of medical records, the degree to which the cost of entering such data into the computer can be traded off against existing costs of dictations, the cost of mass storage hardware and the available capabilities of local computer systems.

Regardless of how much data the clinic chooses to enter, the RMRS data entry program provides a number of features that speed this entry. Conceptually, the entry of each result requires the identification of a patient, a parameter, a date, in some circumstances a time, and the result.

The program takes advantage of natural orderings of data at the time of entry. For example, many of the results for a single patient and results produced on a given day will tend to be clustered together. The program permits the user to take advantage of any such clustering by storing the repeated component of result once for a sequence of results. Alternatively, the user can type in "S" to any given prompt to indicate that the input is the same as that entered for the immediately preceding result.

The program also allows full use of the set definitions mentioned above. For example, to enter the 12 results within an SMA12, one types in SMA12 and the program prompts back with the names of each of the 12 entries. Multiple choice answers may be answered by selecting available responses rather than typing in the name of the response. The program has the same flexibility regarding headers and help messages as the previously mentioned programs.

Between visits, a user can address ad hoc queries to the data base or the medical record file at any time. He can also review the patient's record at any time. In fact, staff physicians currently use portable terminals to review the patient's record at home when they are taking night call.

Between patient visits, statistical and management reports are produced for a number of purposes. For example, clinic workload statistics, by physician, by day of the week, by hour of the day. The program is able to generate charge information for each patient visit. The Regenstrief lab and pharmacy systems currently do generate such data and transfer to a billing system provided by McDonell-Douglas.

Conclusion

As can be seen from the above, the goals of medical record designers should be to achieve maximum generality. This is accomplished by putting into the hands of the users decisions about how the programs should run and what tasks they should perform.

The use of stored data definition and user-oriented command languages described in this report provides some of the needed generality. The provision of generality has two costs. The first is a potential decrease in computer efficiency. While there are always tradeoffs between generality and efficiency, the optimal tradeoff will increasingly favor generality because hardware costs are decreasing so rapidly. The second is management overhead. One could place so much generality into a system that the average practice could not manage the system. In other fields, major progress is usually marked by the invention of a simplifying framework in which to view what had become a very complex system. Thus the degree to which developers will reach the goal of flexible medical practice systems that are easy, both to use and to manage, will depend upon the elegance and simplicity of the conceptual framework on which they are based. There is still much work to be done.

References

[1] Shortliffe, E. H. *Computer-Based Medical Consultations: MYCIN*. New York: Elsevier/North Holland, 1976.

[2] McDonald, C. J. *Action-Oriented Decisions in Ambulatory Medicine*. Chicago: Year Book Medical Publishers, 1981.

[3] McDonald, C. J. "Computer Reminders, the Quality of Care and the Nonperfectability of Man," *N. Engl. J. Med.* **295,** 1976, 1351–1355.

[4] McDonald, C. J., Wilson, G. A., McCabe, G. P. "Physician Response to Computer Reminders," *JAMA* **244,** 1980, pp. 1579–1581.

17

SCAMP System

Alan R. Shapiro, M.D.*

Introduction

The purpose of this report is to provide an overview of the SCAMP clinical information system for ambulatory care. For the past decade, the system had as its goal the acquisition of all clinically pertinent patient information and the processing of this information to enhance the clinical and administrative purposes of multi-physician practices. Over the past four years the system has been totally revised to incorporate modern software techniques for representation of medical knowledge, processing of language, and retrieval of information as well as greatly enhanced ease of overall usage, including modification of system capabilities by clinicians. Over the past three years the system has attained a good deal of practical success in five large family medicine clinics and one private group practice, together comprising over 180 primary care physicians. This chapter highlights the features in SCAMP which have made this system attractive to practicing physicians.

Four principles have guided the evolution of the SCAMP clinic system:

(1) *Ease of use:* a great deal of attention has been devoted to making the user's interaction with SCAMP as effortless as possible. A combination of hierarchially organized menu displays and the capability of the system to understand simple requests expressed in English (natural language interface) make the system unusually easy to learn and operate. The typical orientation time required to teach physicians how to use the system to retrieve patient information is one hour. Clerical personnel without prior exposure to computers can be trained to use the billing, appointment, data entry and editing routines competently in less than 2 days.

*Reprinted with permission from *Journal of Medical Systems* **7**, Plenum Publishing Corp., © 1983; modified from a version previously published in *Proceedings of the Sixth Annual Symposium on Computer Applications in Medical Care*, IEEE, 1982.

(2) *Comprehensiveness:* SCAMP has subsystems for dealing with all information of importance in patient management, including history, physical exams, laboratory information, vital signs, medications, radiology notes, office procedures, hospitalization notes, referrals and therapeutic plans. This patient data is processed by the administrative reporting, practice analysis, clinical reminder and clinical research components of the system and, in some clinics, is integrated with the billing/accounts receivable subsystems.
(3) *Incorporation of medical knowledge:* the ease of use and practical usefulness of the system is largely due to the incorporation of medical or procedural knowledge into the system functions. Spelling correction procedures monitor data entry; drug information can be found by entering only two or three letters of the drug name or of the class of drugs of which the drug is a member; descriptive problem titles can be entered in English and coded by the system; and the problems and drugs are organized in a conceptual hierarchy that matches a clinician's conceptual approach. An extensive set of clinical rules allows the computer to use information to review care and prevent physician errors of omission.
(4) *Ease of modification of system capabilities:* SCAMP features a "fill-in-the-blank" menu system, or *template* system, which allows users to create their own clinical reminder rules, to specify arbitrarily complex retrieval requests and to design ad hoc programs for summarizing patient information without the need for any knowledge of programming. A two-dimensional formatting capability allows users to design the output of reports, again without the need for programming skills. The acceptability of a system and the involvement of physicians in working with the system are enhanced by allowing clinicians to tailor the system's capabilities to their own needs.

Overview

The capabilities handled by SCAMP include patient registration, billing, appointments, entry and printout of the medical record including text of progress notes, patient summaries, letter writing, and extensive retrieval, analysis, clinical reminder and clinical advisor capabilities. These capabilities are modular and many different configurations of these modules have been implemented. In a typical clinic, information is obtained either from an encounter form or directly from the physician's dictation. Clerical personnel enter the data into the computer as a by-product of the transcription of the doctor's dictated progress notes. Information is then available to the clinician either through direct inquiry on a video display terminal or via a multitude of special reports and patient chart printouts. Some major features of the subsystems are indicated below.

Appointments

An on-line interactive appointment system is especially useful in the larger clinics. Patient appointments can be entered, edited or retrieved and special capabilities exist for assigning patients with certain problems or characteristics to specific time periods. The system can automatically generate appointment reminder notes concerning patients with special characteristics selected by clinic physicians. In the larger clinics, the appointment sheet is often used as a communication vehicle to send messages to physicians concerning special events or reminders concerning patients.

Billing

The billing/accounts receivable portion of the system handles Medicare/Medicaid and Blue Cross/Blue Shield, using an A92-R form or superbill. Dictionaries are set up to accept ICHPPC, ICD9 and CPT codes. A large number of administrative reports provide summaries of clinic activities by procedure and by physician. The billing system can interact with the appointment system so that patients with a delinquent account status can be flagged at the time an appointment is scheduled.

Retrieval of Patient Data

On-line examination of patient problem histories, immunization histories, laboratory values, medication histories, etc., proceeds either by use of natural language inquiry or use of an extensive, easy-to-use menu system. We chose this approach because use of high level "query" languages to retrieve information still requires a user to express requests in a restricted unfamiliar grammatical style similar to that required in using programming languages. In practice, these query languages often do represent a valuable convenience for programmers but remained unused by the busy physician for whom the system was intended.

Some of the retrieval capabilities are illustrated in the examples below. In each case, the request specified is exactly what a user might type into the computer to enter a retrieval request.

(1) "HOW MANY VISITS WERE THERE IN THE CLINIC LAST YEAR?"
A number of totals such as number of patients in the clinic, visits to the clinic, physicians etc., are continuously updated and the answer to such a request is immediate.

(2) "HOW MANY VISITS WERE THERE FOR ADVERSE DRUG REACTIONS IN THE PAST 6 MONTHS?"
Again the response to such a request is instantaneous.

(3) "TELL ME THE NAMES OF PATIENTS WHO DEVELOPED ADVERSE DRUG REACTIONS WITHIN ONE MONTH AFTER TAKING ONE OF THE SHORT-ACTING ANTI-ANXIETY AGENTS."
This request illustrates several points. The computer "knows" that benzodiazepine drugs such as lorazepam, temazepam and oxazepam are members of the class of drugs that are short-acting antianxiety agents; the user need only mention them as a group. Similar classes, subclasses and supraclasses exist for problems, lab tests and immunizations. Time relations between events such as the onset of a problem and the prescription of a drug are handled easily by the system.

(4) "LIST MY PATIENTS WHO NEED FLU SHOTS."
Again the clinical conditions indicating eligibility for flu vaccine can be easily entered into the system. Because of the integrated nature of the SCAMP system, immunization histories and laboratory values are easily handled in combination with requests involving problems and drugs. The identity of the physician is known from the logon sequence.

(5) "WHO ARE THE PATIENTS WITH DUODENAL DISEASE WITH OR WITHOUT COMPLICATIONS AND WITH LIVER DISEASE WHO WERE PRESCRIBED B-COMPLEX WITH VITAMIN C AND LIVER?"
Although not a clinically compelling example, this request illustrates the ability of the natural language interface to differentiate multiple senses of a word such as "with" and to distinguish the use of a word such as "liver" when used to designate a problem and when used as part of a drug name such as "B-complex with C and liver."

The above requests illustrate the wide range of syntactic constructions accepted by the SCAMP natural language interface. Of particular importance, the "understanding" capability of the system can be rapidly and easily extended so that the system can be adapted to accept requests phrased in unusual ways.

Still more complex retrieval requests involving laboratory values, immunizations, billing information, etc., are possible through use of a menu system that guides users in forming rules that can be used in retrieval of selected patients, in creation of clinical reminders, or in the development of reports. This menu system makes it possible to locate patients rapidly who had an increase in BUN within 3 weeks after receiving tetracycline or to identify patients whose last hematocrit was less than 33% and who have not had a visit in 3 months. Flow sheets are readily obtainable from the system for the values over time of any particular laboratory result or profile of laboratories, e.g., CBC or urinalysis. Knowledge incorporated into the SCAMP system allows for spelling correction facilities, identification of abnormal laboratory values, and interrelationships between laboratory tests, e.g., an inquiry concerning serum glucose can refer to fasting blood sugar, 2 hour post-prandial blood sugar, glucose (SMA6) or glucose (SMA12).

```
PATIENTS WHO HAVE NOT HAD A VITAL SIGN MEASURED WITHIN 'N' DAYS
BEFORE THEIR APPOINTMENTS WILL BE SELECTED

ENTER THE VITAL SIGN: BP

ENTER THE NUMBER OF DAYS (DEFAULT = PATIENT'S LIFETIME): 365

ENTER  A CATEGORY
       ? FOR A CATEGORY LIST
       'CR' TO FINISH UP THIS RULE
```

Figure 1. The system prompts for the information needed in forming a rule.

Physician-Generated Clinical Reminders

The extensive clinical information stored in the system is processed by a clinical reminder system in an effort to prevent physician errors of omission before they occur. The capability of computers for interpretation of patient data has obvious potential for use in quality assurance and in physician education; nevertheless, often, physicians will not follow the recommendations or reminders offered by a computer [1]. Physicians may disagree with the rules themselves or may simply not trust the operations of a mechanical device whose "reasoning" processes are obscure [2]. SCAMP is designed to provide the physician with a clear statement of exactly what knowledge the computer used in making a recommendation and, more importantly, is designed to let the physicians enter the rules themselves.

Medical knowledge is represented in rules of the form:

IF a patient is over 65
and IF the patient has not been immunized with pneumococcal vaccine in the past five years
THEN print the reminder "Patient is a candidate for pneumococcal vaccine".

The reasoning involved here is straightforward and readily grasped. SCAMP provides physicians with a menu system that lets them enter, edit, delete and review rules in a simple fashion. Entry of a rule takes seconds. Once a menu item is selected, the system prompts the user for the necessary information (Figure 1). The rules can be reviewed as shown in Figure 2. The rules can relate to patient problems, drugs, demographic characteristics, lab values, lab

```
                          RULE #1
THIS RULE WILL SELECT PATIENTS WHO:

(1) HAVE THE PROBLEM 'HYPERTENSION, UNCOMPLICATED':  AND

(2) HAVE NOT HAD A BLOOD PRESSURE MEASUREMENT WITHIN 365 DAYS.

IF A PATIENT MEETS ALL OF THE ABOVE CONDITIONS, HIS APPOINTMENT
WILL BE FLAGGED WITH THE MESSAGE 'CHECK BP: NO BP IN PAST YEAR.'
```

Figure 2. The user can review the reminder rules that are used by the system.

SGOT = 74 IU
ALKALINE PHOSPHATASE = 194 IU
TOTAL BILIRUBIN-2.4 MG %

MS. PRIOLEAU is presently on CIMETIDINE and has evidence of LIVER DYSFUNCTION. CIMETIDINE is largely metabolized in the liver. For this reason consider reducing her dosage of CIMETIDINE to decrease the chances of CNS side effects.

Figure 3. The clinical reminder/clinical advisor system alerts physicians to possible medication errors and to clinically relevant patient characteristics.

trends, vital signs, vital sign trends, billing information, etc. Since the physician took part in the creation of the rule and the reminder that appears when a patient satisfies the rule, he does not feel defensive when he subsequently receives that reminder from a machine. The rules are entered into a physician specific file and tested by a general rule-processing module so that the capabilities of the system are easily modifiable. The reminders can appear on the appointment sheet, on the medical record patient summary sheet, or can appear as the result of a direct inquiry by the user concerning a specific patient or group of patients (Figure 3).

Despite the ease in modification by users, there have been two difficulties with the physician-designed clinical reminder system. The first is that the patient problems which the system currently recognizes are restricted to a subset of those in the ICD-9 coding system. Physicians often wish to make finer clinical distinctions than such codes allow. SCAMP is currently being revised to work with natural language problem descriptions as well as codes to provide still greater flexibility. The second limitation is a relative one in that some physicians use the reminder system more creatively than others. Therefore, we offer physicians the "best" of the rules that have been developed as guides, placing great emphasis on the ease with which these rules can be modified and with which similar additional rules can be created.

Medical Monitor and Drug Advisor

Clinical reminders involving drugs are handled somewhat differently than those involving laboratory tests, vital signs and immunizations. A physician should not have to wait until a patient returns to clinic to learn that he prescribed a drug which interacted seriously with another drug that the patient was taking. Therefore, each night after drug information has been entered, the medication regimen of the patients who were seen is reviewed and compared with over 1000 rules describing drug–drug and, more importantly, drug–problem interactions. Rules are of the format: If the patient is taking <cimetidine> and there is <evidence of liver dysfunction> and the dose is <1200 mg per day or more> then print a warning message suggesting <reduction of

the dose by one third and special monitoring for CNS side effects>. "Evidence of liver dysfunction" is a class of laboratory tests and associated values such as a SGOT greater than 30 I.U., alkaline phosphatase greater than 175 I.U., bilirubin greater than 2.0 mg%, etc.

The drug advisor program has been set up so that a physician can type in a patient's name and a medication that the physician is considering prescribing to the patient; the drug advisor screens the patient's record to see if there are any precautions or contraindications to giving the drug. As the database has grown, physicians with access to the drug advisor program have become increasingly aware that it is worthwhile taking 90 seconds to check with the program before prescribing medications to a patient who has multiple problems and a complicated drug regimen; a potentially serious iatrogenic complication may be avoided and, at worst, the program provides an educational experience.

Use of the Rule Forming System in Patient Selection

Reports from a patient information system typically involve:

(1) selection of criteria to be met by patients appearing on the report;
(2) selection of items of information about these patients; and
(3) selection of a suitable report format.

The menu-based rule formation system described above is also available for selecting patients to appear on reports. All patients of a particular doctor, all heads of household in the practice of a particular doctor, or all patients over 65 with congestive heart failure who are taking digitalis and have not visited in the past nine months can be selected to appear on a report by passing the appropriate rules to the report writing or letter writing modules. Figure 4 illustrates the capabilities of this system, showing a retrieval rule involving time relationships between several events including those based on analysis of the text of the medical record. The rule was used in a study examining the completeness of the medical record, in particular the frequency with which physicians record laboratory abnormalities in their progress notes.

THIS RULE WILL LOCATE THE PATIENTS WHO:

(1) RECEIVED CHLORTHALIDONE FOR TREATMENT OF 'HYPERTENSION': AND

(2) HAD A SERUM POTASSIUM LESS THAN 3.3 WITHIN 8 WEEKS AFTER AN RX FOR CHLORTHALIDONE: AND

(3) HAD THE CONCEPT 'HYPOKALEMIA' MENTIONED IN THEIR RECORD WITHIN 2 WEEKS BEFORE OR 2 WEEKS AFTER THE PRESENCE OF A POTASSIUM LESS THAN 3.3

Figure 4. Printout of a rule involving time relationships between prescriptions, laboratory values, and concepts mentioned in the free text of the medical record.

Report Formatting and Generation

Users of an information system differ greatly in the type of information in which they are interested and in the way they desire the information to be presented. These differences are often legitimate, reflecting differences in practice arrangements as well as in individual preferences. By providing users with control over the information included in the computer output as well as over the formatting of its presentation, SCAMP relieves users of the numbing experience of wading through extensive printouts of irrelevant data.

Selecting the information to appear on a report proceeds using a template system similar to that used in forming clinical rules. The user first selects from a menu the categories of information he wants on his report, e.g., census, problem history, medication, laboratory values, immunization, allergies, billing, etc. As each category is selected, the report facility tells the user which items in that category are available from the patient files for printing. For example, if the problem history category is selected, the system tells the user that items such as the problem title, problem code, number of visits, date of problem definition, and date of last visit for each problem are available for printing. The user can specify whether he wishes the report to be produced for a single patient, a predefined group of patients, patients satisfying certain rules, a particular physician's practice, a predefined group of practices, or all patients seen in a particular clinic.

After the items of information to appear on the report have been selected, the user can design a layout for the report using the report formatter. In using the formatter, the user pretends that his screen is a window looking out on to the printed report page. The user tells the report facility to place an item of information at a point on the page simply by moving the cursor to that point and entering the name of the item. The formatter allows the user to roam around on the screen, placing text and the names of the data items at any location. It supports, among other features, an "offset" facility that lets the user shift the screen "window" to any part of a report "page" that may or may not be larger than the typical video screen of 80 columns and 24 rows. Each of the features of the report formatter is called into play with only a single key stroke. The formatter has been designed to work with many different types of video terminals.

Once the report format has been designed or selected, the user can have the report produced directly or, if the report is intended to be run more than a few times, the system can generate a program from the specifications that were provided. In effect, the user has been led through the process of writing a program to produce a report, without ever being required to use a programming language.

The report facility has been used in tailoring some of the most frequent printouts used on the system. For example, programs that print the patient summary sheet of the patient record have been built using the report facility. Figures 5(a) and (b) show two different formats that were developed to display

PATIENT SUMMARY REPORT

NAME: PELHAM, THELMA OPD #: 241771 AGE: 68 PAGE 1
SEX: F

PERMANENT PROBLEM SUMMARY

ID	TITLE	# VIS	DEFIN. LAST VIS	ID	TITLE	# VIS	DEFIN. LAST VIS
A	PERIODIC HEALTH EXAMINATION	2	11/06/75 11/06/75	B	HYPERTENSION NOS	17	11/06/75 11/06/75
C	DIABETES MELLITUS	47	11/06/75 11/06/75	D	AC MYOCARD INFARCT/SUBAC ISCHEMIA	10	11/06/75 11/06/75
E	OTHER CEREBROVASCULAR DISEASE	3	11/06/75 11/06/75	F	ANXIETY NEUROSIS	3	06/01/72 10/29/79
G	ADVERSE EFFECTS OF MEDICINAL	1	06/01/72 11/30/79	H	PAIN IN JOINT	2	01/18/80 02/07/80

(a)

PATIENT PROFILE FOR PELHAM, THELMA (#241771) AGE: 68 SEX: F

PRB ID	PROBLEM TITLE	PERMANENT PROBLEM SUMMARY DATE OF LAST VISIT	DATE OF DEFINITION	# OF VISITS
H	PAIN IN JOINT	02/07/80	01/18/80	2
G	ADVERSE EFFECTS OF MEDICINAL AGENTS	11/30/79	06/01/72	1
F	ANXIETY NEUROSIS	10/29/79	06/01/72	3
A	PERIODIC HEALTH EXAMINATION	11/06/75	11/06/75	2
B	HYPERTENSION NOS	11/06/75	11/06/75	17
C	DIABETES MELLITUS	11/06/75	11/06/75	47
D	AC MYOCARD INFARCT/SUBAC ISCHEMIA	11/06/75	11/06/75	10
E	OTHER CEREBROVASCULAR DISEASE	11/06/75	11/06/75	3

(b)

Figure 5. Two different formats developed using the report facility for printout of the problem list.

the problem list portion of the patient summary according to the preferences of users at different sites.

Natural Language Processing of the Medical Record

The clinics in Charleston and Spartanburg and the group practice in Mount Pleasant have been set up to use word processing programs incorporated in the SCAMP system for transcribing physicians' dictations and a line printer to print out copies of the medical record. As the text of the record is entered in problem oriented format, the natural language routines process the text to extract drug and problem information. The information is stored both in the patient's computerized record and, simultaneously, in inverted problem and drug files. Programs that are used to edit the text of the medical record simultaneously edit inverted files on-line.

With respect to data entry, the natural language processing programs make the existence of a computer system invisible to practicing physicians. They are not required to fill out lengthy forms in order to record clinical information for further studies [3,4].

We have been experimenting for several years with processing the text of the medical record in order to locate patients having procedures such as performance of a digital exam or having clinical phenomena such as dyspnea on exertion or swelling of the ankles. Depending on the concept to be identified the accuracy of these routines can be very good, with low (often less than 2%) false negative and false positive rates, making them very useful in practice reviews and research involving examination of patient records.

Despite their usefulness in practice review and research, we have not yet incorporated these routines into the clinical reminder system because of a firm policy which maintains that, above all, the computer generated reminders should be accurate. Adherence to this policy has helped gain physician acceptance of the SCAMP system. We do, however, anticipate continuing to extend our use of natural language processing to a variety of clinical applications.

Summary

Acceptance of the SCAMP system by primary care physicians did not arrive until scrupulous attention had been given to improving accessibility and relevance of the information provided. Improvements in these two areas came from incorporating procedural and medical knowledge into the functions of the system. Further advances in gaining physician acceptance have come from including features designed to let users have direct involvement in developing their own clinical reminders as well as in fine tuning the retrieval and report programs according to their individual preferences.

References

[1] McDonald, C. J., Wilson, G. A. and McCabe, G. P., "Physician Response to Computer Reminders," *JAMA* **244;** No. 14, 1980, pp. 1579–1580.

[2] Teach, R. L. and Shortliffe, E. H., "An Analysis of Physicians Attitudes Regarding Computer-based Clinical Consultation Systems," *Computers and Biomedical Research* **14,** 1981, pp. 542–558.

[3] Shapiro, A. R., "A System for Conceptual Analysis of Medical Practices," *Proc. Computer Applications in Medical Care,* November, 1980, pp. 867–872.

[4] Shapiro, A. R., "Exploratory Analysis of the Medical Record, *Proc. Computer Applications in Medical Care,* November 1982, pp. 781–786.

18

Core Record System*

Elizabeth E. McColligan, M.S., M.P.H.

Introduction

The concept of a Core Medical Record was derived as an integral part of a plan for a reorganization of the outpatient medical record system. The Johns Hopkins Hospital currently uses a unit record containing both inpatient and outpatient information organized by a single patient identifier. The reorganized system would establish three types of records for ambulatory care:

The main medical record (Main Record) which is the repository for all medical information. It contains complete records for all inpatient stays plus some information about current outpatient activity. This is the unit record as it currently exists.

The local clinic record (Local Record) which contains complete information about patient care rendered by that clinic. Older information in the Local Record may be retired to the Main Record. Historically, specialty clinics have retained duplicate local records which are used in the absence of an available Main Record. These duplicate clinic records, however, lack information about patient visits outside that clinic.

The automated core medical record (Core Record) which contains a summary of the information included in the various Local Records plus an abstract of recent inpatient admissions. The Core Record is available as a printed record to Medical Records and selected clinics; it is also available at CRTs throughout the institution.

The goal is to have the Core Record contain sufficient information so that it, combined with the Local Record, will eliminate the need for the Main Record, at the time of the outpatient encounter, approximately 90% of the time.

*This chapter was prepared especially for this collection.

Core Record System

The Core Record System [1,2] is an Automated Ambulatory Medical Record System (AAMR) designed for multiclinic use at the Johns Hopkins Hospital (JHH). It provides on-line access to a minimal essential "core" of medical information, prints encounter forms on demand which contain this core record (Figures 1,2), supports an appointment system and is integrated with a registration and charge capture system. The system was developed in an attempt to alleviate the following problems associated with the operations of the Outpatient Department (OPD):

(1) Medical record availability and costs. Because of the heavy request volume, it is difficult and expensive to deliver medical records for all outpatient visits.
(2) Lack of integration with existing hospital administrative systems. A complete patient registration must be performed for each visit, resulting in keystroking data already available in the hospital data base.
(3) Lack of coordination of patient care among clinics. The majority of the patients seen at JHH visit more than one OPD clinic. If a medical record is not available at the time of the visit, the provider of care may be unaware of visits to other clinics and the care which has been provided.

JHH is an 1100-bed teaching hospital where 100,000 patients are seen in 325,000 visits to the Outpatient Department each year. The Core Record System was designed to operate in this environment and perform the following functions:

(1) Maintain a minimal AAMR.
(2) Provide on-line patient registration and processing of charges.
(3) Operate a clinic-oriented appointment system.
(4) Maintain a data base for retrospective analysis and produce management reports.
(5) Integrate with existing hospital administrative systems.

The flow of the Core Record System in the Emergency Department can be seen in Figure 3.

The Core Record System is modeled after the Minirecord System, an automated *mini*mal essential *record* system, used in the Medical Clinic at JHH from 1975 through 1982 [3,4]. The Minirecord contained a list of active and relevant inactive problems plus a current medication list which were maintained by the provider as part of the standard visit note. The minirecord was printed on an encounter form for use at each visit; it was also available at a CRT in appropriate hospital locations. The primary difference between the two systems is that the Core Record System is designed to operate in multiple clinics which share common problem and medication lists. Because the Core Record System is designed to be a hospital-wide system, it also supports additional functions such as charge processing and management reporting.

APPT. TIME:

THE JOHNS HOPKINS HOSPITAL

CORE RECORD (EMERGENCY DEPARTMENT)

IF THERE ARE ANY QUESTIONS REGARDING YOUR TREATMENT OUR STAFF WILL ASSIST YOU - CALL 955-2280. BRING THIS SLIP WITH YOU IF YOU RETURN TO THE EMERGENCY ROOM. ALWAYS BRING YOUR HOSPITAL PLATE WITH YOU WHEN COMING TO THE HOSPITAL.

STAMP PATIENT ID OR PRINT CLEARLY

HISTORY NUMBER

DOCTOR'S INSTRUCTIONS	NURSE'S INSTRUCTIONS
DIAGNOSIS:	
PHYSICIAN'S SIGNATURE:	NURSE'S SIGNATURE:
I ACKNOWLEDGE THAT THE ABOVE INSTRUCTIONS HAVE BEEN EXPLAINED TO ME TO MY SATISFACTION	SIGNATURE:

DISPOSITION

( ) PRN ( ) ADMISSION

( ) ER/PCC RETURN VISIT ( ) JHH CLINIC VISIT

( ) LAB & X RAY TEST ( ) OTHER

( ) COMMUNICABLE DISEASE REPORTED ( ) NOT TREATED

ADDRESS: TIME IN: 1:00 P.M. HISTORY NUMBER: 999

CITY: BALT TIME OUT: NAME:

STATE: MD ZIP: 21213 PHONE: DATE: 02/23/82

ONSET OF ILLNESS - ACCIDENT DATE / / TIME / / COMPLAINT

RACE	SEX	DATE OF BIRTH	MARITAL	RELIGION	NEXT OF KIN NAME	RELATION
B	M	10/29/53				

FATHER'S NAME	MOTHER'S MAIDEN NAME	NEXT OF KIN ADDRESS

CLINIC/CODE	FINANCIAL CODE	HOW ARRIVED	ACCOMPANIED BY
	PPC	☐ WALK ☐ BUS ☐ CAR ☐ TAXI ☐ AMB # ☐ POL # ☐ HEL #	

PERSON RESPONSIBLE FOR BILL/EMPLOYER

NAME/EMPLOYER PT PT

BLUE CROSS PATIENTS

MEMBERSHIP NUMBER POLICYHOLDER NAME: RELATION

MEDICARE PATIENTS **MEDICAID PATIENTS**

MEDICARE NUMBER: MEDICAID NUMBER: DATES TO

PATIENT ORIGINAL MUST BE FILED WITH MEDICAL RECORDS

Figure 1. First page of encounter form—demographic data.

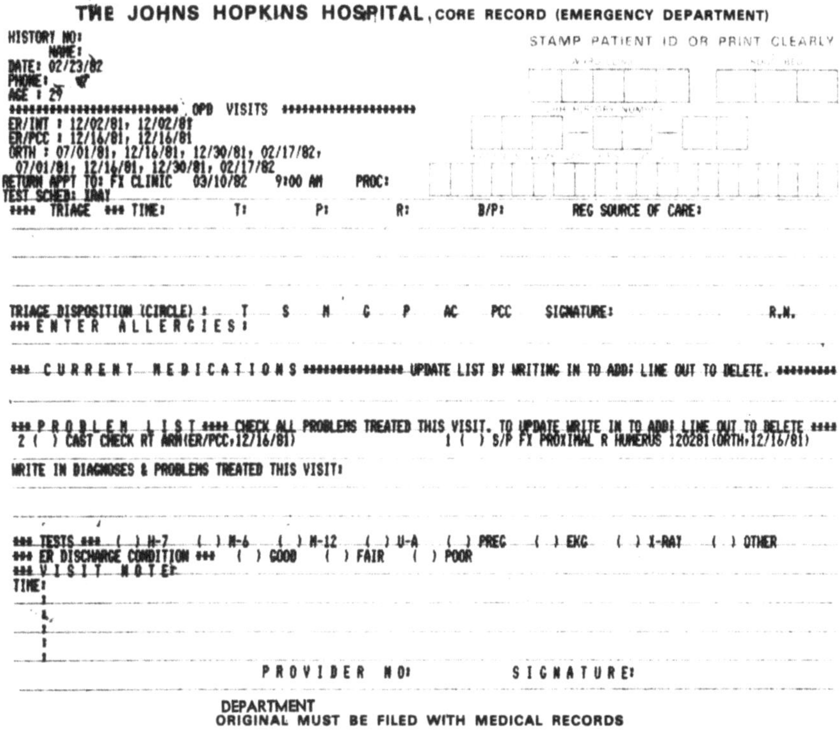

THE JOHNS HOPKINS HOSPITAL, CORE RECORD (EMERGENCY DEPARTMENT)

STAMP PATIENT ID OR PRINT CLEARLY

HISTORY NO:
NAME:
DATE: 02/23/82
PHONE:
AGE : 29
######################## OPD VISITS ####################
ER/INT : 12/02/81, 12/02/81
ER/PCC : 12/16/81, 12/16/81
ORTH : 07/01/81, 12/16/81, 12/30/81, 02/17/82,
07/01/81, 12/16/81, 12/30/81, 02/17/82
RETURN APPT TO: FX CLINIC 03/10/82 9:00 AM PROC:
TEST SCHED: XRAY
TRIAGE ### TIME: T: P: R: B/P: REG SOURCE OF CARE:

TRIAGE DISPOSITION (CIRCLE) : T S M G P AC PCC SIGNATURE: R.N.
E N T E R A L L E R G I E S :

C U R R E N T M E D I C A T I O N S ############## UPDATE LIST BY WRITING IN TO ADD; LINE OUT TO DELETE.

P R O B L E M L I S T #### CHECK ALL PROBLEMS TREATED THIS VISIT. TO UPDATE WRITE IN TO ADD; LINE OUT TO DELETE
2 ( ) CAST CHECK RT ARM(ER/PCC,12/16/81) 1 ( ) S/P FX PROXIMAL R HUMERUS 120281(ORTH,12/16/81)

WRITE IN DIAGNOSES & PROBLEMS TREATED THIS VISIT:

TESTS ### ( ) M-7 ( ) M-6 ( ) M-12 ( ) U-A ( ) PREC ( ) EKG ( ) X-RAY ( ) OTHER
ER DISCHARGE CONDITION ### ( ) GOOD ( ) FAIR ( ) POOR
V I S I T N O T E:
TIME:

P R O V I D E R N O: S I G N A T U R E:

DEPARTMENT
ORIGINAL MUST BE FILED WITH MEDICAL RECORDS

Figure 2. Second page of encounter form—medical information.

The Core Record contains four basic sets of data:

(1) Standard identification and registration information.
(2) Active problem list, inactive problem list and medication summary; all are stored as free text.
(3) Information extracted from other hospital systems; including lists of all outpatient visits captured from the hospital billing system.
(4) Clinic specific data such as the work release information required by the Orthopedic clinics.

In addition to the summary information stored in the Core Record System, which is used at the time of the patient's visit, the system maintains an archival encounter data set which is used for retrospective analysis and management report generation.

Archival encounter data set contains:

Encounter date
Patient history number and name
Patient arrival time
Treatment location

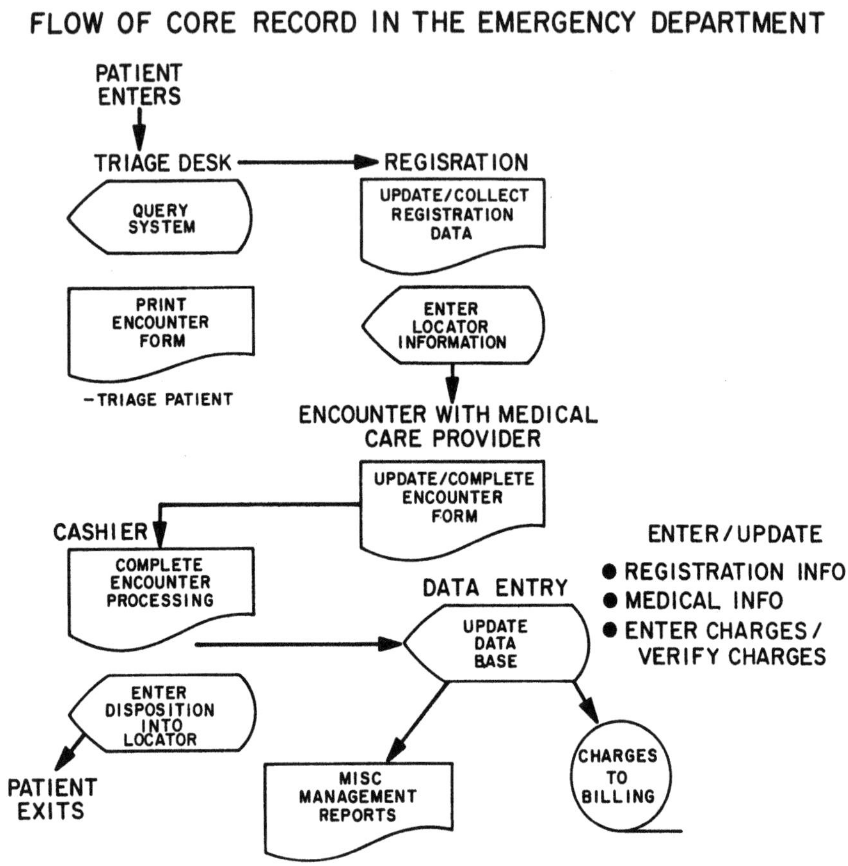

Figure 3. Data flow in Emergency Department.

Type of care rendered
Complaint as verbalized by patient
Date of onset of illness
Method of arrival
Accident information/police notification
  (Data collected by the Emergency Department only)
Provider of care
Visit procedure code
Primary visit diagnosis
Date and time of patient's exit from clinic
Exit disposition
Patient demographics:
  Date of birth
  Race

Sex
Financial code
Zip code

Because the Core Record System was designed to be a prototype system, its installation was limited to three OPD clinics, they are:

(1) *Primary Care Clinic (PCC).* Patients are seen on a "walkin" basis and triaged to other clinics as appropriate, or treated in the PCC if the complaint is minor. PCC was designed to alleviate the overutilization of the Emergency Department by those patients with non-emergency complaints. Prior to January 1981, PCC was a distinct clinic; with the renovation of the Emergency Department it was incorporated into the Emergency Department. In 1977, 14,600 patients were seen in 21,600 PCC visits. Approximately 31% of patients seen in PCC did not visit any other JHH clinics during that year.
(2) *Orthopedic Clinics.* Several different orthopedic specialty clinics with half- or full-day sessions, comprise the Orthopedic Clinics. These include adult and pediatric clinics, fracture clinic, hand clinic, foot clinic, prothesis clinic, etc. In 1977, 4,308 patients were seen in 9,518 visits. Only 10% of the patients seen in Orthopedics did not visit any other JHH clinics during that year.
(3) *Emergency Department.* The Emergency Department is divided into three areas: primary, intermediate and critical care. Since the primary care area functionally replaced the Primary Care Clinic, the primary care area is referred to as the PCC. The intermediate and critical care areas of the department comprise what is referred to as the ER. For the present, the administrative records of the ER and PCC continue to be maintained separately. In 1977, 30,900 patients were seen in 51,500 visits. Approximately 50% of the patients seen in the ER did not visit any other JHH clinic, and had no other ER visits during that year.

In the first six months of 1982 there were 31,000 visits to the Emergency Department and Primary Care Clinics, processed through the Core Record System, and 4,000 visits to the Orthopedic Clinics.

Management Information Component

While most of the necessary financial reports are prepared by the hospital financial system, reports on clinic activity are required for management, assessment, and planning. By utilizing the data recorded in the archival encounter data set it is possible to produce a variety of reports pertaining to clinic activity. Reports on patient arrival times are useful for planning nurse staffing, particularly in the Emergency Department. The report in Figure 4 shows the number of patients arriving in each of the Emergency Department treatment areas by hour of day. This report is available for a single day, a range of dates (e.g., a

08/25/82

EMERGENCY DEPARTMENT MONTHLY REPORT: HOURLY PATIENT COUNTS
01/01/82 TO 06/30/82

TIME	PRIMARY	INTERM	CRITICAL	UNKNOWN	TOTAL
12 AM	48	530	17	87	682
1 AM	34	428	18	63	543
2 AM	32	434	12	61	539
3 AM	19	340	4	58	421
4 AM	12	270	9	28	319
5 AM	12	205	9	27	253
6 AM	17	233	3	42	295
7 AM	562	510	6	254	1332
8 AM	1249	739	10	391	2389
9 AM	1420	1002	10	482	2914
10 AM	1190	1010	11	404	2615
11 AM	969	981	16	376	2342
12 PM	783	914	16	305	2018
1 PM	636	887	16	306	1845
2 PM	514	812	15	319	1660
3 PM	422	656	15	391	1484
4 PM	340	643	21	427	1431
5 PM	280	645	27	394	1346
6 PM	242	709	26	398	1375
7 PM	163	627	15	289	1094
8 PM	159	566	29	264	1018
9 PM	97	527	16	237	877
10 PM	163	389	23	234	809
11 PM	113	408	21	198	740
TOTAL	9476	14465	365	6035	30341

Figure 4. Listing of patient arrival by time of day.

given month) or it is possible to order this report by day of week for a given range of dates. In the latter case, counts for all Mondays occurring in the date range specified are summed together, this allows administrative staff to identify arrival patterns based on the day of the week. The data can also be displayed graphically to permit visual comprehension of the peaks and valleys which occur in the patient arrival patterns (Figure 5).

Figure 6 shows another report that can be produced which displays the distribution of patients by their method of arrival and the type of care received. This information can be useful when analyzing ambulance facilities. The system can also produce reports on patient length of stay in the clinic, the number of patients seen by each provider, and the number of patients seen in each of the visit categories (e.g., initial visit, return visit, etc.). Reports describing the demographic characteristics of the patient population are also available. The number of patients seen by zip code can be useful in determining the geographic distribution of the patients served by the clinic.

It is also possible to search the archival data base for particular groups of patients, such as those patients admitted to the hospital from a particular clinic or all patients with the diagnosis of Asthma. One study, for example, retrieved all information about patients seen in the Emergency Department with the

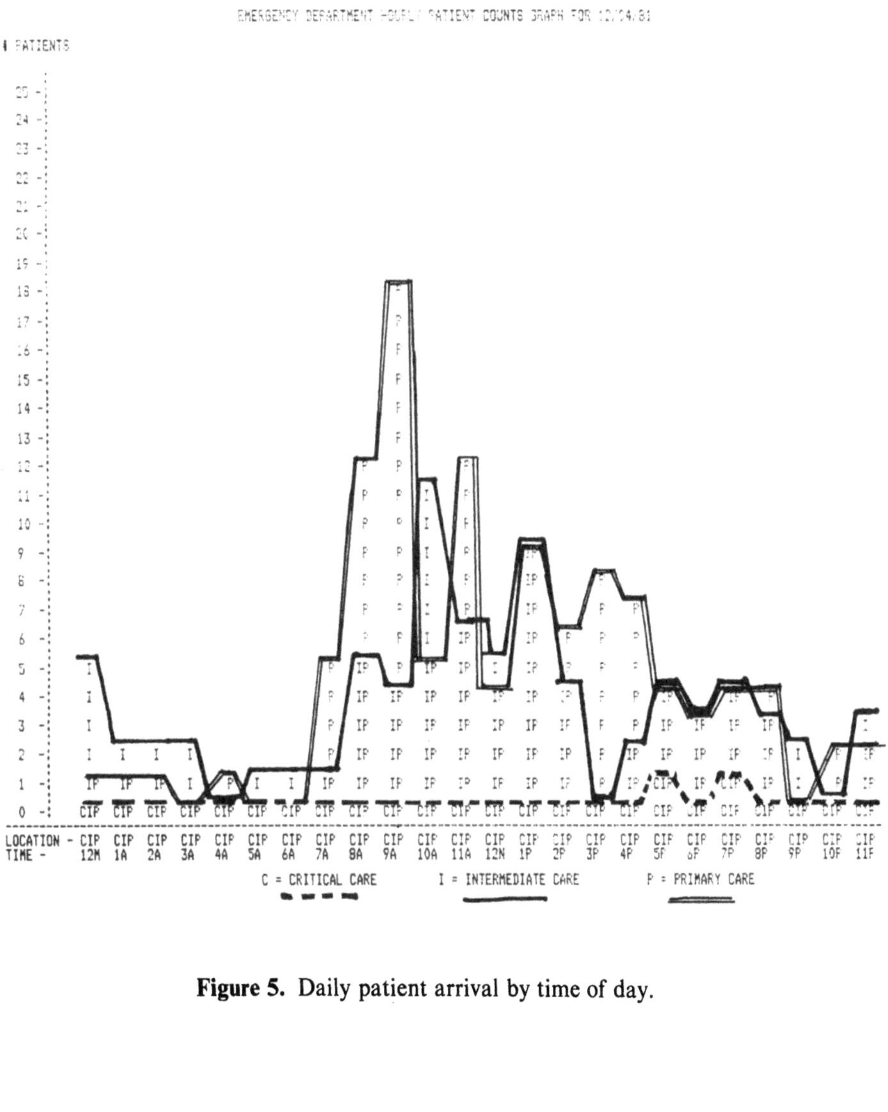

Figure 5. Daily patient arrival by time of day.

02/18/82

EMERGENCY DEPARTMENT MONTHLY REPORT: METHOD OF ARRIVAL

01/01/82 TO 01/31/82

METHOD OF ARRIVAL	PRIMARY	INTERM	CRITICAL	UNKNOWN	TOTAL
AMBULANCE	59	138	12		209
CAR	440	410	2	5	857
POLICE	6	10			16
SELF	774	352	2	5	1133
TAXI	163	134			297
UNKNOWN	419	347	26	34	826
TOTAL	1861	1391	42	44	3338

Figure 6. Distribution of patients by mode of arrival.

complaint or diagnosis of Asthma. With this information it is possible to answer many questions. For example: what is an individual patient's pattern of seeking care for his asthma attack? How many asthmatic patients arrive by ambulance? By providing this information to the medical professionals in the clinic they will have the tools to better understand the ways in which patients seek care. Consequently, they will be able to plan the most effective and efficient way of delivering health care.

Conclusion

The Core Record System has been operational in a prototype mode in three clinics at the Johns Hopkins Hospital for approximately two years. An evaluation of the system performed in January 1982, approximately nine months after implementation, demonstrated that the Core Record System has the ability to communicate useful medical information to the health care provider at the time of the patient encounter at a reasonable cost [5]. The operational cost of the prototype system, without considering the potential savings when the system is completely integrated with other hospital systems, is between $0.75 and $1.00 an encounter.

An informal, subjective survey of medical care providers, conducted at the time of the January 1982 evaluation, indicated a perception that the system led to improved patient care by providing useful medical information at the time of the patient encounter.

The system can also provide administrative personnel and medical professionals with information about clinic activity and the patient population seen in the clinic. Although these management information reports were not included as a specific component of the January 1982 evaluation, and therefore their impact was not measured, they are viewed as quite useful by the administrative staff and are requested on a regular basis. The ability to search the archival data base for particular groups of patients is a relatively recent addition to the system, but it promises to be quite useful from both the administrative and medical perspectives. The fact that new reports continue to be defined by both the administrative and medical staff is viewed as a positive measure of the system's success.

References

(1) Blum, B. I., Johns, R. J., McColligan, E. E. An Approach to Ambulatory Care Medical Records. *Int'l conf. on System Science in Health Care,* 1980, Montreal, Canada.

(2) McColligan, E.E., Blum, B. I. and Brunn, C.W. An Automated Core Medical Record System for Ambulatory Care. *Proceedings of the 1981 Joint Conference of the Society for Computer Medicine and the Society for Advanced Medical Systems,* 1981, Washington, DC.

(3) Johns, C. J., Simborg, D. W., Blum, B. I., and Starfield, B. H. "A Minirecord: An Aid to Continuity of Care." *Johns Hopkins Med. J.* **140,** 1977, pp. 277–284.
(4) Blum, B. I., C. J. Johns, E. E. McColligan, C. R. Smith and D. M. Steinwachs. "A Low Cost Ambulatory Medical Information System." *J. Clinical Engr.* 4, No. 4, Oct–Dec, 1979, pp. 372–284.
(5) McColligan, E. E., Blum, B. I. "Evaluating an Automated Core Medical Record System for Ambulatory Care." *Proceedings of the American Medical Information Association,* San Francisco, CA, May 1982.

19

COSTAR System*†‡

G. Octo Barnett, M.D., Norma S. Justice, Michael E. Somand, J. Barclay Adams, Bruce D. Waxman, Peter D. Beaman, Monica S. Parent, Fredric R. Van Deusen, and Jacqueline K. Greenlie

Introduction

The communication of medical information is a critical element in the practice of high-quality medical care. Traditional recording practices rely almost completely on a manual record folder where physician notes are handwritten or dictated and merged with laboratory data and other-patient material. This manual system has inherent problems of incomplete data and occasional unavailability of the medical record resulting from different physicians recording data at different places at different times. Even when the medical record is available, poor organization and frequent illegibility may make retrieving the desired information a laborious and time-consuming task. It is usually necessary to record data redundantly to provide the necessary information for patient care, scheduling, billing, and management reports. Quality of care studies and clinical investigation research which depend on the aggregation of data from a large number of individual patient records are particularly cumbersome and require the expenditure of many hours of manual searching.

Although many medical practices have begun to use computer technology to assist in information processing, most of the commercial systems are restricted to patient billing, accounts receivable, and preparation of third party insurance claims forms, and have no impact on the information processing related to patient care. There are a number of research projects where the objective is to use the computer to improve availability of medical data to the physician. Most such projects are dependent on research funds and, although

*This work was supported in part by the National Center for Health Services Research, OASH under Grant HS 00240.

‡The authors acknowledge the contributions of all the members of the COSTAR system team in this developmental effort. In the preparation of this manuscript, the editorial criticisms of Ms. R. Zielstorff and the IEEE reviewers were particularly useful.

some of these systems have become operational in their local environment and have achieved significant professional acceptance; none have been nationally disseminated [1]–[5]. There have been two recent state-of-the-art reviews of computer-based medical information systems sponsored by the Department of Health, Education, and Welfare [6], [7], and one review by the Office of Technology Assessment [8]. All three of these reviews emphasized that the different experimental projects were designed to meet the particular and often unique needs of a specific site, rather than being a prototype system that could be easily generalized.

The need to introduce computer technology to facilitate ambulatory practice is becoming increasingly acute, both because of the shift in responsibility for ambulatory care from the solo general practitioner to a loosely coordinated team of medical specialists, and because of the increasing complexity and volume of the medical data recorded on each patient. For example, fifty years ago, many physicians might not even record any information on an ambulatory patient visit, and twenty years ago, the note might be only a few lines long. Now ambulatory records may contain lengthy notes written by a multiplicity of different health care providers (primary care physicians and nurses, dieticians, social workers, etc.), large numbers of different laboratory results, and a diverse set of other data elements such as x-ray and pathology reports, and summaries of hospital admissions. In addition the focus of ambulatory care has changed from preoccupation with treatment of episodic illness to an increased concern with preventive medicine and with ensuring continuity of care in the management of chronic disease.

These changes in patterns of medical care have been associated with changes in the administrative structures of medical practice. The dominant organizational format of ambulatory care has evolved from the solo physician's office to health care organizations of considerable size and complexity. The management of any large health care organization—whether it be a fee-for-service group practice, a health maintenance organization, or a hospital-based clinic—requires the optimal scheduling of visits, the ability to monitor the use of resources and the productivity of providers, and the operation of a billing and accounts receivable operation of considerable magnitude. The manual medical record has proven grossly inadequate to meet the administrative needs of such organizations. Management reports and organizational planning often require duplicate data collection of incomplete and sometimes inaccurate information.

For the past decade this Laboratory has been involved in the development and implementation of a computer-based medical information system COSTAR (Computer-Stored Ambulatory Record) designed to perform the data management functions needed by a group practice in the care of ambulatory patients [9]–[11]. The purpose of the system is to replace the traditional document-based patient medical record with a comprehensive, centralized, and integrated information system that meets both the medical care and financial/administrative needs of either a fee-for-serivce or a prepaid group practice.

COSTAR was originally developed by the Laboratory of Computer Science in collaboration with the Harvard Community Health Plan (HCHP)—a prepaid group practice located in Boston—and has been operational there since 1969. The fourth version of the system, COSTAR 4, is fully integrated into HCHP operations and is supported from the Plan's operational budget. The HCHP management views the system a success based on the criteria of cost acceptability, improved availability of medical information and of management data, reliability of operation, professional acceptability, facilitation of a higher standard of patient care, and stimulation of research in health care delivery by providing a readily accessible database.

However, COSTAR 4 does not have the potential for use by other ambulatory practices, since it incorporates a number of characteristics which were specifically designed for HCHP needs, and has only a limited set of functional capabilities (e.g., there is only a partial scheduling module and no accounts receivable function). In addition, it was obvious that national dissemination of COSTAR could not occur unless there were national support and marketing by private industry. Therefore, a significantly revised and expanded version of the automated information system called COSTAR 5 has been developed in collaboration with the Intramural Division of the National Center for Health Services Research and with Digital Equipment Corporation [12] This paper describes the data management characteristics of COSTAR, and discusses the implementation strategies we have chosen.

Data Management Characteristics of COSTAR 5

Design Goals

The design goals of COSTAR 5 are the following.

(1) Facilitate patient care by improving the availability, accessibility, timeliness of retrieval, legibility, and organization of medical information.
(2) Enhance the financial viability of the medical practice by providing a comprehensive billing system with accompanying accounting reports.
(3) Facilitate medical practice administration by providing the data retrieval and analysis capability required by management for day-to-day operation, budgeting, and planning.
(4) Provide data processing support for administrative and ancillary services (e.g., scheduling, laboratories and planning).
(5) Provide the capability to generate standardized management reports and to support user-specified inquiry and report generation on any elements of the database.
(6) Support programs of quality assurance by monitoring the content of the database according to user-specific rules and reporting automatically any deviations from these standards of care.

Operating System

COSTAR is programmed in Standard MUMPS (Massachusetts General Hospital Utility Multi-Programming System), a compact high-level interpretive data management system particularly suited for interactive applications which require a large shared database and the rapid, efficient manipulation of textual data [13], [14]. The use of MUMPS has greatly facilitated program development and maintenance since it is not necessary to compile or translate a program into another form before it can be executed and debugged. An interpretive language does have decreased execution efficiency, but the adverse effect of this speed reduction is not prohibitive because few COSTAR programs are strictly processor-bound.

A significant feature of MUMPS is its hierarchical data base system. A MUMPS file may be viewed as a spare multidimensional array that is global to all users. The minimal space needed to store each variable-length array element is allocated dynamically as needed; records in the data base are not predimensioned. When sections of the file are deleted or changed, released storage areas are returned to the free pool. Dynamic allocation of storage in MUMPS is an extremely important strategy in implementing a medical information system because the size of any individual record cannot be predicted, and because there is wide variation in the size and speed of growth of the different records. MUMPS does not provide the full degree of data independence of a complete data base management system. However, MUMPS does allow the programmer to refer to information in a symbolic fashion and to deal with the data base in terms of a logically meaningful file structure rather than in terms of the physical layout of the data.

MUMPS programs are written in explicitly defined segments (routines), with only the current segment for the individual user resident in core at any one time. In the COSTAR system there are almost 1000 segments, each of which average over 1000 bytes. In addition, the directories of COSTAR occupy about another million bytes. Because the syntax of MUMPS is very concise, and there is no compilation into machine code, a multi-user COSTAR system can be implemented on a relatively small computer. An equivalent assembly language coded system would have an excess of ten times the amount of MUMPS program code, so that the total COSTAR system in assembly language would be of the order of ten million bytes or more.

Since the original implementation of MUMPS at Massachusetts General Hospital, the language has evolved so that there are currently a number of dialects. Two years ago a standardized version of the language was approved by the American National Standards Institute. Standard MUMPS is now supported by several computer manufacturers. Therefore, COSTAR can be implemented by a number of different computer vendors. In addition, because the MUMPS file structure provides relative independence of the capacity of the physical disk files, COSTAR can be implemented on a wide range of sizes of computer configurations, and can grow modularly via hardware expansion as data processing requirements increase.

Medical Vocabulary—COSTAR Directory

The traditional medical record has a virtually limitless vocabulary and only minimal structure. The only standardization of vocabulary currently used in recording medical data are coding systems for diagnoses and procedures. The *International Classification of Diseases, 9th Edition, Clinical Modification* ICD-9-CM is a recently published disease classification system that is being federally mandated for use by all health care agencies and institutions. The size of this nomenclature is very large, with approximately 13000 primary concepts (many of which have synonyms so that there are over 60000 different terms). This classification system is primarily designed for statistical reporting on disease in a population, and the authors of the classification recognize that there are deficiencies in using the system in the clinical management of individual patients' problems. In particular, the ICD-9-CM is not particularly suited for common health problems, complaints, and ill-defined conditions that are common in primary health care.

There are many different coding systems required by the various third party insurance carriers for reimbursement of common surgical procedures and laboratory tests. A commonly used classification, the Current Procedures Terminology (CPT) published by the American Medical Association, has over 5000 different terms.

There has been little practical success in using computer technology to process the narrative text information of medical records. If a computer system is to be used effectively for automatically selecting and organizing medical information, then it is essential that there be at least a minimum level of standardization of the vocabulary. In COSTAR, standardization of information is provided through the "directory," a dictionary of medical terms that are allowed in patient records. A complete coding system for all the detail of all medical information for all types of specialties would be of enormous size; in addition, there is considerable disagreement within the medical community on appropriate taxonomy systems. The COSTAR directory has a limited taxonomic scheme (e.g., all diagnostic codes relating the cardiovascular system begin with the letter M) and no predetermined or maximal number of terms; each medical group may define the lexicon appropriate for the particular medical specialties involved and the type of practice.

Each element in the directory corresponds to a unique COSTAR "code." This code is a shorthand notation for a medical concept, e.g., MHABI stands for the term hypertension. A code represents the concept rather than the term itself, in that synonymous term (e.g., high blood pressure) are assigned to the same code. In addition, terms which are similar but not identical may be grouped together by using one or more "modifiers' to describe that particular code, e.g., MHABI-A means renovascular hypertension, MHABI-B means secondary hypertension, and so on. Modifiers can also be used to designate various trade names for a drug, so that the specific prescription details can be retained while at the same time medications of the same generic type are grouped together.

The tradition of medical practice requires flexibility in the structure and content of data associated with various terms. For example, weight is merely a single number; a blood pressure, however, may be recorded from various limbs with the patient in different positions, and, therefore, may be associated with two or four or sometimes six numbers; a urinalysis includes a variety of elements such as pH, specific gravity, color, etc.

The directory is the unifying element that allows COSTAR to deal with these varying details, serving a function analogous to a Data Base Management System (DBMS) schema in specifying the structure of the stored data. However, the directory plays a more active role in the processing of data within COSTAR than that of a DBMS schema. Each code in the directory may have associated with it comprehensive validation criteria for input data based on patient parameters such as age and sex. In addition, the protocols defined for the different categories of codes within the directory specify the encoding and physical layout of the associated data within the record, and the appropriate format for reporting the information in output documents. Further, the practice may specify parameters with each term which guide production of reports and statistics. For example, the practice may specify that certain values of a particular laboratory test are considered abnormal, and should be flagged on all medical output; or that certain diagnoses are considered of such important clinical significance that when they occur, the diagnosis should be flagged on all output documents (e.g., penicillin allergy).

The special functional properties of a code can be specified at three levels of detail. First, every code has a "division" attribute that designates one of several broad medical categories, e.g., physical findings, diagnoses/problems, laboratory tests, medications, nonmedication therapies, procedures, and administrative terms. The data stored in the record varies for each division. For example, a medication code usually is associated with related information about dose, route, frequency, quantity, number of refills, and short textual directions, while a diagnosis may have stored with it only textual description or qualification of the patient's condition.

The second level of detail is provided by a set of parameters specific to a particular code that cause specific actions during processing of patient data associated with the code. For example, a particular diagnosis in the directory could specify that output documents produced for any patient having that diagnosis should include a flowchart of data pertinent to the diagnosed disease. In addition, the directory may specify standard fees for procedures or tests, and may also map COSTAR codes to other coding schemes for claims and reports sent to specific insurance carriers.

The third level of capability provided by the COSTAR directory allows MUMPS program code to be executed during data input, output, or analysis processes. Thus virtually unlimited functionality can be implemented by a programmer with a working knowledge of COSTAR. Few directory entries will contain such application program extensions, but the capability ensures sufficient flexibility to accommodate most problems of data input, validation, processing, and output formatting. Such flexibility is required in a medical information system intended for adaption to many styles of practice.

The COSTAR directory shares some of the characteristics of a table-driven information system in providing nonprogrammer users the capability to change the content of the table. This allows local variation in recording data and preparing reports. Although a copy of the COSTAR directory is provided to each site, each practice has the capability to add new codes and to modify any of the characteristics of existing codes. The COSTAR system provides the utility programs required to perform code building and modification without requiring the assistance of a programmer. These directory building programs allow each site to tailor the system to local needs and local practice patterns without adding new functions or changing the computer programs.

The COSTAR directory provides the capability to process and store data which are very nonuniform in nature. Although there is only a finite set of different types of medical information (e.g., diagnoses, medications, laboratory tests, procedures, dispositions, etc.), the information structure associated with each type varies considerably. For example, the minimal information that is required for each code is the name and type of medical information (e.g., the code "MNBC3" may have the name *Hematocrit* and be a laboratory test). However, as Table 1 illustrates, it is also possible to define the parameters for this same code much more extensively.

The number of terms and the complexity of information in the directory varies considerably between different COSTAR sites. A typical multispeciality group practice might have about 4000 different codes (about half of which are diagnoses/problems with most of the others being laboratory tests, medications, procedures, and administrative items). On the average each code usually has three to four different names, synonyms, or abbreviations that can be used to access the same term.

The specific form of the COSTAR code is an arbitrary feature of the system, but it is not essential for the practice to use codes as such, since all interaction with the system can be done using either the full name or the abbreviated name of the code. However, all data are stored internally with the code as an index, and all noncoded information is stored associated with a specific code. For example, any information concerned with the diagnosis of hypertension would be stored with "MHABI" in the patient record. Such information includes each date the diagnosis was used, the clinical status of the diagnosis, the physician who made the diagnosis, and any narrative text associated with the diagnosis.

Modular Construction

In order to facilitate transferability of COSTAR 5, we have adopted an explicit strategy of modular design suitable for phased implementation. We have found that the modular design approach offers several advantages. First, it permits a large complex system to be broken down into a number of smaller, more easily

Table 1.

COSTAR code for Hermatocrit: MNBC3	
Long name of the term: HEMATOCRIT	
Short name of the term: HCT	
DIVISION: LABORATORY RESULT CHECKING	
ALLOWABLE: 1 to 90	
NORMAL:	
FEMALE 15 and up	37 to 42
OTHERWISE, MALE 15 and up	42 to 47
OTHERWISE, up to 0.2	35 to 50
OTHERWISE	38 to 47
FEE: 3.50	
REVENUE CENTER: LABORATORY	
TRANSLATION DIRECTORY: CPT	
MEMBER OF THE PANEL(S): HEMOGRAM MNAS1	

Note: The actual key to the direct entry is MNB3. The *C* is the checkletter of the code and is stored as data in the directory. COSTAR codes have a limited taxonomical meaning. In this example *M* implies hematology, *N*, microscopic, and *B*3, ordinal assignment.

Directory entries always contain the long name of the term. Optionally, short names may be specified. Either can be used for inputting purposes; some output reports will use the short term if available for formatting convenience.

The division assigned to a code is arbitrary. The laboratory division has rules or algorithms specific to it. Input of a code in this division can only be through the encounter input sequence or the laboratory result entry sequence. The clinical statuses which may apply to laboratory codes are "ordered," "pending," "normal," and "abnormal." These status flags affect output formatting and ordering.

The result checks are specified by each practice site through interactive sequences and stored in the directory itself. In this example the acceptable input must be a numeric value between 1 and 90; the normal ranges are based on age and sex of the patient. Laboratory result entry sequences extract these rules from the directory and automatically apply them in checking the test results. A normal or abnormal status is filed with the result data.

The fee is the practice specified amount to be billed to the patient or third party carrier.

The revenue center, in this case "laboratory," is an arbitrary classification of billable items used by management to analyze the practice's income generation.

Entry of this item in the directory provides automatic mapping of this test to the corresponding entry in the *C*urrent *P*rocedural *T*erminology (CPT) translation directory, where the CPT code is specified and optionally a different term and/or fee.

The hemogram represents a set of laboratory codes often ordered together. Generally panel codes are used to facilitate billing, input, and display of related laboratory tests.

handled subsystems and functions. These functions can then be analyzed and specified separately, leading to more rapid implementation. Second, a modular implementation allows the user organization to select a subset of available modules and thus tailor the system to both the requirements of the organization and its current financial resources. Third, a modular organization allows functions to be added or deleted without impacting the integrity of the total system. This facilitates system evolution consistent with an organzation's changing needs, demands, and capabilities. Finally, the modular approach allows flexibility in system software configurations. An organization can take full advantage of technological advances and utilize phased incremental growth which is compatible with their economic constraints.

The basic COSTAR 5 system includes modules for security, registration, and a fundamental medical record system. In addition, there are a number of optional modules which at this time include: expanded medical records, generalized flowcharts, accounts receivable, scheduling, and management report generator. Additional modules under development include: laboratory test reporting, quality assurance, and an information retrieval system.

(1) Security: A computer system concerned with sensitive medical information requires an effective security system. Security relates to two distinct activities:

(1) restricting the use of data to authorized personnel; and
(2) protecting the data against loss by backing up all data to allow reconstruction of the database.

In order to protect confidentiality of medical records, COSTAR currently offers three levels of security. The first level requires all prospective users to identify themselves prior to any terminal use. In addition, identification requirements can be expanded to control access to the various system functions. A second level of security is available through line (or terminal) restriction. Any terminal may be identified in terms of the system options available to it. Thus the practice may specify that no medical data are to be available on the terminals designated for registration, scheduling, or administrative functions. A third level of protection is available through implementation of a password system.

The second portion of the security system—data integrity—is accomplished by copying the entire database at specified intervals onto standard backup storage media and writing each transaction onto a transaction log. If a hardware malfunction necessitating recreation of the database occurs, the copy of the database can be restored and the transaction recorded on the log reprocessed, thus automatically rebuilding the COSTAR files.

(2) Registration: The basic COSTAR registration sequence included routines for both fee-for-service populations and prepaid practices. The programs collect basic demographic data such as patient name, sex, date of birth, guarantor identification, and insurance or prepaid group information. COSTAR also permits the incorporation of locally specified queries into the basic registration sequence.

Unit numbers can be assigned automatically by the system, or they can be entered manually by the clerk. Registration information is available for interactive inquiry and update at any time.

(3) Medical Record Module: The design goal of the COSTAR 5 medical record module is a system that meets the total record-keeping needs of primary health care providers However, adoption of the total system may evolve over time, and individual installations may initially choose to implement only a subset of the functional capabilities such as a computer-based problem list, which would serve as an adjunct to a practice's current medical record-keeping system.

The module has two sections—*data capture* and *medical output*. Data capture is achieved by means of a printed encounter form. Each practice must design forms to meet their individual needs. All items on the encounter form have unique codes in the COSTAR system. Each site will determine the items printed on the form, the appropriate responses to these items, and the arrangement of the items on the form to meet their requirements.

There are three basic formats for displaying medical information:

(1) *The encounter report:* This is a single visit note that includes diagnoses with associated narrative text, objective data, medications, test results, and consultation requests associated with that visit (illustrated in Figure 1).
(2) *The status report:* This is an index to, and summary of, the patient's current medical status illustrated in Figure 2).

```
ENCOUNTER REPORT                  UN: 1Ø36X
DEMO,ABRAHAM LINCOLN (M) 66 YRS     (2/12/12)

3/2/78     SITE OFFICE   TYPE NEW PATIENT   ROUTINE OFFICE VISIT
           BILLED TO GUARANTOR  DOE,JUDY,MD

                            DIAGNOSES/PROBLEMS

BHAE1       COMPLETE HISTORY AND PHYSICAL EXAM [REVIEW, RFV]
              LOOKS REASONABLY WELL CONSIDERING HIS MANY PROBLEMS. IS
              WORRIED ABOUT DOMESTIC STRIFE.
MHAB1       HYPERTENSION [R/O]
              STRONG FH. HAD NEGATIVE WORKUP IN ARMY PHYSICAL TWENTY YEARS
              AGO.  WILL START ON DIURETICS.
QGAA7       HEARTBURN
              PARTICULARLY AGGRAVATING AFTER CABINET MEETINGS

                              PHYSICAL EXAM

CAEF1 **    BLOOD PRESSURE           165/1ØØ RIGHT ARM 175/1Ø5 LEFT ARM
CAKH1       WEIGHT                   165

                               THERAPIES

TTAX1       CHLOROTHIAZIDE
              1ØØMG PO QD TAKE WITH ORANGE JUICE
              QTY: 1ØØ REFILL: 2
CWBB1       SYMPTOMATIC THERAPY
              TRY GLASS OF MILK FOR HEARTBURN
QWDN1       SALT RESTRICTION

                               PROCEDURES

BTTK5-N     SMALLPOX VACCINE INJECTION

                                 TESTS

CNCJ1       BLOOD UREA NITROGEN                        23
KQBA1       STOOL OCCULT BLOOD                         NEG
MNBN3 **    HEMATOCRIT                                 35
```

Figure 1. The encounter report is equivalent to the "progress note" of the manual medical record and reflects the activity at a single patient visit. The report displays in standard format both the data recorded via a single encounter form and the data reflecting laboratory test results associated with that encounter (which may be entered separately).

```
STATUS REPORT                        UN:1036X
DEMO,ABRAHAM LINCOLN (M) 66 YRS     (2/12/12)
PENNSLYVANIA AVE.  WASHINGTON, D.C.  00011
                                                  INS:UNKNOWN

NO PRIMARY PROVIDERS

BHAE1       COMPLETE HISTORY AND PHYSICAL EXAM        DOE,JUDY,MD 3/2/78
              LOOKS REASONABLY WELL CONSIDERING HIS MANY PROBLEMS. IS
              WORRIED ABOUT DOMESTIC STRIFE.

                           --MAJOR PROBLEMS--

YJSN1       DEPRESSION                               DOE,JUDY,MD 4/15/78
              INTERMITTENTLY BECOMES VERY WITHDRAWN AND UNHAPPY

                           --MINOR PROBLEMS--

MHAB1       HYPERTENSION                   DOE,JUDY,MD 3/2/78-2-4/15/78
              BP CONTINUES ELEVATED.  SUGGEST EXERCISE PROGRAM
CGDC2       WEIGHT LOSS                              DOE,JUDY,MD 4/15/78
              20 LBS IN PAST YEAR WITH NO CLEAR CAUSE EXCEPT
              ANXIETY ABOUT COMING ELECTION
QGAA7       HEARTBURN                                DOE,JUDY,MD 3/2/78
              PARTICULARLY AGGRAVATING AFTER CABINET MEETINGS

                           --PHYSICAL EXAM--

CAEF1      BLOOD PRESSURE          * 4/15/78        165/100 RIGHT ARM
CAKH1      WEIGHT                    4/15/78        162

                             --THERAPIES--

TTAX1      CHLOROTHIAZIDE                                        3/2/78
           100MG PO QD TAKE WITH ORANGE JUICE
           QTY:100 REFILL:2
CWMF1      EXERCISE                                             4/15/78
           RUN 2 MILES EACH DAY
CWBB1      SYMPTOMATIC THERAPY                                   3/2/78
           TRY GLASS OF MILK FOR HEARTBURN
QWDN1      SALT RESTRICTION                                      3/2/78

                             --PROCEDURES--

BTTK5-N   SMALLPOX VACCINE INJECTION                             3/2/78

                               --TESTS--

CHEMISTRY
  CNCJ1    BLOOD UREA NITROGEN               3/2/78          23
  FNCW5    FREE THYROXINE [ORDERED]          4/15/78
  KQBA1    STOOL OCCULT BLOOD                3/2/78          NEG
HEMATOLOGY
  MNBN3 *  HEMATOCRIT                        3/2/78        * 35
MISCELLANEOUS·
  WPAN1    ELECTROCARDIOGRAM                 4/15/78
             NSR  LOW T WAVES OVER PRECORDIUM
```

Figure 2. The status report is an up-to-date summary of the diagnoses, medications, and laboratory results which gives the physician an overview of the current state of the patient's medical status. The first part of the Status Report contains identification information, membership enrollment, demographic and personal data of the patient and scheduled appointments. The diagnostic information is organized according to the current status of each problem, with major problems listed first. Each diagnosis is associated with the most recently recorded status and the most recently recorded narrative text as well as the date when the problem was initially recorded, the date when last recorded, and the number of visits at which the problem was recorded. Other sections of the Status Report display information on physical examination, therapy, laboratory test results, and dispositions.

HYPERTENSION FLOWCHART

DATE	WGT	BLOOD PRESSURE	CREA	URIC ACID	K+	MEDICATIONS
3/2/78	165	165/100 RIGHT ARM 175/105 LEFT ARM				CHLOROTHIAZIDE 100MG PO QD TAKE WITH ORANGE JUICE
4/15/78	162	165/100 RIGHT ARM				
6/3/78	160	165/90 STANDING 160/95 LYING			4.3	CHLOROTHIAZIDE 500 MG PO QD
8/9/78		.		* 8.6	3.7	
8/31/78	169	165/90 STANDING 160/95 LYING				CHLOROTHIAZIDE
9/7/78				* 9.2		CHLOROTHIAZIDE 500 MG PO QD QTY 100 TAB X 3 REFILLS METHYLDOPA 250 MG PO BID QTY:60 X 3 REFILLS

Figure 3. The flowchart lists specific parameters or types of data along a horizontal axis with date of events along the vertical axis. The medical practice may create any number of flowchart formats, or "templates" that specify which COSTAR codes are to be displayed, and the output forms of the report. This flowchart would be appropriate for the followup of patients with hypertension. The column marked "medications" includes many antihypertensive drugs, of which only Chlorothiazide and Methydopa have been prescribed for the patient.

(3) *Flowcharts:* A flowchart is a computer-generated display which emphasizes the temporal course of the disease process or the variation in clinical findings over time. The display is a chronological listing, by date, of all occurrences of particular coded items with associated text and/or results. (An example of a flowchart is illustrated in Figure 3.)

These reports can be produced on a scheduled basis or on demand, as printed material, or on the CRT display at the time of provider inquiry.

(4) Scheduling: The on-line scheduling and appointment module finds open appointments, books patients for given times, displays schedules for both providers and patients, and automatically prints support documents for appointments (such as day lists), chart pull lists, and medical records for the scheduled patients. Small practices may not need a complete scheduling function; a subset of the system may be used to enter names or unit numbers of patients one day prior to their scheduled visit in order to trigger the automatic printing of medical records, or to create chart pull lists and routing slips.

(5) Accounts Receivable (A/R): As clerical personnel enter encounter data into the system, the computer automatically files the information for both the med-

ical record and the A/R file, if that is appropriate, thus eliminating duplicate posting. The system prepares bills for individual patients or for third party carriers, translating the COSTAR codes to other coding schemes (to ICDA, for example) when necessary.

Additional A/R functions include entry of payment against accounts, maintenance of balance forward, and preparation of statements and reports such as:

(1) batch verification list;
(2) daily transaction register;
(3) aged trial balance;
(4) outstanding insurance list:
(5) monthly ledger; and
(6) monthly revenue analysis reports by individual provider and for the practice collectively.

All account information is available for on-line inquiry.

Paper Form Data Capture

One of the implementation compromises of the COSTAR strategy is that the physician or nurse does not directly interact with the computer system to enter medical information. Instead, both administrative and medical data for each ambulatory visit are recorded on an encounter form, a preprinted checklist with a defined structure and format (a portion of one encounter form is illustrated in Figure 4). The basic structure consists of classes of recorded items (e.g., diagnoses, medications) and a vocabulary of terms specific to each of those classes. In most of the COSTAR user sites, the majority of information is recorded using a self-encoding format, i.e., there is a unique name and code associated with each item. At a particular encounter, items not covered by codes on the form are recorded in text and later coded by the medical record staff using a master code list (in the HCHP experience, fewer than one in ten encounters require any manual coding). Each form is reviewed periodically and changed to reflect the use of new or different terminology. The coding system used has been developed by the authors to meet the needs of the various specialties and reflects the diseases, problems, and health assessment activities that have been identified in the ambulatory care environment. An encounter form is completed not only for every patient visit in the ambulatory center, but also for many other care activities such as important telephone calls, home visits, emergency department visits, hospital visits, and hospital discharge summaries.

Although for purposes of organization and analysis it would be desirable to collect and store all medical data in a coded form, the enormous variety and richness of medical concepts make this an unobtainable ideal. In COSTAR, the primary topic (e.g., diagnostic name, medication name) is encoded while the bulk of descriptive information is stored in narrative form. Narrative information is recorded either by writing on the encounter form or by using dicta-

DIAGNOSES / SIGNS / SYMPTOMS

Use the column marked "ST" for the following status codes:
✓ = minor I = inactive D = do not display Y = status post
M = major R = rule out P = presumptive H = history of
Place an "S" in the column marked "S?" if the condition is short duration.

S? ST

GENERAL
- ✓ BHAE1 Initial Health Assessment
- BHBK1 Adult Health Review
- BHPQ3 Rx Refill Only
- BHTM3 Immunization Visit
- BWCE1 Health Education

DRUG REACTIONS
- CKPZ3 Drug Allergy - Penicillin
- CKPG4 Drug Allergy - Sulfonamides
- CLTZ3 Drug Reaction:
- CLTJ1 Drug Toxicity:

SYSTEMIC
- CCBP5 Fatigue
- CCCZ6 Fever of Unknown Origin
- DKZW9 Viral Infection

TRAUMA
- CLAY4 Skin Abrasion
- CLCL1 Burn
- CLBJ8 Contusion
- VMZK9 Fracture:
- VLHP1 Head Trauma
- CLDR1 Laceration

RESPIRATORY
- JJEV1 Upper Respiratory Infection
- DKYV3 Influenza
- LJFP3 Acute Bronchitis
- LJKM6 Chronic Bronchitis

S? ST

ENT
- JHJW2 Cerumenosis
- JJGW3 Otitis Externa
- JJGL5 Otitis Media
  - A. Serous
- JLFG1 Rhinitis
- JLFK8 Allergic Rhinitis
- JJDP1 Sinusitis
- JJCJ1 Laryngitis
- JJCM8 Tracheitis
- JJAY1 Pharyngitis
- JJAN3 Strep Pharyngitis
- JGEK8 Epistaxis

EYE
- HJDM1 Conjunctivitis
- HLAC1 Eye Injury
- HJBR3 Hordeolum

MUSCULOSKELETAL
- VGAB6 Back Pain
- VGCZ1 Myalgia
- VJJW2 Gout
- VJME1 Bursitis
- VGFN4 Arthralgia
- VJHH1 Rheumatoid Arthritis
- VLGJ1 Sprain
- VLKA1 Disk Disease
- VJCC1 Osteoarthritis

FREE TEXT FOR DIAGNOSES

CODE
BHAE1 Looks reasonably well considering his many problems. Is worried about domestic status.

Figure 4. The encounter form is a preprinted checklist which is designed by each specialty in each practice to record all medical, administrative, and billing data. Information is recorded by checking the appropriate code, entering the status of the code, selecting any appropriate modifiers, and writing any descriptive information as narrative text adjacent to the coded item. The data on the encounter form are entered into the computer system by clerical personnel.

tion. This text is always associated with a specific item (e.g., a particular diagnosis) and a specific date and provider. Linkage of the narrative information with an encoded item is critical to automatic organization and retrieval of related data. Each clinical code is associated with a "status" at the time of recording to delineate further the clinical relevancy of the particular entity at that time; for example, status flags for diagnostic codes include "major," "minor," "inactive," and "presumptive." The status flag determines how the information is formatted on medical displays (e.g., whether the diagnosis is on the "active" problem list) and is used to suppress routine display when a clinical entity is no longer active or medically significant. However, no medical information is ever deleted from the computer-based file and all data that have been recorded may be reviewed.

After the physician or nurse completes the encounter form, it is sent to the medical record department where clerical personnel enter the data into the computer system via terminals. The interval between the time of physician recording on the encounter form and the time of entry into the computer is dependent on the clerical staff of the record room; at HCHP it currently averages less than 24 hours. The use of a self-encoding form minimizes errors in coding and transcription by clerical personnel. Routine quality control analysis at HCHP indicates that transcription errors of either codes or text occur in less than one of 400 data elements.

Computer-Stored Database

Storage of information in one-line disk files guarantees that it is always available for review either by a printed report or through terminal inquiry. Also, data can be added to the patient's record from multiple sites.

Medical information in the COSTAR database can be automatically manipulated and organized for different styles of retrieval. Because much of the significant information in COSTAR is coded, it is possible to use predefined algorithms to select and organize data for medical record reports. The data are the most current, since the usual strategy in COSTAR operation is to print the record just prior to the patient's visit. COSTAR selects and organizes the information according to the medical significance, so that the content of the medical information reflects the particular medical circumstances of the specific patient.

The size of the patient medical record stored in COSTAR is dominated by three factors: the average number of encounters per year (in most ambulatory practices, this is three to five), the average length of time the patient receives active medical care from the practice, and the amount of narrative text recorded on each visit. There are very large variations in the latter two characteristics: one patient may be a transient who came only for a single episode of care, and another patient may have a major chronic illness lasting a number of years. The limited experience thus far with COSTAR suggests that a visit is associated with an average of about 150–200 characters, and that the average patient record may have on the order of 2500 characters.

Flexibility of Surface Behavior

There are no two medical practices which have identical needs and identical specifications for the functioning of an information system. COSTAR programs are written so that each site can choose among available options that determine certain characteristics (for instance, whether a diagnosis is required as the reason for the visit at each patient encounter). In other instances, site-specified table-stored entries determine the content of an interactive sequence (such as the questions on the registration procedure) or of certain output documents (such as the medical items on a hypertension flowchart). Many of these "switches" can only be set by a COSTAR programmer at the time of installation, but some of the options can be selected or changed at any time by the site personnel.

The encounter form itself may be designed to suit the documentation patterns of each specialty (e.g., the pediatrics encounter form may have a very different appearance from the encounter form used in orthopedics in the same practice). Although each specialty selects the items specific to its encounter form, the system as a whole shares a common set of consistent terms.

To assure that the different needs of the different specialties are met with regard to medical record reports, each specialty in a practice can identify which particular encounter reports should be included in the medical output (e.g., the output for a visit for a surgical consultation could be specified to include all encounter reports which contained the code for which the consultation had been requested) or what flow charts would be useful (e.g., a record of immunizations for a visit to pediatrics for well child care, or a temporal display of blood pressure, weight, and height or uterus for a prenatal visit). The primary objective in the use of different criteria in selecting the content and format of the medical putput is to ensure that the physician can read and evaluate the relevant medical data in a minimal period of time.

Availability of Information

The principal method of providing access to the database is through standardized patient care reports (e.g., status report, encounter report, etc.) as well as through prespecified listings and reports associated with the accounts receivable and scheduling functions (e.g., revenue analysis, lists of currently booked appointments, etc.). However, these reports do not provide the capability for the user to supply the parameters for selective retrieval, analysis, and tabulation of data from the stored records. The effective administration of a complex organization such as a group practice requires the production of a variety of management reports on the characteristics and visit patterns of different patient populations, the types of resources being used by what patients, the costs and income associated with each service provided by the practice, and the workload of different specialties and different providers. These management reports are of increasing importance in facilitating rational administrative decision making, controlling the use of practice resources, appropriate reimbursement of the professional staff, meeting the reporting requirements of third party insurance and governmental regulatory agencies, and planning for growth or expansion of the organization.

In COSTAR 5 there are two different high-level languages that allow access to the database from a symbolic or logical point of view without the need for a programming staff or for the user to have a detailed knowledge of the underlying physical file structure. The first is the COSTAR Report Generator (CRG) which has been developed and is being supported by the Intramural Division of the National Center for Health Services Research. This report generator has been designed to accept commonly understood commands and to permit the user to obtain listings and tabulations associated with any variables indicated by the user. The second retrieval language provides the user with the capability to specify arbitrarily complex retrieval and analysis protocols. Personnel at the local site can use a medically oriented procedural language to specify the parameters for search routines which operate on the database, to format and display patient records selected by the search, and to produce standardized tabulations and cross-tabulations. A further extension of the language will allow the production of user specified reports using the information selected by the information retrieval system.

Implementation Strategies

The prototype COSTAR system at HCHP demonstrates the feasibility and potential effectiveness of automation of information processing at a single ambulatory care site. However, the national dissemination of an ambulatory information system to hundreds of sites requires not only a much more flexible system, but also a significantly different implementation strategy. The four major components the implementation strategy for COSTAR 5 include:

(1) industrial support;
(2) installation planning;
(3) local site modification; and
(4) documentation.

Industrial Support

Any information system that is to be used by a large number of different medical practices must be implemented on current state-of-the-art minicomputer hardware in commercially supported systems and application software. The majority of medical practices do not have, and do not want to have, a programming staff who can develop a sophisticated information system or even modify a system obtained from a research organization such as this laboratory. A few very exceptional medical organizations may have the local talent to undertake the development and installation and support of a computer-based system. However, significant dissemination of COSTAR will depend on commercial organizations providing national marketing, local support for selection of appropriate hardware configuration and installation planning, and local site personnel education in customizing the system for individual practice needs.

COSTAR 5 is written in the ANSI certified Standard MUMPS which is commercially supported by a number of different hardware vendors. All of the developmental activity and programming efforts of this laboratory are supported by a research grant from the National Center for Health Services

Research, and the COSTAR application programs are available to any organization, commercial or private.

Installation Planning

The importance of the installation process cannot be overemphasized, since COSTAR represents an entirely new application area in a marketplace which has little or no experience in automation. Since the installation of COSTAR will affect the methods and procedures not only of providers but of administrators and clerical personnel as well, and since only minimal perturbation of the health care delivery process can be tolerated, the installation process must be carefully planned and executed.

Thus COSTAR installations must:

(1) blend into the existing procedures and behavioral patterns of personnel;
(2) preserve or enhance the efficiency of the normal decision-making processes; and
(3) avoid the maintenance overhead of parallel manual or automated systems. In order to facilitate dissemination of COSTAR 5, an installation guide is being developed to cover the installation process from its early preplanning phase through post installation analysis, and to serve as the operational blueprint for the procedure.

Local Site Modification

At the time of initial installation each user site must make a number of decisions such as which modules to install and the form of unit number identification. In addition, there are a large number of other features which must be initially defined but which can be later modified, such as password conventions for system access, information requirements for each third party insurance carrier, and the revenue centers to be used. However, the major individual site modifications will be made through changes in the COSTAR directory, such as choosing or defining the codes to be used in the practice, defining the translation of these codes into other coding systems, and setting normal value ranges for the laboratory tests.

It is our hope that there will be only minimal need for modification of the computer programs, and that such programming can be done by the industrial field support group. There is a large cost associated both with the creation of site-specific computer programs and more importantly, the support of a large number of different computer software systems on a national basis. Therefore, it is critical that, as much as possible, all necessary changes in COSTAR be made by modification of the site-specific directory of codes. We expect that a major area of continuing development of COSTAR 5 will be adding enhancements to the COSTAR directory functions so that local on-site programming will not be required.

Documentation

The successful transfer of a complex computer-based system depends almost as much on the quality of the associated documentation as it does on the quality of the computer programs. The documentation provided to the user must detail not only how to interact with the system, but also must give the background as to why the different technical decisions were made, and how best to take advantage of the different data recording and display alternatives offered by COSTAR. As much as possible, we have tried to make COSTAR 5 self-documenting in that the system operates in an interactive fashion, wherein the user responds to a series of prompts to control the entry or retrieval of data. To each of these prompts, the user may enter a question mark to learn the meaning of the particular prompt, or to learn the format of the required answer and be given typical examples of possible responses. Ideally, most of the user's documentation needs will be met by this self-documentation so that there will be little need to consult the written user's manual.

There will also be extensive documentation provided to aid in physical education, since it has been our experience that the better a provider's understanding of the different system options, the better the system will be accepted and used. Although we have tried to make the use of COSTAR mimic typical manual recording practices, there are significant differences, and it is essential that the provider appreciate these differences. Changing a physician's recording habits is a challenge; when this must be achieved by an industrial marketing or field support group whose orientation is systems engineering, the task is most difficult and will require an imaginative and comprehensive set of educational materials.

Documentation for the installation and support technical group consists mainly in program and file structure description. The only unique feature of this documentation is the attempt to detail the interaction between the different features of each module and the interaction between modules. We will also try to document explicitly those features and program codes which should not be casually changed because of the possibility of jeopardizing or compromising the integrity of the total system operation.

Technical Evaluation

The evaluation of the impact of all technological innovations in health care has become a topic of great interest to the federal government, primarily because of a concern that many technologies are associated with an increased cost of care without a perceptible equivalent increase in benefit.

The experience of the different sites using COSTAR 5 thus far is so limited that it is impossible to make a definitive evaluation. However we can extrapolate to a certain extent from the experience at HCHP to make certain observations.

Cost

Because of the financial constraints on most medical practices, it is essential that the total direct and indirect costs of COSTAR not be significantly more than the present costs of the manual system and whatever computer support is currently used. The crucial weakness in making such a comparison is the paucity of quantitative information on the costs of information processing in the typical ambulatory practice. Most practices have only a vague idea of the costs of the technology, supplies, and labor directly attributed to the manual medical record and to third party billing, and usually even less data on the other information processing tasks such as scheduling, laboratory reporting, quality assurance, and preparation of management reports.

The relative costs associated with COSTAR are, therefore, difficult to assess, given the degree of uncertainty of comparable costs of manual information processing. It is the judgment of the HCHP administration that the cost of their version of COSTAR is slightly more than the cost that would be incurred in performing the same functions using manual techniques (with the justification for using COSTAR being the much greater effectiveness in performing these functions). It is our expectation that the cost of COSTAR will be comparable to the costs of commercially available systems that deal purely with accounting functions and which have been successfully marketed on a national scale. For a small practice the cost of COSTAR 5 should not be more than \$1000–\$2000/month; for a large practice the cost should be \$3000–\$5000/month.

It is our impression that the monetary costs of the computer system are relatively small compared to the large cash flows of even modest sized group practices, and that there are potential savings in time of both clerical and professional personnel. Given the uncertainty in evaluating cost effectiveness of a computer system versus a manual system, we expect that the main driving force for the decision to acquire COSTAR will not be to save money in information processing, but rather to meet the increasingly critical need for more accurate and complete management and financial information.

Professional Acceptability

The practice of medicine is traditionally conservative, and any radical change in procedures for processing medical information will not be readily accepted. A dominant reason for using a paper encounter form for data entry, and for allowing some information to be recorded in narrative text, is that this technique allows the capture of considerable coded data and yet has at least a superficial resemblance to the classical recording practice of writing in the medical record. The use of direct terminal interaction with the physician recording information through a menu selection process would increase the cost of the system in requiring more powerful computer processing support and more sophisticated terminals. A more important limitation is that many physicians find interacting with a terminal professionally unacceptable. It may be that as more physicians become familiar with using terminals for inquiring about medical data on a specific patient, this restriction will disappear. However, the optimal strategy at present seems to be a system which does not depend on direct physician interaction with the computer terminal.

A second critical charactertistic of COSTAR is that there is no explicit model of medical information recording and no defined data base which the system imposes on the physician. The style and completeness of medical recording is completely determined by local recording practices. It is our belief that the technology should conform to professional standards defined by each site; it is not our objective to reform medical practice by requirements or restrictions built into the technology.

These beliefs have been validated in the ten-year experience with the original version of COSTAR in the HCHP Kenmore Center. A 1975 survey of physicians and nurses at HCHP showed that ninety percent of the providers felt that there was greater record availability in COSTAR, eighty percent felt it required less time to record information in COSTAR, and eighty-seven percent expressed a preference for using COSTAR over a manual record.

Flexibility

As of early 1979, COSTAR 5 is in process of being implemented at seven different sites. In each situation, it has been necessary to make programming changes to meet the specific needs of the practice, although some of these changes were functional extensions that will be useful to other sites. From the beginning, it was planned that the development of COSTAR 5 would be evolutionary, beginning with a basic system (Level One), and progressing with a series of enhancements in later releases. Although the directory functions do support a variety of different surface behaviors at the different sites, the final evaluation of the degree of flexibility inherent in COSTAR 5 cannot be made at this time.

Improvement in Patient Care

Although most medical professionals maintain that increased availability of information is associated with improved medical decision making, and, therefore, with improved patient care, there is little experimental evidence to document this prejudice. In an indirect fashion, a computer-based information system has the potential to improve patient care by providing the opportunity to identify patients at high risk because of a particular diagnosis or patients who are receiving an undesirable set of medications. For example, COSTAR has been used at HCHP for a recall program of women receiving a particular sequential birth control pill and for another program to identify high-risk patients to receive flu vaccination. COSTAR can also be used to monitor the patterns of medical practice according to specific standards of care. For example, COSTAR could be used to support quality assurance programs to ensure that all patients with abnormal Pap smears or elevated blood pressure received the appropriate followup, or that all children received the appropriate immunizations. The standards that are used for the automated surveillance in such quality assurance programs are defined by the local policies of each practice.

The advantage of COSTAR over a manual system is that a computer system can be programmed to be algorithmically sensitive to the implications of the medical content of each patient's record as part of routine system operation. In contrast, a manual medical record is essentially passive and oblivious to the significance of the medical information contained in the document. This use of COSTAR is the engineering equivalent of the control system model where the information about the specific deviation provides the feedback loop by notifying automatically the decision maker in a timely fashion so that action can be taken to correct the deviation. This automatic notification has negligible cost and has been demonstrated at HCHP to improve adherence to the prescribed standards of care [15], [16].

Conclusion

It would be naive to believe that any computer-based information system is completely general purpose or will completely meet the needs of every medical practice. In addition, there is always a cost and efficiency tradeoff between a system that is specifically designed for a well-defined set of needs, and a system designed to have sufficient flexibility to support a broad range of functions and provide a great variety of local options. The dominant objective in the COSTAR 5 development is to provide a system that can meet the information processing needs of a variety of ambulatory practices without requiring excessive programming modifications at each local site. Although the experience thus far in installing COSTAR 5 in the different test sites is encouraging, it is premature to claim that the present implementation is flexible and robust enough to be a completely satisfactory system. We do believe that there is a widespread perception of need for an information system that supports both administrative and medical needs in ambulatory care, and that an information system that met these needs and that could be successfully installed and supported by private industry would be rapidly disseminated. We are optimistic that the basic design strategy we have chosen, and COSTAR 5 as it is presently implemented, provide the system capability that can evolve to meet the diverse requirements for national dissemination.

References

[1] C. J. McDonald, "Protocol-based computer reminders, the quality of care and the non-perfectability of man," *New Eng. J. Med.* **294,** 1976, pp. 1351–1353.

[2] M. L. Braunstein, S. H. Schuman, and H. B. Curry, "An on-line clinical information system in family practice," *J. Fam. Prac.*, Vol. **5,** 1977, pp. 617–625.

[3] J. F. Fries, "Time-oriented patient records and a computer databank," *J. Amer. Med. Ass.* vol. **222,** 1972, pp. 1536–1542.

[4] B. Starfield, D. Simborg, C. Johns, *et al.,* "Coordination of care and its relationship to continuity and medical records," *Med. Care* **15,** 1977, pp. 929–938.

[5] E. C. Schmidt, D. W. Schall, and C. C. Morrison, "Computerized problem-oriented medical record for ambulatory practice," *Med. Care* **12,** 1974, pp. 316–327.

[6] R. R. Henly, "An analysis of automated ambulatory medical record systems," Vols. 1 (Findings) and II (Background Material), AAMRS Study Group, Office of Medical Information Systems, Univ. California, San Francisco Med. Ctr., Tech. Rep. #13 (1) and (2), June 1975.

[7] B. Kerlin and W. Rappaport, "Computer-based health information systems for rural areas: An analysis of demonstration potential," Metrek Division of the MITRE Corporation, MITRE Tech. Rep. MTR-7804, July, 1978.

[8] "Policy implications of medical information systems," Office of Technology Assessment, U.S. Congress, Government Printing Office, Stock No. 052-003-00496-8, Washington, D.C., 1977.

[9] "COSTAR (*CO*mputer-*ST*ored *A*mbulatory *R*ecord)," Laboratory of Computer Science, Massachusetts General Hospital, Harvard Community Health Plan, National Center for Health Services Research, Nat. Tech. Inform. Service (PB-248 314), U.S. Dep. of Commerce, 1975.

[10] J. H. Grossman, G. O. Barnett, T. D. Koepsell, *et al.,* "An automated medical record system." *J. Amer. Med. Ass.* **224,** 1973, pp. 1616–1621.

[11] D. R. Studney, J. B. Adams, A. Gorbach, *et al.,* "A computerized prenatal record," *Obstet. Gyn.* **50,** 1977, pp. 82–87.

[12] COSTAR (*CO*mputer-*ST*ored *A*mbulatory *R*ecord) Functional Specifications (Version 5.5), Laboratory of Computer Science (Massachusetts General Hospital), 1978.

[13] J. Bowie and G. O. Barnett, "MUMPS—An economical and efficient time-sharing system for information management," *Comput. Programs Biomed* **6,** 1976, pp. 11–22.

[14] T. Munnecke, R. F. Walters, J. Bowie, C. B. Lazarus, and D. A. Bridger, "MUMPS—Characteristics and comparisons with other programming systems," *Med. Informatics, 2, 1977, pp. 173–196.*

[15] G. O. Barnett, R. N. Winickoff, J. L. Dorsey, *et al.,* "Quality assurance through automated monitoring and concurrent feedback using a computer-based medical information system," *Med. Care* **16,** 1978, pp. 962–969.

[16] M. M. Morgan, D. R. Studney, G. O. Barnett, *et al.,* "Automated quality assurance in prenatal care," *Qual. Rev. Bul.,* **4,** 1978, pp. 33–36.

[17] Editor's Note. This paper was published in 1979. A more recent description of COSTAR appears in the *SCAMC-6 Proceedings.*

User Satisfaction with COSTAR V*

Barbara D. Kerlin, Ph.D.

Introduction

COSTAR has been evolving over a period of ten years. The earliest and most complete demonstration of COSTAR at the Kenmore Center, Harvard Community Health Plan (HCHP), resulted in [1]

> overwhelming support for COSTAR: ninety percent of the providers felt that there was greater record availability in COSTAR, eighty percent felt it required less time to record information in COSTAR, and eighty-seven percent expressed a preference for using COSTAR over a manual record, . . .

as documented in a 1975 survey of HCHP physicians and nurses.

Since then, the fifth version of COSTAR has been developed and is being demonstrated at a limited number of sites. COSTAR V is substantially different from its earlier versions and the potential marketplace for the system is expanded. The medical objective remains the same, however: "to facilitate patient care by improving the availability, accessibility, timeliness of retrieval, legibility, and organization of medical information" [2]. It is too early to tell whether or not patient care is facilitated with COSTAR V, and quality of care thus improved. No site has fully implemented all of the COSTAR V functions, and there is less than one year's field experience with the automated medical record. Nevertheless, sufficient time has elapsed for a preliminary examination of user response to the system's design and to learn from the users' perception of implementation problems.

In this chapter, "user" is synonymous with medical provider; "satisfaction" is defined as the response of a person to an environment which stimulates valued activities, supports the self concept, and fosters one's special competencies and achievements [3]. A more rigorous study of user satisfaction should be performed which operationally defines and measures satisfaction over time and analyzes data by user groups. To this end, an instrument to measure user satisfaction with computer services [4] is being reformatted and pilot tested at the Baltimore site. Meanwhile, reported medical provider satisfaction—or dissatisfaction—with COSTAR design features, installation, transition and operational phases are summarized below with the hope that they can highlight design and implementation pitfalls and thus contribute positively to the future activities of the developers and installers of COSTAR V.

Design Factors

The designers of COSTAR V made a conscious decision to construct a system which *adapts* to diverse environments, *adjusts* to the idiosyncratic demands of individual providers or provider groups, and *accommodates* the behavioral patterns of personnel. Contrary to the design of PROMIS which requires the user to adjust to the system, COSTAR was structured to involve users in its implementation and to adjust the system to clinic behavior [5]. It does not intend to reform medical practice but rather to enhance its delivery based on the premise that increased availability of data is associated with improved medical decision-making.

This design decision was made at a time when health information systems for ambulatory care lagged behind inpatient applications and information about them was poorly disseminated [6]. However, recent literature and experience in the field support the COSTAR V design decision.

Presentations at past Symposiums on Computer Applications in Medical Care contend that physicians quickly develop their own individual styles of practice [7], prefer technology to be unobtrusive [8] and, in general, find direct physician interaction with the computer terminal to be professionally unacceptable.

A group at the Laboratory of Computer Science, Massachusetts General Hospital, reported in 1977 on a one-year study which tested the acceptability of direct entry (light pen) of narrative made into the medical information record by a nurse practitioner. Lack of acceptability resulted in removal of the system from the clinical setting until technology improves and permits more rapid and more flexible man-machine interaction [9].

A palatable system, therefore, must "have at least superficial resemblance to the classical recording practice of writing in the medical record" [10] and meld into the environment rather than recast it. COSTAR V, therefore, was conceived with flexibility of encounter form design and medical record reports [11]:

> Flexibility . . . is essential for achieving provider acceptance and to ensure that the different needs of the different specialities are met.

Data recording is on paper; data entry is by a trained clerk. A provider, theoretically, can ignore the physical trappings of a technology and continue practicing medicine in his/her established manner aided, unobtrusively, by the superior organizational skills of the computer.

In reality, is this true? Are providers satisfied with the design concept of COSTAR V? Is it truly flexible? Can they ignore the blue/black/tan box in the office? Do they want to? From close observation of one implementation, extensive involvement with a second site, and continuing communications with other major installations of COSTAR V, the following observations are made by the author:

(1) Encounter form design allows vast opportunities for tailoring data capture to the desires of the medical providers.
(2) Installers are taking good advantage of the input design flexibility and providers are becoming more involved in the design tasks (see Installation Process below).
(3) Medical output reports are receiving mixed acceptance. Some providers are not satisfied with the format of the Status Report but are enthusiastic about the flowchart capabilities of COSTAR V.
(4) Interfacing with a terminal appears to be a personal response. Some providers are resigned to their use; some excitedly access them; others ignore them to the greatest extent possible.

COSTAR V can accommodate the personal preference of the medical user or group without its being the sole criterion in the success or failure of the system on site. The COSTAR V site at Vancouver, Canada reports a strong commitment to use of CRTs rather than printers. Additionally, they indicate high user satisfaction with the COSTAR system. It should be noted, however, that they have just begun to automate the medical record, having already automated registration and scheduling activities.

Lack of enthusiasm for the Status Report, the summary medical record, has been observed at different sites. First, the S.O.A.P. format is preferred. Secondly, the quantity and mixture of data can become overwhelming. Lastly, important trends in physical exam, chief complaint and history data are lost in the Status Report. These observations may reflect unrealistic expectations for the Status Report on the part of the user. In any case, the use of flowcharts to rearrange the output into S.O.A.P. format and to present a longitudinal view of selected items has satisfied the medical user. In fact, flowcharts are being enthusiastically received at the two federal demonstration sites.

Clearly, the design of COSTAR V is flexible. In fact, the danger lies in the other extreme. COSTAR V may be too flexible and encourage excessive accommodation to the whims and wishes of providers. For example, a recent addition to the COSTAR software permits the automatic creation of a lab results encounter whenever no encounter is found with that test ordered (and unresulted), and no encounter is shown for the date on the lab test results sheet. In an operating environment, this can cause havoc: lab test results may get input prior to the patient's medical encounter data, two encounters for the same day are then recorded; and the Status Report shows the test as being unresulted. In brief, it is *too easy* to create a "lab results only" encounter. Secondly, in encounter form design, the initial tendency in installation is to bend over backwards to satisfy individual providers. Flagrant violations of flexibility in design, however, may result in:

(1) too many forms with overlapping content but different COSTAR codes clogging up the Status Report;
(2) too much free text encouraging illegibility, and resulting in backlog of data input and high disk usage; and
(3) long lists of data elements on the encounter form which are seldom referenced and could just as easily be written in.

To summarize, flexibility of input and output design is a major factor in COSTAR's acceptability. However, restraint is required to balance user satisfaction with a viable, timely, operational system.

Installation Process

Again, the literature is replete with consistent reference to the importance of user involvement in system design and installation. Dr. Sobolewski, in an excellent paper presented at the First Symposium of Computer Applications in Medical Care, and reprinted here as Chapter 11, wrote [12]:

> . . . problems (in applying computers in the area of patient care) include failure to adequately . . . communicate with the end users, involve them more in the design process and evaluate the impact of the computer application on their environment.

Zimmerman and Rector, in a recently published book, discuss human engineering problems in medical computer applications, contend that new systems engender natural opposition, and suggest that introduction of a system should be preceded by discussions with users [13]. Similarly, Giebink and Hurst reviewed 29 medical computer application projects and reported that "highly supportive physicians have generally been those who participated actively in the development of the system by working directly with the developers or actually leading the project [14].

What has been the experience with installation of COSTAR V? Is the level of provider satisfaction related to the degree of involvement in the planning stages? To date, informed reports from COSTAR V installations generally support the findings in the literature and in other projects. In fact, field experience at COSTAR sites strongly suggests that earlier involvement, i.e., participating in the decision-making process for *selecting* COSTAR V, increases user acceptance and satisfaction with COSTAR V.

However, the statement that user involvement and early exposure to COS-

TAR V has a positive effect on user satisfaction is both naive and superficial. The picture is more complex—a splash of contrasting colors rather than stark black and white. This is especially true in larger installations where heterogeneous groups of providers—physicians, nurses, physician assistants, health educators, etc.—come from various training backgrounds and espouse differing philosophies of medical care, provider role, and medical data requirements. COSTAR V, unobtrusive as it may want to be, still imposes structure where there may not have been any before, requires cooperation when it could have been avoided in the past, and more easily exposes medical care delivery practices to the blaring lights of a CRT and the inquisitive eyes of other providers. This says nothing about the potential friction between non-medical users (data entry clerks, front desk personnel) and medical users who, involved in the encounter design process, must weigh ease of data entry against ease of data recording.

This multi-faceted issue of user involvement reflects, perhaps, complexity of process rather than of outcome. A more rigorous study of user satisfaction may well indicate that satisfaction is not a direct function of level of involvement, i.e., the more actively involved in the early installation phase, the greater is the user satisfaction with COSTAR V. Rather, the level of involvement required for high user satisfaction is more likely variable and dependent upon personality structure (rigid or flexible), perceived importance of data needs, and congruence with the majority ruling group.

A corollary to user involvement is that of peer acceptance. Installation experience with COSTAR V supports the theory that medical providers respond more positively to advice and guidance from other medical providers than from "outsiders" [15,16]. Enthusiasm is contagious, especially so when it is shown by a person whom you know and accept and who speaks your language. At the Vancouver, Canada, COSTAR V site, inhouse training is a key factor in acceptance of the system. Likewise, at San Marcos, California, the reassignment of two nurse practitioners to provide technical assistance and training at the five clinics is contributing significantly to the installation effort.

Transition Phase

Ideally, for each module of COSTAR V, there should be planning, installation, transition and operational phases. A COSTAR V site could very likely be in different phases at the same time, i.e., "live" with registration, testing medical encounter forms on a selected basis, while planning for automated scheduling. The issue of user satisfaction is especially relevant to the transition phase. Documented experience in other systems strongly suggests that premature release of an application can lead to serious problems [17]:

> Since first impressions are sometimes difficult to overcome, every effort should be made to ensure that the application performs satisfactorily before it is made operational. Not only is it necessary to ensure that the programs implemented

are fully debugged and are working correctly but it is necessary to ensure that the application does solve the actual problem and meets all of its stated goals and objectives.

Yet, Dr. Barnett, developer of COSTAR, writes: "To be successful, COSTAR installations must . . . avoid the maintenance overhead of parallel manual or automated systems" [18]. Maintenance overhead, in Barnett's statement, can refer to the difficulty and strain of dual entry in the medical record as well as to the high costs and heavy data entry workload in supporting two accounting systems. The Catch-22 element of the situation is only too apparent. Premature release of the medical record module coupled with no entries in the paper record and immediate withdrawal of the previous system can result in chaos, potentially lower quality of care, and disillusioned providers. On the other hand, writing in a medical record and filling out a billing or encounter form in addition to completing a COSTAR encounter form quickly discourages the providers and slows down the delivery of care.

Different COSTAR installations have addressed this problem with varying techniques and varying levels of user satisfaction. The gradual approach, e.g., one or two pro-COSTAR providers testing it on their patients, a random subset or an easily identified subset of their patients, appears to result in fewer reported crises of conscience or critical cases of absent data. In this way, the more robust provider is able to survive the inevitable initial changes to forms, procedures, etc., without adversely affecting the group's morale. Furthermore, if positive experiences are reported by the provider to the group, the snowball effect occurs, increasing user acceptance of the system.

The San Maros installation, on the other hand, felt it necessary financially and timewise to switch all clinics and all providers to COSTAR V on the same day. Swings in user satisfaction and dissatisfaction have been extreme for a multitude of reasons. Although a transition period may have mitigated some anxiety on the part of the providers, it was not a viable alternative from the viewpoint of the administration.

Operational Phase

The core of a provider's dissatisfaction appears to be rooted in the quality of his/her delivery of patient care. Presence or absence of a computer system is of secondary importance and is judged principally on whether or not it works, is reliable, makes patient care easier to provide, and enhances the quality of patient care [19,20,21,22]. Status of the medical record is the key element in the operational phase: Is it available, easily retrieved, complete and accurate? Several factors may interfere with the desired status, the most important among them being:

(1) reliability of equipment;
(2) data entry workload;
(3) clarity, conciseness, legibility of the encounter form.

Computer downtime and malfunctioning communications can clearly lower user satisfaction with COSTAR V apart from any consideration of system design, objectives, etc. The other two factors are interrelated and hard to resolve. For instance,

(1) Accuracy of the automated medical record may require sending the written encounter form back to the provider for clarification, thus adversely affecting the timeliness of retrieval.
(2) Or, the site may elect to enter the readable sections of the form, thus making part but not all of the medical record available on retrieval.
(3) Free text may satisfy the providers' perceived data needs, but excessive free text results in a heavy input workload, thus adversely affecting the timeliness of retrieval.
(4) Free text offers opportunity for illegibility, unknown abbreviations and sometimes requires innovative data inputting to condense the text.

In brief, without *enough* free text, the providers feel constrained by COSTAR V. With *too much* free text, the system becomes bogged down—data entry is slow, disk space disappears rapidly, data base searches become time consuming and cumbersome—and, from the provider's viewpoint, the quality of care then suffers due to the lack of an updated, quickly retrieved medical record.

Summary

The issue of user satisfaction with COSTAR V is clearly complex. There is not one single answer and never will be. What is becoming apparent at site installations, however, is that the system design incorporates the means to satisfy the diverse data needs of the various provider groups; the installation and transition phases can contribute to user satisfaction, or dissatisfaction as the case may be, dependent upon a multitude of interrelated decisions made within the specific environment; and the operational phase depends upon earlier decisions as well as upon the actions of the providers themselves. COSTAR V continues to be an evolving system. With feedback from the users at the demonstration sites, the developers and installers have increased opportunities to improve the system and its dissemination process.

References

[1] Barnett, G. O., "A Computer-Based Medical Information System for Ambulatory Care," Draft Paper for Publication in IEEE Proceedings, January 1979, p. 19. [Reprinted as Chapter 19.]
[2] *Ibid.*, p. 5.
[3] Holland, John L., *Making Vocational Choices: A Theory of Careers*, Englewood Cliffs, New Jersey: Prentice-Hall, 1973.
[4] Pearson, S. and Bailey, J., "User Satisfaction, Measurements of Computer User Satisfaction," *Computopic*, Volume XII, No. 6, March 1979, pp. 12–18 and No. 7, April 1979, pp. 17–22.

[5] "Policy Implications of Medical Information Systems," Office of Technology Assessment, U.S. Congress, November 1977, p. 55.
[6] Henley, Ronald R. and Wiederhold, Gio, *An Analysis on Automated Ambulatory Medical Record Systems,* Volume 1, NCHSR, June 1975, p. 16.
[7] Ronis, Norman, "My Girl Sends Out My Bills—Doctors and Computers," *Proceedings: The Second Annual Symposium on Computer Application in Medical Care,* IEEE, 1977, 491.
[8] *Ibid.,* p. 492.
[9] Zielstroff, R. D., Rugliei, J. L., Marble, K. D., Poitras, J. W., Van Duesen, F., Follayttar, S. M., and Barnett, G. O., "Experience with a Computer-Based Medical Record for Nurse Practitioner in Ambulatory Care." *Computers and Biomedical Research,* Volume 10, Number 1, February 1977, pp. 61–73.
[10] Barnett, G. O., *Op. cit.,* p. 19.
[11] *Ibid.,* p. 8.
[12] Sobolewski, J. S., " Towards the Implentation of Successful Medical Computer Applications," *Proceedings: The First Annual Symposium on Computer Applications in Medical Care,* 1977, p. 84. [Reprinted as Chapter 11.]
[13] Zimmerman, Joan and Rector, Alan, *Computers for the Physician's Office,* Forest Grove, Oregon: Research Studies Press, 1978, p. 6.
[14] Giebink, G. A., and Hurst, L. L., *Computer Projects in Health Care,* Ann Arbor: Health Administration Press, 1975, p. 39.
[15] Ronis, N., *Op. cit.,* p. 494.
[16] Reps, D. V., "Performance Requirements for Data Processing Systems in Health Care Delivery," *Proceedings: The Second Annual Symposium on Computer Applications in Medical Care,* 1978, pp. 661–663.
[17] Sobolewski, J. B., *Op. cit.,* p. 87.
[18] Barnett, G. O., *Op. cit.,* p. 16.
[19] Office of Technology Assessment, *Op. cit.,* p. 56.
[19] Barnett, G. O. *Op. cit.,*p. 5.
[21] Reps, D. N., *Op. Cit.,* p. 661.
[22] Sobolewski, J. S., *Op. cit.,* p. 86.

21

Proliferation of COSTAR—A Status Report*

Richard H. Fiddleman

Introduction

The National Center for Health Services Research (NCHSR) began funding the development of the Computer-Stored Ambulatory Record (COSTAR) system by the Laboratory of Computer Science (LCS) of the Massachusetts General Hospital (MGH) in 1968. In 1975, the Digital Equipment Corporation (DEC) joined the development effort, and by 1978 an improved and more generalized version of the software was ready for widespread use. However, when DEC decided in late 1978 not to market COSTAR, NCHSR faced the problem of how to transfer the system from a research setting to a widespread operational setting. In early 1979, NCHSR developed a dissemination strategy that was based on having the dissemination effort primarily accomplished by commercial firms. These "COSTAR vendors" would perform installation and programming services using either the government, e.g., the public domain version of COSTAR, or their own enhanced proprietary version of the system. NCHSR recognized that commercial firms would need support while they contemplated venturing into the COSTAR marketplace and during the period when they first installed and enhanced/debugged the software. Documentation, technical assistance, and information dissemenation was needed; and the MITRE Corporation was selected to provide this assistance.

Since September 1979, MITRE has been performing these dissemination support tasks. The number of COSTAR installations has grown steadily, as has the number of COSTAR vendors. While the dissemination strategy laid down by NCHSR has worked well, a number of unexpected events have also occurred. This paper summarizes the current state of the COSTAR community, and presents MITRE's reflection on certain aspects of the dissemination strategy.

Current Status of COSTAR'S Dissemination

Since the objective of the dissemination effort is the proliferation of COSTAR, one measure of the success of the effort is the number of COSTAR installations. A COSTAR installation is defined as the operational or planned use of the system by an organization to meet a specific set of objectives. By this definition, one organization can have more than one COSTAR installation if each installation is directed at attaining different objectives. A COSTAR time-sharing company, of necessity, supports numerous installations, e.g., each of its customers is treated as a separate installation. And, one organization operating multiple facilities, but using COSTAR to meet uniform objectives in these facilities, is treated as one installation.

Installations can be further subdivided into operational and planned installations. Operational installations are those that are, at a minimum, registering patients in COSTAR; they usually are using other major COSTAR modules. Planned installations are those that have signed a contract with a COSTAR vendor, or if they are not using a vendor, have obtained the software from MITRE and have committed resources (e.g., hardware, staff, consultants) to the installation effort.

The first tally of COSTAR installations using these definitions was completed in February 1981. At that time, there were twenty-seven operational and fifteen planned COSTAR installations in the U.S., Canada, United Kingdom, and Japan. The U.S. accounted for twenty-six of the operational installations and eleven of the planned installations.

Data on the number of installations (and their characteristics) was compiled again in July 1982. The data was obtained from the COSTAR Users' Group, conversations with COSTAR vendors, and MITRE's knowledge of the COSTAR community. The counts presented below should be considered as minimum values, for it is likely that a small number of COSTAR installations were unknown to the author when the installation data was compiled.

As of July 30, 1982 there were 54 operational installations and 19 planned installations worldwide. The location of these installations is shown in Table 1.

In 15 months the number of COSTAR installations has grown by 74%. The actual growth percentage is somewhat higher in that ten of the installations reported in the February 1981 figures are no longer in existence. It should also be noted that there are known COSTAR installations in Finland and Australia, but data on these installations is not available.

Another measure of the success of the dissemination effort is the use of the system by a wide variety of ambulatory care organizations. As shown in Table 2, virtually every type of ambulatory care organization, with the exception of large HMOs, are or will be using COSTAR (an early version of COSTAR is used in a large HMO but this installation is not treated as a COSTAR installation in this report).

The use of the various COSTAR modules is also of interest when reviewing the results of the dissemination strategy. COSTAR contains six major modules

Table 1. Number of Installations

Location	Number of Installations Operational	Planned	Total
Arizona		1	1
California	18	1	19
Connecticut	1	1	2
Colorado	2		2
Delaware	1	1	2
Florida		2	2
Illinois	6	3	9
Maine	1		1
Maryland	2	2	4
Massachusetts	4	1	5
Minnesota		1	1
Mississippi	1		1
Nebraska	1	1	2
New York	1	2	3
New Hampshire	1		1
Ohio		1	1
Oklahoma	1		1
Pennsylvania	1		1
Texas	1		1
Virginia	3		3
Washington, D.C.	1		1
U.S. Total	*46*	*17*	*63*
Canada	2		2
Colombia	1	1	2
United Kingdom	2		2
Germany		1	1
Sweden	3		3
Foreign Total	*8*	*2*	*10*
Grand Total	**54**	**19**	**73**

Table 2. Types of Organizations Using Costar

Organization Type	Number of Installations	Percentage
Hospital Clinics/Departments	18	25%
Small Group Practices	16	22%
Solo Practitioners	8	11%
Public Health Departments	7	9%
Community Clinics	7	9%
Large Group Practices	5	7%
Other Clinics	4	6%
Organizational Health Clinics	3	4%
Research Units	2	3%
Unknown or Other	3	4%
Total	**73**	**100%**

Table 3. Installation Module Usage

Type of Installation	Operational	Planned
Medical Records	8	4
Medical Records/Accounts Receivable	14	3
Medical Records/Accounts Receivable/Scheduling	12	6
Medical Records/Scheduling	6	3
Accounts Receivable	14	3
Total	**54**	**19**

three of which—Registration, Report Generator, and System Maintenance—are used by all COSTAR installations. The use of any of the other three modules—Medical Data, Accounts Receivable, and Scheduling—is optional, and an installation is free to use any combination of these modules. It was expected that most installations would use the Medical Data Module, for this module, which enables the user to automate the organization's medical record data, is unique to COSTAR. (There are numerous data systems that contain accounts receivable and scheduling capabilities.)

There are eight possible combinations of the Medical Data, Accounts Receivable, and Scheduling modules. The number of installations using each combination of these modules, is presented in Table 3.

The data in Table 3 indicates that 77% of the operational and planned COSTAR installations use the Medical Data Module. Vendors have indicated informally that many of their installations currently using only the accounts receivable module will ultimately also use the Medical Data module.

Unfortunately, reliable patient registration and annual encounter load data is not available at this time. However, it is MITRE's belief, based on conversations with COSTAR users and vendors, that the smallest installation is a sole practitioner who handles approximately 5,000 visits per year, while the largest is a Canadian twenty physician group practice that handles 130,000 visits per year. Ultimately, the largest COSTAR installation appears to be the inpatient and ambulatory installation currently being implemented in Ventura County, California. This installation already has over 250,000 patients registered in COSTAR.

One item of data of interest to prospective users of COSTAR is the number of installations using various models of minicomputers and microcomputers. This data is presented in Table 4. Obviously, DEC, with its extensive MUMPS experience, is the most commonly used hardware vendors.

Characteristics of COSTAR Vendors

The dissemination scenario envisioned that prospective new COSTAR users would obtain software and support services by using commercial COSTAR vendors. It was recognized that some organizations would decide to implement

Table 4. Installation Usage of Computer Hardware

Computer	Number of Installations		
	U.S.	Foreign	Total
DEC PDP			
11/23	7	1	8
11/24	1		1
11/34	23	5	28
11/40	2		2
11/44	2	1	3
11/60	1		1
11/70	15	2	17
VAX	1		1
HARRIS	7		7
TANDEM	2	1	3
MICROS	1		1
UNKNOWN	2		2

COSTAR on their own, and that some organizations would use nonprofit support groups such as MITRE and LCS for installation support. The actual current sources of installation support to the operational COSTAR installations is summarized in Table 5.

The table clearly shows that most new users do employ COSTAR vendors or support groups.

The number of commercial COSTAR vendors constantly changes as new firms enter and existing firms drop out of the COSTAR marketplace. In February 1981 there were seventeen commercial COSTAR vendors. As of July 1982 there were 24 commercial COSTAR vendors.

Observations of the Activities in the COSTAR Community

The original dissemination scenario was that commercial organizations of differing sizes, and in differing locations, would modify and install COSTAR in a variety of ambulatory organizations that had medical record problems. From the preceding material it can be seen that many elements of the scenario have occurred. There have been some surprises, however.

Table 5. Sources of Support

Source	Number	Percent
Commercial Vendor	41	65%
Support Group	13	21%
In-house	8	13%
None		
Unknown	1	1%

The major element in the scenario that has occurred is that COSTAR vendors, as envisioned, have made or are making creative enhancements to the base public domain system. A few of these enhancements are:

modifying the Medical Data Module so that data is stored and displayed in a problem-oriented medical record format;
converting COSTAR into an inpatient system;
adding an order entry/results reporting and a problem specific encounter form to the Medical Data Module;
implementing COSTAR on a microcomputer;
adding a forms driver to the system;
adding a variety of additional capabilities to the accounts receivable module.

Also it was anticipated that small firms would initially become vendors, for it was felt that such firms would have less of a problem in committing resources to such a novel undertaking. It was expected that larger firms would enter the COSTAR marketplace when it became clear that COSTAR was a viable product. This is exactly what has occurred. All existing COSTAR vendors are small or modest size companies; however, it is expected that in the next six months at least one major corporation will announce its entry into the COSTAR marketplace.

The one real surprise is the number of organizations who have decided to implement COSTAR themselves or through the limited use of consultants. As shown in Table 5, 13% of the COSTAR installations have decided not use a COSTAR vendor. This decision has usually been made in the face of contrary advice from MITRE. The reasons that organizations have given MITRE on why they decided against the use of commercial vendors are that either they couldn't afford a vendor or they felt they would lose control if they had to rely on a vendor for software support.

Another surprising aspect of the COSTAR dissemination process is the length of time required for an organization to reach a decision on using COSTAR. Vendors report that it is not unusual for a prospective customer to take over a year to reach a decision. The usual reason given for the delay is that COSTAR is used by many units in a organization and, therefore, it takes longer for an organization to reach a consensus on the use of the system.

Finally, there is no discernible pattern of installation size, location, etc., except as can be seen in Table 1, the large number of installations in California. The growth of COSTAR installations in California can partially be traced to the highly successful use of the system by a community health center located in North San Diego County, and the joint NCHSR/San Diego County Department of Health Services COSTAR demonstration effort. These efforts have made it easy for Californians to visit successful operational installations. This factor, continued with the State of California's decision to help fund public health agencies' development of data systems, has given COSTAR a considerable impetus in this state. It will be interesting to see if a similar "clustering effect" occurs in other parts of the country.

Summary

The major objective of the COSTAR dissemination effort is to achieve widespread proliferation of the software. The dissemination strategy as postulated by NCHSR is based on the use of commercial COSTAR vendors as the main dissemination agents. These vendors are expected to enhance the software, as well as install and support it. This has occurred. The number of COSTAR installations is growing, and most installations use COSTAR vendors for support. The vendors, as hoped, have made numerous creative changes to the software.

It is unclear at this time if the system (in its current form) will be widely disseminated. The 74% growth rate in COSTAR installations over the last fifteen months is encouraging, especially considering that the rate of system dissemination is typically slow during the initial dissemination period, and then rapidly increases once the system's benefits become well known. The actual number of installations is small, however, in comparison to the number of potential users. Only time will tell if the major objective of the dissemination effort, the widespread proliferation of COSTAR, will be met.

A COSTAR Quality Assurance Program in Private Practice*

D. R. Studney, M.D., and R. Kunstaetter, M.D.

Introduction

Computerized medical records are an attractive alternative to conventional paper charts because of their potential for improved information access, legibility and timeliness over manual systems. Additionally, where the computer record is highly encoded, the possibility arises of automated analysis of the patient data base to provide information for assuring or enhancing the quality of care. One system, COSTAR [1], (Computer Stored Ambulatory Record) developed by the Laboratory of Computer Science at the Massachusetts General Hospital has been successfully implemented in a variety of practice sites world-wide.

Work by Barnett [2] and his coworkers [3] at the Harvard Community Health Plan where COSTAR was first implemented, demonstrated that automatically-derived quality assurance information from a COSTAR data base could have meaningful impact on provider behaviour in specific care processes. To make this capability more widely useable at diverse user sites, effort has been directed to develop software which would permit the implementation of these protocols without the need for specialized applications programming. One strategy involved developing predefined, preprogrammed modules [4] which could be implemented and used by providers using COSTAR with little or no requirement for applications programming. We report here our observations of the transportability, acceptability and efficacy of one such acceptability and efficacy of one such protocol, the COSTAR Hypertension module [5].

*This chapter has been modified from a version previously published in *Proceedings of the Fifth Annual Symposium on Computer Applications in Medical Care,* IEEE, 1981.

Method

COSTAR had been implemented in 1978 at a 17 doctor multispecialty private practice in British Columbia, Canada, and by early 1981 35,000 patients had been registered, including 20,000 who were primary-care patients of the Clinic's 12 family doctors. All hardware, software and support costs at this site were met by and continue to be borne by the clinic physicians solely from fee-for-service revenue. All modules of COSTAR were in use including the full medical record which had replaced the paper chart as the repository for medical data. A unique feature of this implementation was the installation of video terminals in each physician's desk as the principal means of data retrieval, which has successfully eliminated the routine need to print medical records.

All blood pressures (BP) were measured and recorded by physicians who made one or more determinations and recorded the results on encounter forms for subsequent input into the COSTAR data base. The frequency of blood pressure measurements ranged from 14 to 57% of patients seen, depending on patient age.

Prior to proposing any plans for automated quality assurance to Clinic physicians, the BP recording tenets and follow-up objectives of these physicians were elicited by means of a questionnaire, and compared with published recommendations [6]. Next a meeting was held with the doctors to present the results of the questionnaire, comparison with the literature, and proposals for computerized surveillance and feedback.

A tape copy of the Hypertension Quality Assurance program designed for integration into COSTAR was requested from the Laboratory of Computer Science and installed at the British Columbia site. To obtain the largest possible sample, the data base was searched to identify all patients having any BP greater than 90 mm Hg. The physicians were divided (without their knowledge) into control and study groups, matched with respect to time since graduation. The two groups were further structured so as to include within the same group physicians who customarily paired with each other for purposes of providing patient coverage during days off and vacation.

The feedback consisted of reports (Table 1) supplied to the study but not the control group, initially weekly and later biweekly.

The physicians were not instructed or advised to follow a particular course of action regarding the reports, but rather were informed that these were for information only. After seven weeks feedback was then supplied to both control and study groups as it appeared that :

(1) in about 12% of patient-visits crossover between physician groups was noted;
(2) all physicians had become aware of the attention being paid to patients with unconfirmed elevated BP.

Table 1. Example of Feedback Provided to Doctors

COSTAR HYPERTENSION QUALITY ASSURANCE REPORT....... 4/27/81

INTERVAL FROM: 1/1/79 to 4/27/81 AGES: 0–150 SAMPLE SIZE = 100%

ELEVATED BP DEFINED BY: HIGHEST MEASURED BP
CONSIDERING DIASTOLIC ONLY
WITH DIASTOLIC LIMIT = 91

DIAGNOSIS OF HYPERTENSION INCLUDES: MAJOR, MINOR, PRESUMPTIVE, RULE OUT, STATUS POST, HISTORY OF, INACTIVE

RESULTS FOR: Example, P. B., MD

INDIVIDUALIZED FEEDBACK—PATIENT WITH ELEVATED BP AND NO DX

UNIT NUMBER	NAME	BP
62Z	SAMPLE, SUSAN	180/100
3543P	SAMPLE, DONALD	145/95
46621	SAMPLE, SHERYL	190/110
12720A	SAMPLE, BILL	175/110
24575B	SAMPLE, TERRY	140/95

Results

Installation and site specific modification of the Hypertension Quality Assurance module required one half-day of programmer time to achieve compatibility with a COSTAR system that had undergone some degree of local modification during the three year period since initial implementation. To search the entire 35,000 patient data base required 2 hours and 50 minutes during the quiet hours of operation. Training of a non-technical clinic staff person to run the Quality Assurance program required one quarter hour.

The blood pressure recording and follow-up practices of the clinic physicians conformed closely with published recommendations, and at a subsequent Quality Assurance meeting all physicians indicated a desire for feedback on patients who had abnormal blood pressures and who had not been given a diagnosis of hypertension. A diastolic blood pressure of 91 mm Hg or higher was agreed to be the level which should trigger feedbsck.

The results of the first data base search are tabulated in Table 2 where we note a rather low (34.7%) frequency of BP recording. Analysis of representative records indicated that blood pressure data was generally recorded as a "physical finding" (and hence was accessible to the search program) and only occasionally contained in free text associated with diagnostic codes.

The cohort of "undiagnosed" hypertensives was that group identified on 3/16/81 and was followed for 6 weeks prior to and 6 weeks following feedback.

Table 2. Results of Database Search: Hypertension. Time Interval = 28 months; Elevated BP defined by any diastolic being greater than or equal to 91; Clinic's Primary-Care Population: N = 20186

	Patients with a Visit	Patients with a BP	Patients with Elevated BP	Elevated BP Plus Diagnosis	Elevated BP and No Diagnosis	Normal BP Plus Diagnosis	No BP and Diagnosis
Number	16532	5735	1009	817	192	511	15
Denominator	20186	16532	5735	1009	1009	4726	10797
Proportion	(81.9%)	(34.7%)	(17.6%)	(81.0%)	(19.0%)	(10.8%)	(.14%)

The observations shown in Table 3 suggest that the rate of making the diagnosis of hypertension in the study cohort may have more than doubled.

The effect of the feedback was difficult to judge (Table 3) partly because of the small number of patients seen in the postfeedback period and also because of the patient crossover previously described.

Discussion

We have demonstrated that a preprogrammed Quality Assurance module is readily transferable at minimal cost between COSTAR systems despite site-specific modifications that had been made locally over three years. The standard COSTAR directory appeared to be one factor in the successful transfer of the programs to a site which supports no full time programming staff.

Provider interest in Hypertension Quality Assurance is high and avidity for feedback appears to exist among full time nonacademic physicians. During the 6 weeks following the beginning of feedback, several of the physicians indicated to the investigators that they had called in patients, reviewed them and in those cases where appropriate, revised the diagnosis as a consequence of the feedback

Table 3. Observations on Undiagnosed Hypertensives During Two Six-Week Intervals.

	Before Feedback 3/16/81 to 4/27/81	After Feedback[a] 4/27/81 to 6/5/81
Number of undiagnosed "hypertensive" patients in primary-care population	192	204[b]
Number of above diagnosed in six-week interval by all providers	4 (2%)	10 (5%)

[a]Feedback was begun 4/27/81.

[b]This includes 188 of the original (3/16/81) cohort who remained undiagnosed after six weeks of observation.

that had been provided. Our observations noted in Table 3 support this impression, and it should be emphasized that only a fraction (34%) of the undiagnosed cohort were actually seen by the physicians in the 6 week pre-feedback period and a similar proportion (27%) in the post-feedback interval, recalling that no specific recommendations had been made to call patients in as a consequence of feedback. On the average, blood pressures were taken on these patients in only 26% of visits prior to the beginning of feedback and in 33% of visits following feedback despite the fact that in many patients repeat determinations were indicated, because of previously abnormal blood pressures. This strongly suggests that a lack of followup or the failure to determine blood pressure at a subsequent visit may be a much more important deviation than the recording of a diagnosis.

Of great interest were several related observations during the study:

(1) Physicians not receiving feedback approached the authors and asked that they too be provided with feedback as soon as possible and;
(2) physicians also requested provision of computer generated information to assist both the investigation and management of newly-discovered and long term hypertensives.

It may prove possible to effect more appropriate blood pressure determination and diagnostic confirmation by providing reminders automatically on the physician's desktop video screen at the time of a patient visit, recalling that computer inquiry into the patient record is routinely made at virtually every visit. It also appears that a limited amount of feedback regarding recommendations for followup investigation, and management would be well-received and these modifications are being planned.

In summary, we found that a preprogrammed COSTAR Quality Assurance module was indeed transportable, well-received in a busy private practice and appeared to affect physician behaviour favourably. Further, this program stimulated physician interest in and a request for more timely and extensive automated Quality Assurance information, with the potential for improving both the process and outcome of ambulatory care.

References

[1] Barnett, G. O., Justice, N. S., Somand, M. E., Adams, J. B., Waxman, B. C., Beaman, P. D. et al "COSTAR—A Computer-Based Medical Information System for Ambulatory Care", *Proc. IEEE* **67,** No. 91. September 1979. [Reprinted as Chapter 19.]

[2] Barnett, G. O., Winickoff, R. N., Dorsey, J. L., Morgan, M. M. and Lurie, R. S.: "Quality Assurance Through Automated Monitoring and Concurrent Feedback Using a Computer-Based Medical Information System", *Medical Care* **16:** 1978, pp 962–970.

[3] Morgan, M. M., Studney, D. R., Barnett G. O. and Winickoff R. N.: "Automated Quality Assurance in Prenatal Care", *Quality Review Bull.* **4** 1978, pp. 33–36.

[4] Colloff, E., Morgan, M. M., Beaman, P., Justice, N., Kunstaetter, R., and Barnett, G. O.: "Quality Assurnace in a Computer-Based Outpatient Record", *Proc. of Fourth Annual Symposium on Computer Applications in Medical Care,* Washington, D.C. November 1980, pp. 1375–1376.

[5] Laboratory of Computer Science: "COSTAR Quality Assurrance User's Manual," October 1, 1980.

[6] "The 1980 Report of the Joint National Committee on Detection, Evaluation, and Treatment of High Blood Pressure," *Arch. Internal Medicine.* **140,** October 1980, pp. 1280–1285.

23

Enhancement to COSTAR*

A. Eugene Osburn, D. O., Norman M. Neches, M.D., G. Edward Shissler, M.D., and Diane Kittredge, M.D.

Introduction

Medical care has become progressively more complex. Multiple providers may be involved in management of a patient and increasing numbers of diagnostic and therapeutic options are available to providers. The need for identification and communication of details of care under such conditions has contributed to the development of the Problem Oriented Medical Record. The Problem Oriented approach to health care management involves construction of lists of problems which group a patient's clinical findings into discrete problems. Management plans are then formulated for each identified problem. This paper is a description of additions made to the COSTAR package which provide automatic maintenance of problem lists and organization of clinical data into Problem Status Summaries. Decision support options have also been added to the COSTAR-based package to facilitate utilization of the problem oriented approach to health care. The described enhancements are operational and are being evaluated in a clinic in the Department of Pediatrics at the University of Oklahoma Health Sciences Center.

Structure of Problem Oriented Record

Data captured from an encounter is stored in a patient's clinical record in one of the following divisions in COSTAR: Diagnosis/Problems, Objective Findings, Tests, Procedures, Medications, NonMedications; Therapies, and Dispositions. All the COSTAR clinical divisions are also used for the problem oriented implemention; however, additional divisions have been added to allow

TRANSIENT/EPISODIC (MINOR) PROBLEM LIST 8/27/81

NONAME, HECTOR
DATE OF BIRTH: 3/11/78 AGE: 3 YRS 5 MOS 2 WKS

===

DATE OF ONSET	DATE OF ENCOUNTER	PROBLEM LETTER	PROBLEM TITLE
8/1/81	7/21/81	A	INCOMPLETE DATA BASE
	8/5/81	B	UPPER RESPIRATORY INFECTION

Figure 1. Sample problem list.

discrete storage of symptoms elicited during an encounter and also of symptoms confirmed absent—i.e., significant negatives. The Problem Oriented Record Structure provides a means of depicting a patient's problems and all clinical entries related to each problem (e.g., symptoms, physical findings, laboratory tests, etc.). This grouping of data provides a context within which one may interpret clinical information.

As a summary of a patient's health care history, the Problem Oriented Record contains the following problem lists:

Established/permanent (major) problem lists;
Transient/episodic (minor) problem lists;
Resolved problem list;

Figure 1 is an example of a problem list. In addition to identification data the problem list indicates the onset date of each problem, the date of the last encounter addressing the problem, the problem number (if a major problem) or letter (if a minor problem), and the name of the problem. During entry of problems, if a problem is selected which the patient had before but has become inactive, then the previously used problem number or letter for that problem is retained and a recurrence number is incremented. Problems may be resolved to other problems or resolved without sequela. The Resolved Problems List reflects a patient's currently inactive problems. When a problem is resolved an audit trail is automatically maintained of the resolution of the problem by the processing routines. All updating and associated "housekeeping" required to maintain each of the problem lists is done automatically by the system from information captured about the problem during data entry. The various Problem Oriented display options will be shown in a later section.

Problem Oriented Data Entry

COSTAR was designed to facilitate entry of encounter form encoded data by a transcriptionist regardless of the specific clinical or administrative division to which the data belongs. The clinical data entry routines of COSTAR have been rewritten for the Problem Oriented implementation to allow entry of clinical data in three different formats:

Entry of Abstraction of Medical Record;
User Defined Problem Oriented Record;
Problem Oriented Record with Automatic Consultation.

For brevity, this discussion uses the word "user" to refer to anyone whose job classification in the Security Module allows use of the option being described. The user is usually a provider.

One of the powerful features designed into COSTAR is the capture and storage of primary terms which are expandable by the use of modifiers. The primary terms and associated modifiers are defined in a directory which point to an internal code used for storage. The use of a vocabulary provides a well structured format which is important when searching stored data. It is also an effective method of forcing semantic consistency. Free text and statuses may be used with the encoded terms to allow unformatted qualification of a patient's specific instantiation of a given encoded term. Segregation of data in this manner allows separate capture of clinical data which has global import over the entire patient population from data which serves solely to uniquely qualify a given term for a specific patient. Another advantage of the use of primary terms and modifiers is that, since processing routines can easily collate modifiers chosen by a user, the order in which modifiers are chosen is not critical for recognition of synonymous terms. Thus, if a primary term such as "Otitis Media" is modified by "Acute" and "Bilateral" it does not matter whether the person entering the data chooses to use the phrase "Acute Bilateral Otitis Media" or "Bilateral Acute Otitis Media." This capability is invaluable when the medical data is searched by the report generator, when problem lists are updated and for other functions. An additional benefit is that data capture can be done with only a minimal amount of typing required of a user.

Although directory defined coded terms and modifiers are used internally for storage and retrieval of clinical data by the Problem Oriented Module, their use has been made transparent to the user. An example of entry of a diagnosis is given in Figure 2. Entry of all types of clinical data follow the sequence of primary name, optional modifiers, optional onset date and time (if applicable) and optional qualification of the primary term. An alphabetic lookup feature facilitates selection of primary terms and obviates the necessity of knowing how to correctly spell primary terms.

```
ENTER DIAGNOSIS OR PROBLEM TITLE > HEADACHE

 HEADACHE MODIFIER:
           1  CEPHALGIA
           S  SINUS
           T  TENSION

 SELECT MODIFIER > T

 STATUS OF TENSION HEADACHE > PM
   STATUS:  PRESUMPTIVE MAJOR

 ONSET DATE OF TENSION HEADACHE > 3 (8/3/81)

 FREE TEXT PROBLEM TITLE QUALIFICATION:
 1>
```

Figure 2. Sample diagnosis entry.

The option to enter an abstraction of medical record data and items is designed to allow capture of a practice-specified minimal data set from a patient's chart. Since this minimal data set is usually abstracted by someone other than the provider who made the chart entry, it is designed to avoid decisions about the data being made at the time of abstraction. Use of this option also establishes flags which are used by the display output routines to indicate additional data concerning the encounter may be in the patient's chart. Data entry via the other two options requires decisions by the user at the time of data entry. The User Defined Problem Oriented Entry option is designed to expedite capture of the results of provider's decisions in a straightforward manner. The Problem Oriented Entry With Automatic Consultation option functions to provide decision making support during data entry.

All three problem oriented options for data entry use the COSTAR medical record Header information. This identifies the provider, date of visit, and so on. The COSTAR Header entry has been minimally changed to be consistent with the style of the problem-oriented entries. For both an abstraction of the medical record and a user defined problem-oriented entry, the user is expected first to specify a diagnosis or problem. After a diagnosis is entered any of the following types of clinical data may be entered in any order and they are automatically linked to the diagnosis by the processing routines:

Symptoms Elicited
Absent Symptoms
Examination Findings
Tests
Procedures
Medications
Non-Medication Therapies

The entry of all types of clinical data follow a format similar to that used for entry of diagnosis. For symptoms, the entry of a primary term and/or modifiers is followed by entry of an optional onset data and time and an optional status or statuses. If an abstraction of a medical record is being entered qualification of symptoms is done entirely by free text. However, for the other two entry options symptom qualification may be entered from a screen display of multiple choice questions such as:

A = Since Onset it has
  1 = Slowly Worsened
  2 = Rapidly Worsened
  3 = Remained Unchanged
  4 = Slowly Improved
  5 = Rapidly Improved

The screen display of multiple choice questions for a given symptom are specified and/or updated by a practice through routines which have been added to the Systems Maintenance module of COSTAR. Updating or changing the screen display for a given symptom does not effect prior clinical entries. A user, when selecting choices, simply enters the letter and number corresponding to each of the questions he wishes reflected in the qualification of the symptom; this is converted to phrases for display. Multiple choice responses are given to afford a means of decreasing user typing and system storage requirements, as well as provide prompts of most likely events associated with given symptoms. The user is also able to interspace free text among the selected multiple choice responses during data entry. With both the User Defined Problem Oriented Entry and the Problem Oriented Entry with Automatic Consultation the user is warned of medication interactions, incompatibilities and previously known allergies, before the medication is accepted for entry into the patient's record.

Implementation of the Clinical Data Entry with Automatic Consultation option is made possible by relational Knowledge Content Files which have been added to the COSTAR package. These are files which contain "knowledge" or problems and their associated clinical features. Creation and maintenance of these files is discussed below.

Instead of passively recording information relative to the selected clinical divisions, the option for entry of data with automatic consultation is designed to provide decision making support at the time the data is captured, i.e., in the examination room. This option should not be confused with other approaches which attempt to simulate an experienced physician's traversal through decision modes during the diagnostic process. As used here, decision-making support refers to the automatic presentation of information which a user can consider, together with his personal knowledge in making decisions about what features a patient has, what diagnostic approaches should be taken, and what therapeutic intervention is required. Entry of symptoms in this option triggers routines which automatically list possible diagnoses containing the symptoms entered, as well as the additional symptoms of those diagnoses which have not yet been specified as present in the patient. For example, if the symptoms headache and fever are entered the user would see the display shown in Figure 3. Additional symptoms can then be entered and the cycle is repeated. This allows the user to rule in, or out, additional symptoms before deciding upon a specific diagnosis. During entry of physical examination findings, the user is automatically shown the last results for the findings being entered. In order to focus on identification of potential complications of problems identified, an assessment option is used. When making decisions about whether to treat a given diagnosis

```
PURSUING THE POSSIBILITY OF:
MENINGITIS, OTITIS MEDIA, ROCKY MOUNTAIN SPOTTED FEVER

SYMPTOMS TO RULE OUT:

          1.  EAR PAIN
          2.  HEARING DEFICIT
          3.  LETHARGY
          4.  MYALGIA
          5.  NECK PAIN
          6.  NECK STIFFNESS
          7.  PETECHIAL RASH
          8.  RUBELLIFORM RASH

Indicate additional symptoms elicited:
>
```

Figure 3. Sample entry of symptoms.

or order a test in order to determine whether treatment should be given or withheld a decision analysis can be accomplished by use of routines which are based on the work of Pauker and Kassier [1]. Another method of supporting decision making is the availability of situation-action rules (which can be added by the practice via an extension of the COSTAR System Maintenance module) to list choices of diagnostic tests and treatments for diagnoses under variable situations. For example if the patient is under two months of age and has a fever above 38.5°C the practice may have specified that the options to do cultures and start antibiotics be given; and if so, will be presented to the user as choices at the appropriate time.

Requirements of the COSTAR Billing Module are satisfied by the Problem Oriented Entry options. The Problem Oriented enhancements are also compatible with an expanded version of the COSTAR Billing Module developed by Neches.

Display of Clinical Data

The COSTAR Status Report and Encounter Report have been expanded to include the capability to display symptoms separately from diagnoses and problems. All the COSTAR display options have been retained and the following have been added:

Problem Status Summary
Medication History Graph
Iatratropic Data
Transient/Episodic (minor) Problem List
Established/Permanent (major) Problem List
Resolved Problem List
Audit Trail of Resolution of Problems

The Problem Status Summary compliments the COSTAR Status Report and Encounter Report. The Status Report allows review of the most recent status of all prior clinical entries. The Encounter Report reflects data in a non-problem oriented format and has the advantage of showing all diagnoses and clinical features addressed during an encounter even if incorrect associations were made by the user during creation of a problem oriented record. The Problem Status Summary allows review of chronological clinical details for each of a patient's problems. An example of a Problem Status Summary is given in Figure 4.

Discussion

It is obvious that the source of information contained in the Knowledge Content Files is critical for successful utilization of the described system. Data for the Knowledge Content File can be entered by using routines added to the

```
                         PROBLEM STATUS SUMMARY          8/27/81

               NONAME, HECTOR
               DATE OF BIRTH:  3/11/78     AGE:  3 YRS 5 MOS 2 WKS
==========================================================================================
//////////////////////////////////////////////////////////////////////////////////////////
               PROBLEM LETTER:  B    PROBLEM TITLE:  UPPER RESPIRATORY INFECTION
//////////////////////////////////////////////////////////////////////////////////////////

VISIT SITE:  PPM                                 DATE OF PATIENT CONTACT:  08/04/81
PROVIDER:  OSBURN, A  EUGENE, DO                 TIME OF PATIENT CONTACT:  UNSPECIFIED

SUBJECTIVE DATA:
  REASON FOR VISIT:  HISTORY OF COLD SYMPTOMS

  Symptoms supporting Problem Title:
    INTERMITTENT FEVER:  ONSET ON MON AUG 3, 1981 AT 2:00 AM, AT FIRST MOTHER THOUGHT WAS FROM PLAYING
    IN THE SUN TOO LONG.  BECAME WORRIED WHEN IT DIDN'T RESPOND TO ASPIRIN.
    RHINORRHEA:  CAUSES HIM PROBLEMS WHEN HE TRIES TO WATCH TV.
    HEADACHE:  BEGAN IN FRONTAL REGION, NOW INVOLVED ENTIRE HEAD, KEPT HIM AWAKE LAST NIGHT.

  Symptoms confirmed absent:                              TREATMENT PLAN:
    COUGH, MYALGIA, NECK PAIN.
                                                             DIMETAPP ELIXER 1 TSP
  Symptoms still unexplained when Problem Title was chosen:  Q4H
    INTERMITTENT ABDOMINAL PAIN                              QUANTITY:  120 CC     REFILLS:  2
                                                             TO BE TAKEN FOR:  RHINORRHEA.
OBJECTIVE DATA:
  Physical Findings supporting Problem Title:
    TEMPERATURE 102.6.

  Physical Findings confirmed absent:
    NONE.

  Findings still unexplained when Problem Title was chosen:
    NONE.

ASSESSMENT:
  Because of the presence of:  FEVER, HEADACHE,
  The possibility of:  ROCKY MOUNTAIN SPOTTED FEVER
  Still needs to be ruled out.

  CONDITION DUE TO UPPER RESPIRATORY INFECTION:  GOOD
  EXPECTED OUTCOME:  RESOLVE WITHOUT SEQUELAE

DIAGNOSTIC PLAN:

  FEBRILE AGGLUTININS to rule out ROCKY MOUNTAIN SPOTTED FEVER.
```

Figure 4. Problem status summary.

Systems Maintenance module. Also, the information is updated automatically by background routines which use the chi-square statistic to evaluate associations between clinical entries and problems. Statistically significant associations are added automatically to the Knowledge Content File by these background routines.

One potential problem which could result from updating the Knowledge Content Files from a user defined data base arises from cases in which users supplying data incorrectly associate clinical features with diagnoses. To circumvent this possibility, the "knowledge" is not abstracted from the problem oriented relationship specified by users. Instead, clinical features are added to the expected features of given problems by background programs automatically if a given feature is occurring more often than statistically by chance in the same encounter regardless of whether such associations have ever been made by users in the Problem Oriented Record entry or not. This still leaves the possibility that inadequate clincal skills by the user could result in incorrect knowledge eventually being placed in the Knowledge Content Files. This possibility however is viewed as an asset rather than a liability. It provides a means for self-audit by the practice. This allows the user group to investigate those instances in which the accumulated expected features for specific problems differ from associations reported in the medical literature. Such investigation is a basis for correction of practice habits or the literature, as appropriate. One should also remember that data from the Knowledge Content File is used only to present knowledge about problems. The decision support capability does not make diagnoses nor prescribe treatments. It is used only to support decisions about diagnosis and treatment made by the provider.

At present, a simple chi-square statistic is the only automated method used for deciding what information from encounter data should be associated in the Knowledge Content Files. The framework for implementation of more sophisticated multi-varient discriminate analysis techniques has been accomplished. Current efforts are directed toward implementation of such advanced techniques within this framework. Development is also under way to enhance utilization of information from the Knowledge Content File during decision making support activities by incorporation of multiple techniques for feature selection including: Statistical pattern recognition and other decision analysis approaches, symbolic reasoning approaches, and artificial intelligence approaches. It is important to note incorporation of such techniques should be guided by an appreciation of the significance of possibilities and probabilities in clinical medicine. Simply being able to list relative probability with statistical techniques must often be overridden by the need for diagnostic and/or therapeutic intervention because of a less likely but still present possibility of a specific disorder and the consequences to the patient if the disorder should be present.

Implementation of the described package requires considerable attention to each Systems Maintenance Module option in order to specify actions to be taken under given situations. We are finding it beneficial to utilize the requirement of explicitly given categorical rules for various clinical situations, an aid for guiding Housestaff Education. A similar benefit for continuing education activities of other practices is expected.

Conclusion

An expansion of COSTAR to include a Problem Oriented Record Structure and decision support capability has been described. While it is felt that the development thus far is significant, it is also felt that the most important feature of the system is the framework which it provides as a foundation for future development.

References

[1] Pauker, S G; Kassirer, J P., "The Threshold Approach to Clinical Decision Making," *N. Engl. J. Med.* **302,** 1980, pp. 1109–111.

24

Impact of an On-Line Information System in the Medical Office*

Frederick R. Jelovsek, M.D., Barbara P. Deason, M. S., and Hope Richard

Introduction

Most automated computer systems for the physician's office available today are designed to be used in a "batch" mode. Under this arrangement, data entry is not performed until after the patient encounter has taken place either later in the same day or as much as several days later. As a result, determination of what took place is always retrospective and subject to uncertainty. Data is only as up-to-date as performance of the last "batch". Since the source of most office data is either the patient, who is no longer in the office, or the physician, whose memory for detail decreases with elapsed time, any question about activity in the past many require extensive personnel time to answer.

Since September 1980, we have utilized an on-line, ambulatory medical information system—TMR—in the Woman's Clinic at Duke University Medical Center. This paper describes our experience and concentrates on the on-line and realtime aspects of TMR versus our previous off-line, batch system. The impact on both patient and office personnel over the last three years is emphasized.

System Description

Although there are many features available, the major interfaces of the information system with patients, office support personnel and physicians are as follows:

Patients arrive at a registration area in the clinic where one of three registration clerks will greet them, retrieve their record on a video terminal, update any demographic data, order any laboratory studies as indicated by the patient type (e.g., obstetric patients get a hemoglobin and urine test at each visit) or by M.D. indication at the patient's last visit. These recommended or pre-ordered studies appear automatically on the video terminal. Following this process, a route sheet (Figure 1) and pre-encounter summary (Figure 2) are computer-produced and studies requisitions are printed in the laboratory. After pre-ordered studies are obtained, the patient waits until called to see the physician. The route sheet is attached to the patient's medical record and taken to the area where she will be seen by the physician. After the doctor/patient encounter takes place, the physician fills out the route sheet, indicating the type of visit, additional studies to be obtained, the diagnoses/problems determined for that visit, medications given, and disposition, usually in the form of the next appointment. The patient then takes this route sheet to a "checkout clerk" who enters the data via video terminal and orders studies. The patient then goes to

```
WOMAN'S CLINIC                                              * H.
CATEGORY: PVT/LROB PVT                                 RV   *  E
 PRIMARY MD: JELOVSEK                                   DB  *  829 CHALICE ST
 OCCUPATION:                                                *  DURHAM, NC  27705
   EMPLOYER:                                                *919-383-5174/919-737-2581
                                                            *  BIRTHDATE: 03/01/49
                                                            * N239         323-58-06
APPOINTMENT:  3:30 - JELOVSEK                               *
                                                            *  33 YO ORIENTAL MAR FEMALE
03/11/82 10:52                                              ****************************

TYPE VISIT:  LROB RV   HROB RV   LAB ONLY

DATA:        WEIGHT-LBS __________ BP(SYS-DI) __________

WAITING STUDIES ROUTINE                        XRAY                    SUPPLIES
  2-PRO, SUG, KE            43-HGB
113-GTT 3H                 119-GLU 2H
185-US INIT                186-US  REPEAT




  ONSET  RESOLVED SEEN CODE      ACTIVE PROBLEMS
12/10/81 ________ __           1 PREGNANCY, INTRAUTERINE-18 WKS

________ ________ __ ____  _______________________________________________

________ ________ __ ____  __________  ___________________________________

       MEDICATION            DOSE      FORM   NO.     FREQ      # DISP REFILLS DAW

_________________________ __________ _____ _____ _________ _______ _______

_________________________ __________ _____ _____ _________ _______ _______

_________________________ __________ ___   _____ ___ ___ _______ _______

          PATIENT DISPOSITION                          NONE      PRN

RETURN VISIT ___________________________ CLINIC/MD: ______________________

STUDIES NV: __________________ ADMIT __________ FOR ________ # DAYS ________

          SIGNATURE: ________________________________________
```

Figure 1. Computer printed route sheet.

```
WOMAN'S CLINIC                                               * h
CATEGORY: PVT/GYN PVT                                    RV  *  E
 PRIMARY MD: JELOVSEK                                    DB  *  829 CHALICE ST
 OCCUPATION:                                                 *  DURHAM, NC   27705
   EMPLOYER:                                                 *919-383-5174/919-737-2581
                                                             *  BIRTHDATE: 03/01/49
                                                             * N239        323-58-06
APPOINTMENT:  3:30 - JELOVSEK                                *
                                                             *  33 YO ORIENTAL MAR FEMALE
03/11/82 10:52                                               ****************************

  VISITS: 82 (3): 81 (6)
---------------------------VISIT PROFILE ------------------------------
                                   YEAR: (--82---)(-------81-------)
                                  MONTH:   3  2  1 12 11  8  7  7  6
CODE         PROBLEM                DAY:  11 11 14 10 24  5 24 23  8
----------------------------------------------------------------------
   1 PREGNANCY, INTRAUTERINE-18 WKS        8  8  8
 235 AMENORRHEA                                    51
 273 INFERTILITY-SECONDARY                            8     8  8
 274 INFERTILITY-NOS                                     8
----------------------------------------------------------------------------
                06/08/81 07/23/81 07/24/81 08/05/81 11/24/81 12/10/81 01/14/82 02/11/82 03/11/82
----------------------------------------------------------------------------
HGB G/DL                                                     13.0     12.0
HCT %                                                        37.4
VDRL                                                          NEG
B TYPE                                                       B EG
ABS                                                          NEG.
PAP             TEXT1
UCG                                                 POS
HSG                                        TEXT2
WBC CMM                                                      9,800
RBC THOU/CMM                                                 4.29
MCV CU MIC                                                     90
MCHC %                                                       34.8
COLOR                                                        YELLOW
CLARITY                                                      HAZY
SP/GR                                                        1.022
PH                                                              6
PROT                                                            N        N        T
GLUC                                                            N        N        N
BLOOD                                                           N
KETO                                                            N        N        N
WBC                                                           OCC
RBC                                                             N
EPITH                                                           1
BACT                                                            1
CASTS /LPF                                                      N
TEXT1 = CL II INFLAMMATION
TEXT2 = NORMAL UTERUS &R. TUBE,NONFILLING L. TUBE SEE REPORT
```

Figure 2. Preencounter summary sheet.

a different station where appointments and cashiering functions are performed. Patients receive a computer produced "superbill" at time of leaving the clinic and a computer printed appointment reminder letter about four days prior to their next scheduled visit.

Findings

There are many office functions and activities which have been changed or influenced due to the real time nature of the TMR Ambulatory Medical Information System in our setting. Some of the advantages are listed in Table 1.

Data is much more current and complete since the software forces clerks to collect and update demographic and financial data in a systematic fashion. If

Table 1. Advantages of the Real Time Aspect of Ambulatory Medical Information System

A. Data Quality
  1. Less missing data
  2. Improved accuracy of existing data
  3. Timeliness—data is current
  4. More pertinent data due to customization of data collection
B. Clerical Work Volume
  1. Less repetitive transcriptions
  2. Replacement of manual reports by computer printed reports
  3. Elimination of manual control process
C. Financial
  1. Less missed charges
  2. Improved cash flow

there is a need to know a patient's address or charge status, this is instantly available by video terminal and contains the latest information. Once data is entered into the system, it is printed on all documents without error that is usually attendant a manual transcription. Clerks do not have to remember the code or price for studies, such as a hematocrit, to write on the bill or claim, since this is resident in the software. There are no missed charges since studies cannot be performed without a requisition, and existence of the requisition is evidence it has been recorded in the information system. The fact that a bill is ready when the patient leaves the clinic and can be verified by the patient, means less contested bills after the visit as well as more timely payment. Producing bills and claims on a daily basis speeds up collection of accounts outstanding and improves cash flow.

The major disadvantage of a real time system (Table 2) is increased equipment needs such as terminals and printers as well as having hardware which provides a rapid response time even at peak service times. With a batch system an entire day's work can usually be entered in much less than an eight hour period of time. The number of video terminals and geographic locations necessary for data entry are minimized. In a real time system, however, data are

Table 2. Disadvantages of Real Time Aspect of Ambulatory Medical Information System

A. Equipment Requirements
  1. Number of data entry terminals and printers is increased versus that needed for batch system.
  2. CPU capacity and speed and disk access time directly affect patient flow and an undersized system (for the patient volume) will slow patient flow and increase reliance on system availability.
B. System implementation requires modification of behavior on the part of all office personnel and physicians.
C. Established procedures and logistical flow have to be followed with more disicipline to take advantage of the real time information system benefits.

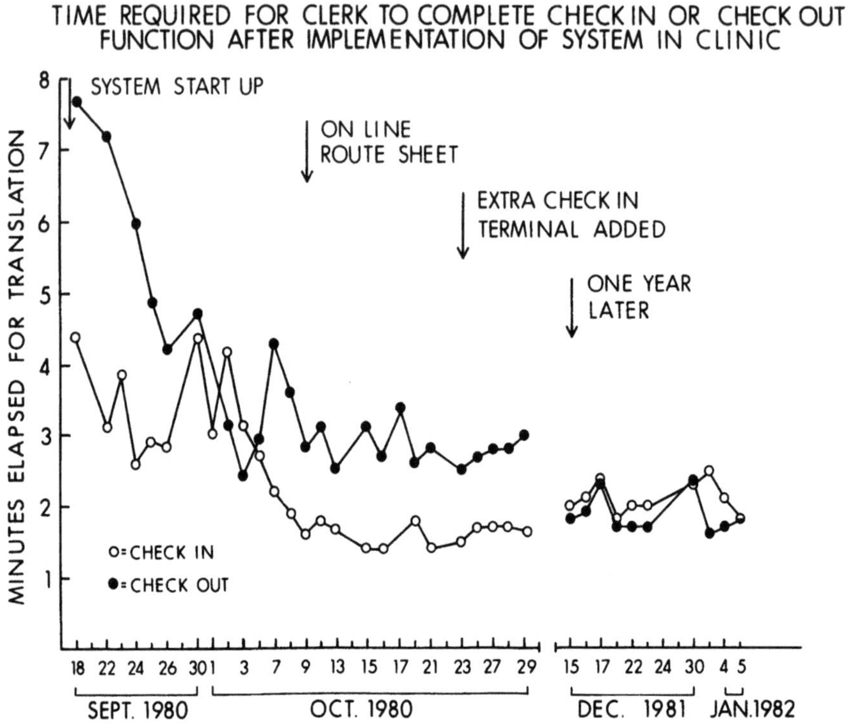

Figure 3. Learning curve for realtime checkin and checkout functions.

entered when the patient is standing in front of the clerk. More video terminals are required since patient arrival and processing times are not evenly distributed throughout a day, and each point of patient interaction requires a data port.

Any major change to establish procedure in a medical office will have a disorganizing effect until system learning and acceptance of change on the part of all people involved takes place. Figure 3 illustrates the learning curve we observed for two clerical functions performed in real time on the video terminals.

There is an increased rigidity to a real time system in that all patients must "check in" and "check out" and not bypass any of the functional clerical stations; otherwise the books will not be balanced at the end of the working day. This adherence to rules is quite necessary in order to provide some of the benefits of a real time system. In most offices there is significant manpower allocation to correcting and completing paperwork from prior day's work when physicians, staff or patients bypassed the normal office procedures. Although listed as a disadvantage here, the lack of following established rules and procedures is really a deficiency in most offices and should not exist in any information system whether manual, batch or real time.

Discussion

When patients present to the clinic for a physician visit, their demographic and financial data require verification and update if necessary. Prior to the medical information system (MIS), this update process took place to a very limited extent. Transcription of the patient's name and address onto the printed encounter sheet (superbill) took place; however, updating of the patient's master record occurred infrequently. As a result, old addresses, employers, telephone numbers and name changes were not corrected as often as they should have been. Since approximately 20% of all our patient's have some change to their demographic and financial data at each visit, the pre-MIS system had many errors.

In the original manual system, sensitive data items such as marital status and age were not confirmed for accuracy. We initially had some difficulty upon implementation, since the clerks would ask these questions of the patient in the presence of other patients waiting to register. Office personnel did this because they had a video screen in front of them with a data element which needed to be filled out. After familiarity with the system increased, our personnel learned to be more discreet about collecting this information. Also, in the first 6 months of system implementation there seemed to be less eye-to-eye contact between registration personnel and patients, due to newness of the video terminal requiring full attention of the registration clerk. Patients felt that employees were exhibiting a lack of care and concern for them due to the amount of time pushing "buttons on a scope".

Although personnel had the new duty of entering data items via a video terminal, they were relieved of writing the patient's demographic information onto the encounter sheet, since the computer automatically printed it. Initially, the route sheets were printed the evening before a patient's visit in order to save the time required for the printer to fill in the blanks, thus reducing the patient's wait. It was immediately observed, however, that it took the registration clerk as long to find the printed encounter sheet from a stack which contained those for all patients to be seen that day, than it did to produce a computer printout with the latest demographic and financial information on it.

Order entry and results reporting of studies are major office functions which were implemented from the start. One of the TMR laboratory options is to print requisitions each time a test is ordered. A printer was set up in the clinic laboratory so that our registrations would be filled out by the computer and transmitted directly. Laboratory technicians had to take an extra step to tear requisitions off the printer instead of having them directly presented from the patient. Although this extra step seemed not to be a major effort, it initially disrupted the work flow pattern of laboratory technicians with resultant complaints and lack of commitment to helping the system run smoothly. Lines of patients formed waiting to get their lab studies done and often patients bypassed the lab altogether since they had no requisition in hand to remind them about the study. Laboratory requisitions printed out in the order in which they were entered at various terminals in the clinic rather than the order in which patients arrived at the clinic. Occasionally this resulted in the conveyance of an impression to the patient that a later arrival was being serviced before them. Some patients believed that their blood and/or urine specimens were being "mixed up" because they no longer carried a requisition with them to the lab.

It is important to note that the laboratory personnel were not under our direct supervisory control. It was almost as if lack of cooperation on their part would make the "changed system" disappear. Once it became clear that the new computerized system was not going away, acceptance upon the part of lab personnel took place and the patient flow and perception greatly improved.

Physicians have benefited significantly from the on-line system because the encounter sheets now include the patient's permanent problem list and the laboratory results from the last ten patient visits. This available data has decreased the frequency of phone calls to the laboratory to find out a result which did not make it to the medical record in a timely fashion. Conversely, physicians occasionally have been irritated either by their own inconsistency in diagnostic terminology or by mistakes of the clerks in interpreting what the physician wrote as a diagnosis at the last visit for that patient.

It is difficult to determine what impact the computerized system has had on the patient's perception of the physician encounter. There have been many positive comments from the patients about the medical information system, but perhaps computer printouts imply a precision and accuracy about the doctor encounter beyond that which is warranted.

Prior to the computer system, there was a tendency on the part of check out clerks and physicians to allow patients to leave the clinic and expect to be billed. Often there was a line of people waiting to be serviced at that station and bypassing was encouraged in an effort not to inconvenience patients. Billing, however, suffered from this practice, and with the advent of computer terminals at checkout stations, patients were not allowed to leave the clinic without "checking out". Both patients and some physicians initially felt inconvenienced by this change. After approximately six months of operation however, a greater acceptance took place. Flow was smoothed out enough so that there were not excessive waiting times to checkout. The ability of the patient to have an accurate bill in her hand at the time of leaving the office benefited cash flow significantly, and we think left a better understanding with the patient as to what was done and what the exact financial obligations would be.

Several physicians have changed their habits to complete the encounter sheet and present it to the clerks so that check-out data processing can take place while the physician finishes talking with the patient. This procedure enables the checkout process to be completed as soon as the patient has finished seeing the physician and decreases the total amount of time the patient spends in the clinic.

There has been a significant time saving at the cashier desk by allowing the computer to provide coding of procedures and diagnoses on the route sheets. The improved accuracy of coding has also helped the reimbursement process.

Conclusion

In summary, the on-line, real time medical information system has made an impact both on the office personnel and on the patient flow through the practice. We feel that most of the changes have been beneficial, although there were some difficult times at the initial implementation of the computer system.

The attitude of the clerical and financial staff has changed from reluctance to total support, and there is no evidence of desire to return to the previous method of operations.

Lessons we have learned from this process can be used to enhance the implementation of on-line, real time medical information systems such as the TMR database management system in physician's offices.

25

A Computer-Assisted Problem-Oriented Medical Record System for Office Use*

Daniel J. Forward, M.D.

Introduction

The purpose of this chapter is to describe initial results in the development of a computer-assisted office medical record system in a small medical group practice setting. Rapid access to medical information is particularly desirable in primary care practice because of the large volume of telephone calls for which chart information is needed. Approximately 32,000 such calls per year are handled by Harwood Medical Associates, a group of eleven internists. Each call requires that the medical record be removed from file, reviewed by the physician, and then refiled. This paper will describe the development of a computer data base containing the information normally required by the primary physician in providing telephone consultation. The following aspects of the development of this data base will be discussed:

(1) Definition of the record;
(2) Physician communication with the computer;
(3) Information retrieval using a coding system;
(4) Justification of physician time and effort in establishing the data base.

Methods

An IBM system 34 computer was installed in July, 1980 for use in billing and insurance applications. Prior to installation, computer software was developed, using the RPG II computer language, for establishing and maintaining a base of patient information accessable by physicians using the computer display stations. Basic demographic information is repeated as heading information on each of the three computer display screens comprising the medical record abstract. Date of last visit is included as well as a frequency code indicating the frequency of visits in months as recommended by the physician. The field labeled "Last Rx" indicates the date on which the medication list was last updated.

Figure 1 indicates the information on page 1 of the 3-page computer display. A problem list serves as the foundation for the record summary. Spaces for two single character codes are displayed with each problem. The code labeled "C" allows definition of previous surgical or radiation therapy for the problem using the letter "S" or "R" respectively. The code labeled "A" is a single numeric assessment: 1 indicating a problem under ideal control; 2 a problem under borderline control; 3 a poorly controlled problem. The "X" indicates an inactive problem. A listing of therapy is also included on page 1 with fields oriented toward listing of medication, dosage and frequency. Dietary and other forms of therapy may also be listed. A single character code labeled "D" is provided with each therapy record to identify the department within the clinic prescribing the drug. An "X" in this field indicates that the drug is prescribed outside the clinic.

Figure 2 indicates page 2 of the 3-page display. Drug allergies are listed with a code "I" for drug intolerance. Chronologic past medical history is also listed with codes indicating past surgical or radiation therapy or the code "H" for hospitalization. Pertinent nonmedical history may also be included in this chronology. An additional code labeled "R" is included here for the patient recall. The number listed in this column indicates the recall interval in years recommended by the physician for the condition listed. Computer generated recall lists may be produced using this information.

Figure 3 indicates the final page of the computer summary which is a listing of intercurrent or minor illnesses. Date, mode of contact, a statement of the problem and a brief statement of plan are all listed in this summary. This 3-page summary serves as a table of contents for the medical record.

```
         Patient ID No  100033  Physician No  00098       Home 414-771-5565
PROBLEX  Name WILSON                  MICHAEL      P  Sex M   Work 000-000-0000
INQUIRY  Addr 7744 W.VLIET ST.                            Phar 414-258-1000
PAGE  1  Last Visit 04/13/83  Freq 02  Last RX 04/13/83   Birth 09/25/06

PROBLEM LIST (X = INACTIVE)        C A THERAPY             DOSAGE    FREQUENCY D
01 ASHD                              2 COUMADIN            5 MG      AS DIR
02 ANGINA PECTORIS                   2 LANOXIN             0.25 MG   1 QD
03 DIABETES MELLITUS                 2 ISORDIL             2.5 MG    1SL ACHS
04 GRAVES' DISEASE                 R 1 INDERAL             10 MG     2 BID
05 ATRIAL FIBRILLATION               2 NITROGLYCERIN       GR 1/150  1 SL PRN
06 THYROTOXIC EXOPHTHALMOS           2 SINEQUAN            50 MG     1 HS        *
07 PULMONARY EMBOLISM                1
08 NEUTROPENIA                       2
09 COLON CARCINOMA                 S X
11 DEPRESSION                        2


Enter # 1 - 6:     1-Page 1  2-Page 2  3-Page 3  4-New Inq  5-Overflow  6-Menu
```

Figure 1. Page 1 of display.

```
        Patient ID No  100033  Physician No  00098        Home  414-771-5565
PROBLEX  Name WILSON               MICHAEL      P  Sex M   Work  000-000-0000
INQUIRY  Addr 7744 W.VLIET ST.                             Phar  414-258-1000
PAGE  2  Last Visit 04/13/83  Freq 02  Last RX 04/13/83    Birth  09/25/06

DRUG ALLERGY-INTOLERANCE (I)     YEAR    CHRONOLOGY                C   R
  PENICILLIN                     1906  BIRTH
                                 1912  MEASLES
                                 1913  MUMPS
                                 1929  T&A
                                 1956  COLON CARCINOMA             S
                                 1967  RECTAL POLYPS               S   1
                                 1972  CARDIAC CATHETERIZATION
                                 1973  ACUTE MI                    H
                                 1977  GRAVES' DISEASE             R
                                 1978  PNEUMONIA                   H


  ter # 1 - 6:    1-Page 1  2-Page 2  3-Page 3  4-New Inq  5-Overflow  6-Menu
```

Figure 2. Page 2 of display.

```
        Patient ID No  100033   Physician No  00098        Home  414-771-5565
PROBLEX  Name WILSON                MICHAEL      P  Sex M   Work  000-000-0000
INQUIRY  Addr 7744 W.VLIET ST.                              Phar  414-258-1000
PAGE  3  Last Visit 04/13/83  Freq 02  Last RX 04/13/83     Birth  09/25/06

                          INTERCURRENT ILLNESS - THERAPY
    DATE    CONTACT                PROBLEM                        PLAN
  11/22/79    OC    VIRAL URI                             SYMPTOMATIC
  01/12/80    PH    PROSTATITIS                           BACTRIM-DS
  02/20/80    PH    PROSTATITIS                           BACTRIM-DS
  05/30/80    PH    STREP THROAT                          ERYTHROMYCIN
  03/20/80    OC    OTITIS MEDIA                          ERYTHROMYCIN
  11/10/80    PH    SINUSITIS                             ERYTHROMYCIN


Enter # 1 - 6:    1-Page 1  2-Page 2  3-Page 3  4-New Inq  5-Overflow  6-Menu
```

Figure 3. Page 3 of display.

```
^^^^^^^^^
*PROBLEX* Patient ID No. 100033  Physician No. 00099       Home 414-771-5565
*INQUIRY* Name WILSON              MICHAEL      P  Sex M    Work 000-000-0000
* CHART * Addr 7744 W.VLIET ST.                             Phar 414-258-1000
*  COPY * Last Visit 02/22/81  Freq 02  Last RX 02/22/81    Birth 09/25/06
*********

 PROBLEM LIST  (X = INACTIVE)      C A   THERAPY             DOSAGE    FREQ    PROB

01 ASHD                              3 COUMADIN             5 MG      AS DIR     7
02 ANGINA PECTORIS                   3 LANOXIN              0.25 MG   1 QD       5
03 DIABETES MELLITUS                 2 ISORDIL              2.5 MG    1SL AC&HS  2
04 GRAVES' DISEASE                 R 1 INDERAL              40 MG     2 BID      2
05 ATRIAL FIBRILLATION               2 NITROGLYCERIN        GR 1/150  1 SL PRN   2
06 THYROTOXIC EXOPHTHALMOS           2 SINEQUAN             50 MG     1 HS      11
07 PULMONARY EMBOLISM                1
08 NEUTROPENIA                       2
09 COLON CARCINOMA                 S X
11 DEPRESSION                        2

 DRUG ALLERGY-INTOLER (I)   DRUG ALLERGY-INTOLER (I)   DRUG ALLERGY-INTOLER (I)

 PENICILLIN

 RX REFILL
```

Figure 4. This medical record copy is produced for permanent chart entry after each patient visit. Space is allowed for written notes. A second copy is edited by the physician and used by medical transcription for updating the computer record.

Physician communication with the computer occurs initially when the physician fills out a form indicating the information which is to be filed in the computer. This information is updated through use of a chart copy of a portion of the data base which is printed after initial entry. This is shown in Figure 4. Note that an indication for each therapy given may be stated as a problem number. Two copies are placed in the chart. One is a permanent chart entry used by the physician for note keeping. The second is used for communication with the clinic transcription department. This information is edited by the physician and sent to medical transcription for use in updating the computer files. Previous information on file is then displayed and edited by the transcriptionist and the computer files are updated with current information after each patient visit.

Information can be viewed by the physician using a display station by entering the patient's medical record number, last name or home telephone number.

The design of the system includes automated coding of information by the computer. Patient problems, previous surgical or other diagnostic procedures are coded using ICD-9 codes in a master file. A method of coding medications had to be devised and is outlined in Figure 5. Up to four codes indicating generic components of medications are allowed. In addition, an excretion code is allowed for each generic substance. Codes for identification of generic substances, contolled substances and an administration code are also included. Coding may be bypassed for entries in the chronology file.

The coding system allows retrieval of information from the computer data base in spite of the wide vocabulary of medical terms which may refer to a single disease entity. Similarly, patients taking given generic substances may be retrieved from the data base without regard for the multiple drug names which may be used to identify a single substance or combination of substances.

A daily listing is produced of diagnosis or therapy records entered by the transcriptionist but not found in the master coding file. These are coded manually and entered into the master file. At the same time, patient records are updated with missing codes.

A variety of written reports can be generated using this data base. Prescrip-

PMS140 THERAPY TABLE LIST 7/31/81 PAGE

THERAPY DESCRIPTOR	ADMIN CODE	CTL SUB	GENERIC CODE	DRUG #1 CODE	E	DRUG #2 CODE	E	DRUG #3 CODE	E	DRUG #4 CODE	E
DIETHYLSTILBESTROL			G	4765							
DIGITOXIN			G	1530	H						
DIGOXIN			G	1540	R						
DIHYDROERGOTOXINE			G	1255							
DIHYDROTACHYSTEROL			G	5010							
DILANTIN				2420	H						
DILAUDID		C		2080							
DILAUDID INJ	P	C		2080							
DIMETANE				0010							
DIMETAPP				0010		1230		1235			
DIMETAPP EXTENTABS				0010		1230		1235			

Figure 5. Drug coding structure.

```
                         Prescription Request
 7/30/81          Harwood Medical Associates, S.C.             414-771-5300
                         7400 Harwood Avenue
                         Wauwatosa, WI 53213                 BNDD # AF6146953

Patient Name: MICHAEL       P WILSON                         Record # 100033

  Medication:             Dosage:           SIG:        Refill:       (PHAR. USE)

01 INDERAL                40 MG      #___   2 BID       ___ Times
02 NITROGLYCERIN          GR 1/150   #___   1 SL PRN    ___ Times

  Additional Directions:



                                   ---------------------------------------
                                   DANIEL J. FORWARD                 M.D.
```

Figure 6. Prescription rewrite.

tions may be rewritten as shown in Figure 6 and a medical report can be generated (see also Figures 8 and 9). Codes used in the data base to indicate problem assessment are translated by the computer into words for use outside the clinic. A computer generated hospital admission note similar to the medical report may be printed on demand. These reports are produced with simple processing of the information in the medical record abstract.

The computer also finds all patients who are one month or more overdue for office followup and retrieves patients due for followup from the chronology listing. A sample patient recall report is shown in Figure 7.

Results

To date, this system has been implemented on a limited basis with approximately 635 patient records now summarized in the computer. Some initial difficulty was encountered as physicians became accustomed to eliminating details from descriptions of patient problems so that a codable diagnosis could be entered. At present four of eleven clinic physicians have begun to use the

Harwood Medical Associates, S.C.
Patient Recall Report - Dr. DANIEL J. FORWARD 7/30/81

The following patients are one month or more overdue for office followup:

Name			Last Visit	F Code	Record #	Phone #
WILLIAMS	ROBIN	J	8/25/80	04	000030	414-771-0022
BROWN	GERALD	T	6/20/80	06	000040	414-258-1410
JONES	CHERYL	P	3/12/80	06	000050	414-771-2660
JONES	DIANNE	M	8/25/80	06	000060	414-321-5477
JONES	JOHN	S	8/20/80	06	100001	414-453-2144
SMITH	SARA	J	4/20/80	03	100002	414-784-1112
SMITH	MARY	E	7/30/80	06	100003	414-453-6541
ADAMS	JOSEPH	P	1/10/79	03	100004	414-453-9801
ADAMS	JAMES	S	4/12/79	06	100005	414-453-9801
JONES	MARY	P	8/04/80	06	100008	414-786-3214
MASON	JOHN	S	8/20/80	03	100030	414-771-0331
SULLIVAN	FRANK	J	8/20/80	03	100031	414-453-4411
WILSON	MICHAEL	P	2/22/81	02	100033	414-771-5565
JONES	JAMES	R	12/08/80	04	100045	414-111-1111
JONES	JOE	P	0/00/00	02	100050	414-258-1212

Figure 7. Patient recall report.

```
Page 1                      Medical Report                          7/31/81
                     Harwood Medical Associates,S.C.
                          7400 Harwood Avenue
                          Wauwatosa, WI 53213
                             414-771-5300

Patient:  MICHAEL       P WILSON                              Sex M
          7744 W.VLIET ST.                                    Birth date 09/25/06
          WAUWATOSA                  WI  53213                Last visit 02/22/81
          Home Phone  414-771-5565                            Weight 164
          Record # 100033                                     BP 140/070
```

Problems	ICD-9-CM	Assessment
ASHD	414.0	PROBLEM CONTROL
ANGINA PECTORIS	413.9	PROBLEM CONTROL
DIABETES MELLITUS	250.0	BORDERLINE CONTROL
GRAVES' DISEASE	242.0	IDEAL CONTROL
ATRIAL FIBRILLATION	427.3	BORDERLINE CONTROL
THYROTOXIC EXOPHTHALMOS	376.21	BORDERLINE CONTROL
PULMONARY EMBOLISM	415.1	IDEAL CONTROL
NEUTROPENIA	288.0	BORDERLINE CONTROL
COLON CARCINOMA	153.9	INACTIVE PROBLEM
DEPRESSION	311	BORDERLINE CONTROL

Therapy	Dosage	Frequency	Drug Allergy or Intolerance
COUMADIN	5 MG	AS DIR	PENICILLIN
LANOXIN	0.25 MG	1 QD	
ISORDIL	2.5 MG	1SL AC&HS	
INDERAL	40 MG	2 BID	
NITROGLYCERIN	GR 1/150	1 SL PRN	
SINEQUAN	50 MG	1 HS	

Figure 8. Sample entry for a complex case.

system. Other clinic personnel adapted quickly to the use of the computer for maintaining and updating patient records. Extensive training in the use of the computer was not required. At present, approximately 3,000 procedures and diagnoses are contained in the master list for coding of problems and procedures.

Initial experience with the system has involved entries for the most complex cases with multiple problems. These patients are seen at more frequent intervals and present the greatest difficulty in adequate medical record keeping. In addition this population of patients is more frequently admitted to the hospital requiring transfer of office record information to a new file. Record keeping is simplified and clarified with computer generated chart entries. At present, the system has received only minimal use in handling of patient phone calls and appears even more useful in organizing and updating of the paper medical record. A computer display station is not required for each doctor in order to use the system.

Conclusion

The use of a brief computerized medical record abstract appears to be feasible in medical practice. The computer record, even for complicated cases, does not exceed 25 to 50 words in length and therefore requires a reasonable expendi-

Page 2 MICHAEL P WILSON Record # 100033

Chronology

1906	BIRTH	
1912	MEASLES	
1913	MUMPS	
1929	T&A	
1956	COLON CARCINOMA	Surgically Treated
1967	RECTAL POLYPS	Surgically Treated
1972	CARDIAC CATHETERIZATION	
1973	ACUTE MI	Hospitalized
1977	GRAVES' DISEASE	Radiation RX
1978	PNEUMONIA	Hospitalized

Intercurrent illnesses / Followup Parameters

Date	Contact	Problem	Plan / Parameter
11/22/79	OC	VIRAL URI	SYMPTOMATIC
1/12/80	PH	PROSTATITIS	BACTRIM-DS
2/20/80	PH	PROSTATITIS	BACTRIM-DS
5/30/80	PH	STREP THROAT	ERYTHROMYCIN
3/20/80	OC	OTITIS MEDIA	ERYTHROMYCIN
11/10/80	PH	SINUSITIS	ERYTHROMYCIN

Comments:

This information is confidential and may not be released to other parties without written patient authorization.

________________________________M.D.
DANIEL J. FORWARD M.D.

Figure 9. Sample entry continued.

ture of effort by the physician. The physician time and effort in establishing the computer data base is compensated by the unique capability of the computer to rapidly format, display and print information. Generation of medical reports for insurance companies and other requests, printing of admission notes and rewriting of prescriptions consume considerable time in daily practice. These tasks are quickly handled by the computer. Computer generation of chart entries improves legibility and organization of medical records and provides an easy method for physician communication with the computer through the medical transcription department. Repetitive typing of chart entries such as medication lists is eliminated since these lists can be updated for reprinting as a chart entry.

Discussion

Initial experience with a brief computerized medical record summary would suggest significant advantages to the physician in reduction of paper work as compensation for effort required in establishing a computer data base. The system also offers considerable potential for documentation of quality of care. Coding of diagnoses and medications essentially offers full medical record room function to the small clinic records department. It is anticipated that the computer data base will become the basis for documentation of quality of care in this small clinic setting. Retrieval of information from this base can document drug interaction or drug diagnosis incompatibility and can review patterns of drug use in controversial areas such as the use of controlled substances. In addition, problem areas in practice can be easily identified since assessment of each problem is included in the data base.

It is expected that computer display stations will be provided for each physician using the system as the data base grows. On-line access to a patient summary may provide an efficient system for correlation of clinical and laboratory findings. A physician reading an x-ray or electrocardiogram, for instance, could have instant access to a record summary indicating the presence or absence of valvular heart disease. Easy rapid correlations of this type are not possible in a purely paper based record system.

It is also anticipated that this data base will have considerable scientific value and will allow our clinic physicians to speak with greater confidence about their clinical experience.

26

A Comprehensive Patient Care System for the Family Practice*

David D. Schlager

Introduction

A primary goal of the health care system is to optimize the health of individuals by providing high quality care that is complete in scope and reasonable in cost. For this goal to be achieved, medical information systems must be modernized and brought to the same levels of efficiency as the advanced medical services to which they relate [1]. In the field of family and community medicine, the medical information system must have the ability to integrate data which are brief in nature and vital statistics that are traditionally unorganized. This system will have to demonstrate that the operating costs are comparable, if not less than, the current medical information system. And ultimately, this system must demonstrate an improvement in the quality of care provided.

Currently, the family physician is assuming a greater responsibility for the provision of comprehensive health care. The physician can no longer rely upon memory or a looseleaf patient record as the primary means of managing the large amounts of data being produced. Computerized medical information systems could provide a viable alternative to the conventional systems. The insurgence of the minicomputer and the microcomputer with on-line and interactive capabilities, has eliminated frequently discouraging situations, such as time-sharing and data entry in batch mode. Computers have been improved to store sufficient volumes of prose to satisfy the needs for episodic care and continuing care. The succss of computerized medical information systems lies in the ability of current users to demonstrate to other health care providers the cost benefits and the quality of care improvements realized through the use of these systems.

This chapter will provide information regarding the design and the function of a computerized medical information system currently used by this family practice medical group. The results of a cost per record study and the results

*Reprinted with permission from *Journal of Medical Systems* 7, Plenum Publishing Corp., © 1983; modified from a version previously published in *Proceedings of the Sixth Annual Symposium on Computer Applications in Medical Care,* IEEE, 1982.

of two studies which evaluate the system's effect on the quality of care will be presented.

Practice Demographics

Family and Community Health Associates (FACHA) has incorporated an in-house, on-line computer medical management system into a three physician, two office, family practice medicine group. These three physicians have practiced 25 years, 4 years and 1 year, respectively. The offices are located five miles apart and have access to the same data files by using dedicated telephone lines and modems. Two clinical pharmacists operate a licensed pharmacy and work in close association with the physicians. They utilize the same medical records and interact with the patient in the same physical setting.

The practice encounter profile and age/sex profile is presented in Table 1. The total practice experience parallels the findings in the Robert Wood Johnson Special Report of Medical Practice in the United States [2]. The differences between male and female physicians' age/sex profiles and encounter frequency related to practice experience is demonstrated by the data. The practice data was collected for a one year period from July 1, 1981 through June 30, 1982.

System Demographics

All financial, pharmaceutical, scheduling and medical data are stored in a Basic Four System 610 minicomputer. The system consists of seven video display terminals, two dot matrix printers and two letter quality printers. It has 150M bytes of hard disc storage and 224K bytes of RAM memory. The programs are written in Business Basic II. The video display terminals at the main site transmit data at a rate of 9600 bits/second and the video display terminals at the remote site transmit data at a rate of 1200 bits/second. Video display terminals are located in the laboratory, business office and data entry center in

Table 1. Encounter Profile and Age/Sex Profile

Phys.	# Enc.	% M/F	% >65	%65–45	%44–20	19–0
1[(a)]	8,226	48/52	21	38	30	11
2[(b)]	5,409	30/70	10	21	41	28
3[(c)]	3,792	41/59	10	19	45	26
Total	17,427	41/59	15	29	36	20
RWJ[(d)]	6,714	42/58	21	22	33	24

[(a)]Male physician, 25 years of practice experience.
[(b)]Female physician, 4 years of practice experience.
[(c)]Male physician, 1 year of practice experience.
[(d)]Robert Wood Johnson Foundation Special Report.

both offices. These locations were selected so as to maintain an isolation of the patient from the instrumentation. This arrangement of hardware has served to maintain the traditional physician-patient interaction.

The hardware is operated by entry level personnel previously not trained in the use of computers, but with proficient typing skills. The system is integrated and uses full duplex communication. The only restriction applied to the operators is that they may not use the same patient's records at more than one video display terminal. The system has been in use since September, 1980, and 29,000 patient encounters have been stored in the computer. The software is a reproduction of the manual system previously used by the physicians. The components of the software can be divided into five modules:

(1) medical module;
(2) financial module;
(3) pharmacy module;
(4) scheduling module; and
(5) codes maintenance module.

The software is currently maintained and improved by an in-house computer programmer. This relationship is an important part of the system since changes in practice style and standards can be integrated immediately into the software.

Data Input and Output Mechanisms

Patient encounter data are recorded on a standard SOAP formatted encounter form. Specific fields are designated for height, weight, blood pressure, pulse and temperature. Symptoms, diagnoses and laboratory data are coded using Social Security Administration codes, ICHPPC codes and practice developed laboratory codes respectively. Each encounter has over 1300 characters available for free text. Data relating to the patient's medical history are recorded using the following categories:

(1) prior medical history;
(2) prior drug history;
(3) allergy history;
(4) immunization history;
(5) social history; and
(6) family history.

The data updates are performed in a real time mode by a data entry person, so that the updated patient records are immediately available.

Prior to each patient encounter, the following medical reports are routinely generated.

(1) Patient Progress Notes
    a. Last five encounters
    b. Concurrent laboratory data and reports
(2) Chart Summary
    a. Diagnosis list
    b. Medication list
(3) Health Maintenance Report
(4) Patient History
(5) Physical Parameters Flow Chart

These reports form the medical record that the physician uses when seeing the patient.

System security and patient privacy are provided by shredding these medical reports after the data updates are performed. The encounter form is shredded after a complete system backup is performed as a precaution against the loss of patient encounter data. Original copies of resource material, such as discharge reports, referral letters and consultation letters are maintained in hard copy filing system after they are summarized and transcribed into the computer.

Cost

Cost containment is a stated goal of most systems, as it is for this computer system [3]. Such things as improved utilization of health manpower, the reduction of the duplication of medical data, the increase in patient record retrievability and the use of previously archived medical data to improve the quality of care can provide possible savings. This practice documented the cost of computer patient record maintenance and the cost of manual patient record maintenance to analyze the cost savings generated through the use of the computerized medical information system.

The initial cost of this computer system hardware was $80,000.00 and the software development cost $20,000.00. Annually, the computer system costs $29,000.00 to operate. This includes programmer's salary ($10,000.00), hardware depreciation ($7,500.00), hardware maintenance ($10,000.00) and paper and supplies ($1,500.00). The annual cost of the previous manual filing system was $7,200.00, which included filing equipment depreciation ($3,200.00) and paper and supplies ($4,000.00). Table 2 displays the patient record maintenance costs associated with an average thirty patient per day work load for this family practice medicine group that averages 16,500 patient encounters per year. Since the annual operating cost of the computer system is four times greater than the manual filing system ($29,000 ÷ $7,200 = 4), the maintenance cost per computer stored record was increased to reflect this difference.

The reduction in record retrieval time is due to the integration of the sched-

Table 2. The Patient Record Maintenance Cost for an Average 30 Patient Per Day Work Load

Manual Activity	Time	Cost
Retrieve 30 medical and financial records	Receptionist 40 min., $5.00/hour	$3.33
Pre-audit 30 medical records	Nurse-auditor 60 min., $8.00/hour	8.00
Data transcription including dictation, typing and chart summarizing	Physician 30 min., $100.00/hour. Nurse-auditor 60 min., $8.00/hour Receptionist 60 min., $5.00/hour	50.00 8.00 5.00
Refile 30 medical and financial records	Receptionist 30 min., $5.00/hour	2.50
Total		**76.83**[a]
Computer Activity	**Time**	**Cost**
Retrieve and collate 30 medical records	Receptionist 15 min., $5.00/hour	$1.25
Pre-audit medical records	Nurse-auditor 40 min., $8.00/hour	5.32
Data transcription	Receptionist 40 min., $5.00/hour	3.32
Total		**9.89**[b]

[a]Each patient record costs $2.56 to maintain.
[b]Each patient record costs $0.33 to maintain. Accounting for the difference in system operating cost, each patient record costs $1.32 to maintain.

uling and the medical modules which eliminates the need for a receptionist to retrieve medical records. The fifteen minutes associated with the automated retrieval of records results from the collating of the computer printed medical reports. A reduction in the time used to pre-audit the medical records is attributed to the records' clarity, organization, and, in the case of the health maintenance report, computer-assisted audit. The cost associated with data transcription is drastically reduced because the physician no longer dictates the patient encounter. The encounter form serves as the source of data used by the data entry person to update the patient records. The responsibilities of the nurse-auditor are eliminated because problem lists, medication lists, immunization histories and health maintenance records are updated by the data entry person following the patient visit. The data transcription time required to record the data is reduced because the encounter form is duplicated on the video display terminal, codes are used for symptoms, diagnoses and laboratory data and data is entered in a single entry mode. Costs associated with refiling patient records are eliminated because the computer automatically refiles patient records. The maintenance cost per record using the manual filing system was $2.56. The maintenance cost per record using the computer system is $1.32 per record. The record maintenance cost using the computer system is below the cost of $2.04 calculated by the users of the Regenstrief Medical Record system [4]. It is comparable to the computer-augmented record audit cost of $1.26 or 7% of the charge calculated by the Duke University Health Services Clinic [5].

Nonquantitated savings associated with traditional office procedures are also realized. Computer-generated receipts eliminate the process of completing insurance forms. Off-hour duties can be accomplished with fewer personnel. No longer is the patient record sequestered because it is being used for dictation, for health maintenance auditing, for insurance form processing or for use during the next office period. Individuals can retrieve the part or parts of the medical record which is relevant to their task. And finally, this practice was able to expand from a one location, one physician, one pharmacist practice to a two location, three physician, two pharmacist practice while continuing to employ only four nurses and four receptionists.

Quality of Care

Tangible benefits, such as improved manpower utlization and increased productivity do not guarantee that the quality of care will improve [6]. This practice studied this computer system's effect on the quality of care using two different evaluations.

Computer-Assisted Medical Record Audit

The patient record provides the basic data upon which decisions regarding preventive medicine and health risk management are made. Through the use of a computerized medical information system, the encounter data can be used to create information summaries which can be used for more efficient and accurate medical record audit. A pre-audit of the five medical reports is routinely performed. Remarks relating to the patient's health maintenance needs are made on the encounter form.

An evaluation of the effects on the quality of care of the use of these computer-generated reports in the auditing process was performed. A group of one hundred patients was analyzed for compliance to health maintenance suggestions. The practice was divided into two groups of patients. The first group was

Table 3. Percentage of 50 Patients Documented as Having a Current Health Maintenance Status in These Categories

	Immun.	BP	Wt.	Hx	Hclt.	Phys. Ex
Group I:	32	88	88	14	16	38
Group II:	56	100	100	90	36	76

composed of patients who were seen at least five times during a two-year period from July, 1972 through June, 1974. The second group of patients was selected from a group of individuals who were seen at least five times during an eighteen month period from September, 1980 through March, 1982. The former group was treated during a period prior to the installation of the computerized medical information system. The patient record during this period consisted of a narrative of the examination findings presented in chronological order. The latter group was treated following the introduction of the computerized medical information system. The areas of investigation for this study included:

(1) immunization updates;
(2) blood pressure recordings;
(3) weight measurements;
(4) historical updates;
(5) hemoccult testing of stool specimens; and
(6) routine health maintenance physical examinations.

Table 3 represents the comparison of the two test groups. A marked improvement is noted in the percentage of patients who accepted immunizations, hemoccult testing and health maintenance physical examinations. The increase in the updated histories is exceptional and can be attributed to a system which enabled data to be collected in an organized fashion.

A revenue analysis shows that gross revenue was increased by $12,443.00 by using a computerized medical information system. These results are presented in Table 4. The expenses for this practice are seventy-five percent of the total gross revenue. The profit realized through the use of this computer system was $3,111.00.

The simplicity of a single entry system that can produce a multiplicity of medical reports which can be used by the entire health team has increased the quality of care in this practice. The medical record audit has also increased the gross revenue of this practice.

Table 4. Increase in Gross Revenue Generated from the Use of Computerized Medical Information System in the Auditing Process

Immun.	BP	Wt.	Hx	Hclt.	Phys. Ex.
$4,411.00	0	0	0	$757.00	$7,275.00

Computer-Generated Flow Chart

Another study that evaluated the use and the benefits of a computer-generated flow chart as a tool for quality of care review and as a tool for evaluating system effectiveness was performed. All patients scheduled for an appointment have a graphical flow chart printed prior to their visit, displaying weight, systolic blood pressure, diastolic blood pressure, pulse, blood sugar and urine sugar. The date and value of each data point are noted on the graph. This report is presented by the physician to the patient during the encounter. Progress or regression are clearly demonstrated. This flow chart is given to the patients for their personal records. An evaluation of the use and the benefits of flow charts relating to the management of chronic hypertension was performed. The study period encompassed a period of ten weeks from May through July, 1982.

Flow charts were produced for each visit for only one of the three physicians in the group. The patients cared for by the remaining two physician had no flow charts produced. At the conclusion of the study period, the computer was programmed to search for all patients coded with ICHPPC code designation 401 (hypertension) during the study period. The computer grouped the patients by examining physician. All patients who were not seen at least twice during the study period were excluded from the study. The diastolic blood pressure value associated with the first visit during the study period was compared to the diastolic blood pressure value at the conclusion of the study period. A diastolic blood pressure value of 90mm of mercury or below was considered to be normotensive. Four categories of results were developed;

(1) values that became normotensive;
(2) values that became hypertensive;
(3) values that remained hypertensive; and
(4) values that remained normotensive.

The Student's Distribution for three variates was used to test the significance of the results [7].

Table 5 presents the encounters for hypertension expressed as a percentage

Table 5. Percentage of Hypertensive Encounters Sex Distribution and Median Age

Phys.	# Cases	% Enc.	% M/F	Median Age
1[(a)]	129	24.8	47/53	56
2[(b)]	56	12.9	16/84	55
3[(c)]	20	8.2	50/50	47
Total	**205**	**17.2**	**39/61**	**55**
RWJ[(d)]	—	5.1	41/59	58

[(a)]Physician used flow charts.
[(b)]Female physician.
[(c)]Physician with one year of practice experience.
[(d)]Robert Wood Johnson Foundation Special Report.

Table 6. Change in the Diastolic Blood Pressure Value

Phys.[a]	# Cases	Became Normotensive	Became Hypertensive	Stayed Hypertensive	Stayed Normotensive
1	129	16 (73%)	5 (23%)	1 (5%)	107
2	56	16 (59%)	7 (26%)	4 (15%)	29
3	20	3 (38%)	0 (0%)	5 (63%)	12
Total (2&3)	**76**	**19 (54%)**	**7 (20%)**	**9 (26%)**	**41**

[a]See Table 5 for physician information.

of the total encounters for the practice. It also presents the sex distribution and median age for the cases examined during the study period. The table presents the total number of patients seen at least twice during the study. Comparable data from the Robert Wood Johnson Foundation Special Report on Medical Practice in the United States is also presented [8]. The total percentage of hypertensive encounters for the practice for the period July 1, 1981 through June 30, 1982, was 13.9%.

Table 6 presents the change in diastolic blood pressure values between the initial visit and the visit at the conclusion of the study period. It was assumed that the maintenance of a normal diastolic blood pressure value could not be attributed to the use of flow charts, since previous methods had achieved the normal state. Therefore, this group was not included in the analysis of the effect of the flow charting process. As demonstrated in Table 6, 73% of the patients cared for by physician 1 became normotensive and comparable results for physicians 2 and 3 were 59% and 38% respectively. These results did not reveal a statistically significant improvement in the achievement of a normotensive state.

The analysis of the data indicates that computer-generated flow charts employed in the management of chronic hypertension did not improve the quality of care when judged by the lowering of the diastolic blood pressure. Jonathan Rodnick, M.D., also observed that quality of care improvements from automated medical record systems are limited and that the relationship between the quality of care and the recording of data has not been proven [9]. The use of a computerized medical record management system does however offer important advantages in the process of monitoring care. Patient acceptance of physical parameter flow charting was mostly supportive. Many patients were provided with a better understanding of the control of their disease, hypertension. The physician who used flow charts was given a quick and an accurate overview of the patient's progress toward control.

Summary

The results presented indicate some of the unique benefits realized following the installation of this computerized medical information system. The financial, medical and quality of care reviews presented, detail the applications of this system. They address one of the major shortcomings, stated by Ingeborg Kuhn in her review of patient medical record systems, that is, the lack of formal evaluation of the effects of these systems on the quality of clinical practice [10]. These studies represent an attempt at the design and completion of prospective and retrospective studies relating to the use and management of medical record data. The advancement of family medicine into a more traditional research study mode was suggested by the Family Practice Residency at Madigan Army Medical Center in a 1978 research paper [11].

Today's health care system, its providers and users, are demanding more data, better documentation and more flexibility from the medical record. Family physicians will be concerned with the task of maintaining these records at a reasonable cost and demonstrating the improvements in quality of care. The computerized medical information system with its reduction in maintenance cost and its contributions to the improvement of the quality of care is a viable alternative to the current medical information systems.

References

[1] Levinson, D., "Information Management in Clinical Practice," *J. Fam. Prac.* **7** (No. 4), October, 1978, p 799.

[2] Medical practice in the United States. Special Report Robert Wood Foundation, Division of Research in Medical Education. University of Southern California School of Medicine, 1981, pp. 11–19.

[3] Rodnick, J. E., "The use of automated ambulatory medical records," *J. Fam. Prac.* **5** (No. 2), August, 1978, p. 255.

[4] McDonald, C. J., *et al.*, "A Computer-based Record and Clinical Monitoring System for Ambulatory Care," *Am. J. Pub. Health* **67** (No. 3), March, 1977, p. 245.

[5] Ahlers, P., *et al. al.*, "The Cost of Auditing Outpatient Records," *South Med. J.* **69** (No. 10), October, 1976, pp. 1329–1330.

[6] Rodnick, J. E., "The Use of Automated Ambulatory Medical Records," *J. Fam. Prac.* **5** (No. 2), August, 1978, p. 255.

[7] Alder, H. C., Roessler, E. B. *Introduction to Probability and Statistics,* Sixth ed. San Francisco: W. H. Freeman and Co., 1977, pp. 171–185.

[8] Medical practice in the United States. Special Report Robert Wood Johnson Foundation, Division of Research in Medical Education. University of Southern California School of Medicine, 1981, p. 61.

[9] Rodnick, J. E. "The use of Automated Ambulatory Medical Records," *J. Fam. Prac.* **5** (No. 2), August, 1978, p. 264.

[10] Kuhn, I. M., "Looking at Outpatient Medical Record Systems," *Hospitals* **51** (No. 20), October, 16, 1977, p. 152.

[11] Hollison, Jr., R. V., *et al.*, "A Medical Information System for Ambulatory Care, Research, and Curriculum in an Army Family Practice Residency: 51,113 Patient Problems," *J. Fam. Prac.* **7** (No. 4), October, 1978, p. 794.

27

Medical Computing, Realities for the Private Practitioner*

David M. Brickman, M.D., John M. Ashton, M.D., and Jeff Balsam

I will start by telling you this chapter is intended for solo practitioners or those of you in groups of up to 8–10 doctors. This is my area of expertise and experience. It is my desire to give you some basic tools with which to analyze, collate and, in some cases eliminate what you hear in these SCAMC sessions.

My business associates and I have been involved in medical office computing for over 5 years now. We started by researching and analyzing what was available for the private practitioner. Simply put, we found there were essentially no systems at a reasonable cost available to meet the needs of the vast majority of physicians. This was a clear result of the cost factors involved.

Despite our desires to find more efficient ways to maintain and manage the vast amount of patient data we collect, it just is not cost-effective for the average practitioner to spend the $35,000 to $100,000 or more necessary to purchase a computer system.

What we did then was to design a complete system to meet these needs using a relatively low cost microcomputer. We now have over 150 installations nationwide. We continue to work closely with our many clients and we typically talk with approximately 1,000 physicians a year at medical conferences such as this.

It is from this practical base of experience that this talk has evolved. This session is for those of you who have come to this meeting seeking BASIC knowledge about how computers can be applied to your own practices.

While I don't pretend to have all the answers, I do have extensive real world experience in medical computing with private practitioners.

Many of you are looking for *practical* approaches and *specific* applications to help improve the quality of patient care and ease the burdens which assualt you every day as you practice medicine.

That brings us to the question of what applications are practical. In my experience there are six primary applications. These are:

(1) Patient billing;
(2) Clinical data and data management;
(3) Communications with national data bases;
(4) Word processing;
(5) Appointment scheduling;
(6) General accounting.

Before we discuss each of these applications in detail, let us first consider costs.

My experience after many meetings suggests that most physicians would like to spend $10,000 on the purchase of their computer system.

Let me tell you what $10,000 will buy today: a good hardware system including a small, fixed (or hard) winchester disk plus a good quality printer. This is the *minimum* system to be recommended for all but the smallest practices.

This doesn't even consider the software. Unfortunately, what few people really appreciate, is that the computer is just a fancy box without the software . . . and good reliable software is darned difficult to find.

There is a general misconception both in and out of the computer industry, especially in what is called the low end—that is systems priced in the range we are talking about—that software is cheap. This misconception came about because personnel computer software which performs single functions is quite inexpensive. For instance game software can be purchased for about $30.

But medicine is not a game. We are talking about very sophisticated software that is debugged, reliable, well documented, well supported and easy to use. We are talking about installation, training and total on-site hardware and software support . . . because you can not afford to have your medical practice shut down for a couple of weeks when you have a computer problem. As with any mechanical device computers are going to need service from time to time. There are also going to be times when your office staff will have procedural questions. For these reasons it is absolutely essential that you deal with a knowledgeable, reliable and *responsive* supplier. You need to have somebody who is interested and supportive when you need help.

So, even though $10,000 is certainly the ideal price-point for physicians who want to purchase their own systems, it really is not enough money to provide the quality of hardware, software and initial support necessary for a medical practice. What is possible, however, is a price around $15,000. For this amount, if you are careful in you selection, you can purchase an excellent system.

With this $15,000 base figure in mind let us look at specific applications. Although most of the lectures in medical meetings revolve around medical data systems, I know for a fact that medical data comes as a very distant second for private practitioners as a group. What comes first? Quite simple . . . medical billing. We get more than 50 medical billing requests for each medical data request. This is a natural and logical way to begin medical computing. The cost justification for the computer is clearly evident in the paperwork bound billing department which is still using manual methods that have not changed significantly in more than two decades. There also are growing demands from government and other areas which are increasing the paperwork.

What the computer can do for your billing department is nothing short of revolutionary. Using a good medical billing system in-house is on the order of 5 to 20 times faster than antiquated manual methods. The time savings and reduction of human error alone will typically pay the entire cost of the system in 3 to 12 months for most practices.

By taking only a minute to register the patient and less than a minute to add charges and post payments, the computer will automatically do everything else related to patient billing. This includes maintaining patient ledgers, printing insurance forms and patient statements, giving you aged accounts receivables, practice management reports, a daily transaction journal and even a check register and deposit slip. So patient billing is clearly the most practical initial application; and despite everything you are likely to hear, I can tell you that most of us in private practice want a billing system first.

Of course many physicians who are primarily interested in billing also mention clinical data as a secondary interest. And there is really no reason you can not have both. In most cases the same system that is used for billing can also be used for clinical data. In fact several of the things we want most are already in the billing system and may be easily retrieved.

For instance, the most common request from physicians is to get a listing of all of their patients with a specific diagnosis or set of diagnoses. The next most frequent request is to locate all patients upon whom we performed a specific procedure. Frequently we want to locate patients with a combination of diagnosis and procedure. Retrieving this information is easy because it has already been entered into the computer for billing purposes as well as other information such as age, sex, zip codes, referring provider, etc.

Another practical clinical application is recalling patient by medication as in the case of a drug recall. Certainly this is an easy program for the computer. It does require, however, additional and somewhat time consuming data entry on the part of the office staff. It is, nevertheless, a realistic option for your in-house system and may be implemented at any time you desire.

As for patient history and physicals, the computer certainly is able to ask the questions and store the answers for you. The negative, however, is that this is very memory intensive requiring a tremendous amount of disk space if you wish to keep it all on-line. The same is true for patient charting. So, if you wish your computer to maintain all of this information on your computer you must be willing to spend additional money to buy the necessary storage as well as paying someone to *enter* all of this extra data into the system.

I think what *is* practical is to keep specific types of data on patients such as diagnosis, medications, allergies, operations and pertinent lab and clinical data. By doing this you cut down the amount of time spent inputting the data to a manageable amount, tying someone up for a few hours a week rather than full time. In the final analysis, you must decide for yourself what information is pertinent and worth the time and expense to enter into the computer.

Other Applications

One of the most mature applications available for the in-house system is word processing. Using a word processor your staff can save amazing amounts of time each week if you write many reports, consultations or other correspondence. Since reports are largely made up of standard paragraphs a word processor will allow you to store these paragraphs and quickly put together an almost unlimited number of customized letters in a fraction of the time it takes to do it manually.

Yet another existing application is communications with national data bases. This is accomplished quite simply using an inexpensive device called a modem which connects your computer via telephone to large computer systems throughout the country. With this devise you can take Category I CME Courses in your office, access medical literature, communicate to other offices, the stock market and just about anything else you can imagine. In fact the AMA in conjunction with GTE is developing a national medical information data base.

The remaining most requested application is general accounting. Your office system is capable of cutting your accounting bills dramatically by using accounting software to perform many of the tasks you now pay your accountant to do. Of course your accountant still will have to look over and reconcile the data, but this is less expensive than having him (or her) do *all* of the work for you. Accounting Systems typically include General Ledger, Accounts Payable, Payroll as well Inventory for those dispensing medications. The accounts payable and payroll will even write the checks for you.

Appointment scheduling is also a handy application for your in-house system. It can locate open times, print out daily appointmnt sheets for each doctor and/or room as well as sending reminder notices in advance to your patients.

Many other applications are also available for your in-house system including financial spreadsheets, statistical programs, property management and a host of other useful tools.

So when should you purchase your system?

While its true that the industry is changing rapidly and prices trend downward, it is also true that any any point in time there will always be exciting future developments.

I recommend getting in as soon as you find a system that reasonably fulfills your *primary* computing goal. Whether that primary goal is billing, word processing, patient data, national data links, accounting or whatever, check out what's available and make your choice. In all likelihood any future reduction in price will be offset by additional revenues the system would have generated by savings in staff time, better practice management and better patient care.

Be reasonable—don't expect customized software unless you are willing to pay the price and spend a great deal of time necessary to create custom programs. In fact, unless you have extremely unusual needs custom programs are not necessary and are definitely not recommended.

Don't go in with the mentality that if you've seen one system you've seen them all. The quality of software varies incredibly and there is no reason for you to settle for anything less than first-quality. Remember you are going to be putting important practice records on that computer.

So how do you choose the right computer with the proper software from a quality supplier? How do you know that the system will do all the necessary tasks that you want and need? How do you know that it will be easy enough for your office staff to operate? Remember you do not want to hire a computer programmer to operate the system.

To help answer these questions to your satisfaction I have prepared a list of questions to ask any vendor you are considering. This will help you to qualify them. By far the most important of these questions is will they provide a list of 15–20 names of current users. If they have a satisfied users base that you can speak with personnally, they will be more than happy to give you their names. Otherwise forget them.

Here is the list of questions:

(1) Will they provide names of current users? (If not forget about them. Be sure to check out the referrals. They will tell you a lot more than the salesman.)
(2) How easy-to-use is the system?
(3) Will they train your office staff as much as necessary and is this training included in the cost of the system?
(4) How many systems are currently in the field?
(5) How long has the company been in business?
(6) How much training will be necessary before the system may ben operated by your staff?
(7) Can *you* understand what the salesperson is showing you? (If not, the office staff will probably have even more trouble.)
(8) Can you enter your own data into the system and see the reports, or are you only seeing a pre-arranged demonstration?
(9) Will they provide on-site maintenance and what is the cost? Without on-site maintenance, don't purchase the system.
(10) How are software updates handled? Every system needs periodic updates.
(11) Do they offer one-stop service or will you have to deal with two, three or more vendors to maintain the computer? With more than one vendor, it is possible no one will accept responsibility for the problem.
(12) How thorough and well written is the user's manual?
(13) What enhancements are currently in the works?
(14) What are each of the reports the computer supplies?
(15) How are you protected in case of power failure or system malfunction?

(16) Are hard copy print-outs always readily available for all transactions and other valuable information?
(17) Who makes the hardware and how reliable is it?
(18) How many patients will the system handle with complete ledger detail?
(19) How many doctors will the system handle?
(20) How many transactions will the system handle daily?
(21) How many patients may be maintained with complete ledger detail for a year?
(22) Does the system maintain a permanent record of all patients or are they removed every time their balance is 0?
(23) Does the system support the CP/M operating system? (There is more software available for this operating system than any other.)
(24) Is it a hard disk or floppy system? A floppy system is definitely not recommended except for exceptionally small practices.
(25) Can the system be easily and cost-effectively expanded and/or upgraded?

IV. OTHER APPLICATION AREAS

The previous two parts focused on information systems in hospital and ambulatory care settings. The chapters in this section provide different perspectives for automated applications. The first chapter is by Collen and contains an introduction to automated multiphasic health testing (AMHT). While the AMHT concept was one of the first computer applications to be articulated, it failed to achieve the early success of other experiments. To a large part, this was a function of its early start and the associated equipment costs. With new generations of inexpensive and powerful computers, one can be reasonably certain that AMHT will be re-examined with a renewed interest.

Another area that received early attention was the use of computers in mental health. The chapter by Greist and his colleagues contains a historical overview and then goes on to examine why there have been delays in the development and implementation of systems. The basic functions of a medical health system—patient interviewing, clinical consultation, etc.—are reviewed, and changes to facilitate their use are considered.

The next chapter, by Zielstorff, presents a survey of the automated tools that impact nursing administration. It should be noted that, in most institutions, nurses are the primary users of the HIS. Indeed, a reduction in hours devoted to nonnursing tasks is a common justification for the installation of a HIS. The catalog of functions presented in this chapter is largely derived as a byproduct of clinically oriented systems. That is, data gathered by clinical systems can be used for administrative purposes (*cf.*, chapter 18 on the Core Record System). Unfortunately, the converse is not true; systems designed for financial or administrative functions will have no positive impact on the delivery of care.

The next two chapters address the use of computers in a research environment. The chapter by Lenhard and his associates describes a system for cancer patients. The system provides both HIS and ambulatory care support. It combines and retains all clinical data in a single, integrated data base. The focus is patient care in a terciary care setting. The existance of a complete and comprehensive data base is also an important resource for research.

The chapter by King and his colleagues is concerned only with the analysis of clinical data. With the growth in power of the small computer, it is now possible to use an inexpensive system for the manipulation and analysis of research data. In this chapter the functions which such a system must provide are described, and a survey of eleven systems is presented.

The final chapter in this section, by Maturi and DuBois, contains a survey of systems used by hospital ancillary functions. It will be remembered that some of the first successful applications were implemented for the ancillary functions. In the context of the current interest in systems networking (*c.f.*, Chapter 9 describing the UCSF hospital system), a renewed interest in stand-alone capabilities can be expected.

In concluding this review of information systems in patient care, we note that many systems have a narrow set of objectives. The Oncology Clinical Information System described in Chapter 31, for example, has a limited potential set of users. But the tools it provides overlap with those described in the HIS and Ambulatory Care sections as well as with chapter on research data bases. The conclusion is, therefore, that as we learn to build integrated systems, we will find that functions transfer across organizational boundaries. Furthermore, as we demonstrate a capability in one area, we should be able to transfer it to other applications. And when we do this, medical informatics will become a true discipline.

28

Automated Multiphasic Health Testing*

Morris F. Collen, B.E.E., M.D.

Introduction

Automatic multiphasic health testing (AMHT) is a systemized process for efficiently providing health examinations to large numbers of persons. The concept and practice of periodic health examinations has been generally recommended for decades. Attempts to decrease the costs and improve the quality of these examinations has resulted in the historical evolution of screening, multiphasic testing and automated multiphasic health testing.

Screening persons for a disease (or a condition) refers to the performance of a test that is usually simple and quick but has sufficient sensitivity and specificity to detect a disease, if present.

Multiphasic screening combines several individual screening tests into an integrated test battery to screen for a number of diseases.

Multiphasic testing has evolved from multiphasic screening by the employment of modern measurement technology; it uses semi-automated and automated instrumentation to provide accurate quantitative test measurements. Whereas "screening" sorts persons into two groups wherein one is "positive" and the other is "negative", "testing" sorts persons into three groups, "positive", "ngative" and an uncertain group usually called "borderline".

Automated multiphasic health testing is the expanded systems concept of utilizing automated equipment and a computer to determine automatically whether there is sufficient likelihood of disease being present to warrant further specific diagnostic testing. Such programs employ computerized decision (advice) rules that automatically sort out those patients whose test results exceed prescribed limits. The mere employment of automated equipment or a computer in a multiphasic testing program is not sufficient to satisfy the concept of "automated" multiphasic testing. AMHT is a total systems approach

*This chapter was prepared especially for this collection.

in which history and test data are entered online into a computer, which, by using programmed decision rules, automatically prints out advice in realtime (that is, before the patient leaves the multiphasic laboratory) as to appropriate followup tests and referrals to medical care and health maintenance.

In considering an AMHT service or program, it is very important to clearly distinguish between the systemized multiphasic process for providing the test procedures from the outcome effects upon patients from the selected battery of tests. The consideration of specific tests for health maintenance and disease detection is not a consideration of this paper. The criteria for the selection of cost-effective tests has been previously reported; and, if it is decided to provide periodic health examinations to large numbers of persons, then the AMHT approach is the most efficient [1,2]. In this chapter, only the computer supported AMHT process will be considered.

AMHT Functional Requirements

If it is decided by a medical care program, by an industry or by a government agency to provide health examinations to many persons, then the goal of the AMHT information system will be to satisfy all of the information processing requirements necessary for the specific health examinations provided. The AMHT system can then be considered as a clinical departmental information system having broad requirements for processing patient identification data, patient history information, test procedure results including clinical laboratory, x-rays and ECG's, physical examination findings, clinical diagnoses, and appointment and referral arrangements [3–5].

To implement successfully an AMHT information system, one must first define its functional and technical requirements, select competent personnel, obtain funding, acquire an appropriate hardware configuration; develop or acquire necessary software, install the system, orient and train the users, and finally operate, evaluate, revise and improve the AMHT.

Although these functional requirements will vary from one AMHT to another, they will all usually include the following:

(1) *Patient Identification.* Uniform patient identification will be required to provide positive identification of the patient at the time of registration.

(2) *AMHT Data Base.* It will be necessary to decide what patient data are to be stored and for how long. Usually, clinicians will want to collect and store all AMHT data for each individual patient over the continuing period of patient care by the AMHT program, which may involve many periodic AMHT examinations over many years.

Data will need to be collated and processed on all patients' specimens collected in AMHT or elsewhere.

(3) *Computer Medical Records.* AMHT computer printout of the patient

medical record will require data retrieval of an overall integratable computer data base so that the final report can present the results of the current health examination compared to the results from prior health examinations. The final report should have flagged abnormalities, individualized patient normal values and a summary of high priority information.

(4) *Data Communication.* To functioning support systems, such as the clinical laboratory, radiology, ECG, etc., AMHT will need a results reporting application to permit the physicians to order specific AMHT tests for their patients and obtain a printout of the results incorporated in the AMHT final report. AMHT scheduling will need to be sufficiently flexible to permit selective batteries of tests in accordance with a patient's age, sex and prior AMHT results.

(5) *Special Data Sets and Knowledge Bases.* Inverted files for important diagnoses, and patient registries will probably be desired; and the use of patient data to support clinical research may be a requirement. It will be increasingly desired that an AMHT system be established with or interfaced to computer knowledge bases to support clinical decision making for diagnosis and followup for detected abnormalities. Normal limits for AMHT test procedures will need to be established by age, sex, time of day, etc.

(6) *Terminals.* It will be necessary to consider carefully the kind, number and location of data acquisition terminals for AMHT, since this will significantly effect patient and technician user satisfaction. An acceptable data entry terminal and response time will vary with the application. Technicians and clerks will accept key board terminals but patients providing history information will require visual display terminals.

(7) *Support Functions.* Additional applications of the AMHT computer, such as a patient appointment system, may be requested, or a patient "tracking" system for followup and recall of patients with specific positive tests. Administration will request periodic usage, statistical and financial reports.

(8) *Reliability and Security.* Functional requirements for AMHT will include good system's reliability and backup. Security and confidentiality of patient data must be carefully considered in order to meet legal and ethical requirements as for any other patient record system.

(9) *Flexibility.* Since AMHT is a very dynamic system and tests are constantly being changed, an ideal requirement for AMHT is to allow change and growth. AMHT must allow both sequential testing with all tests provided and selective testing for small test groups. AMHT must satisfy all users: patients, physicians, operational personnel, nurses, technicians and clerks and administration.

(10) *Cost.* In the end, affordability and cost are usually the factors limiting what one will ask the AMHT to do.

AMHT Technical Requirements

To satisfy the functional requirements specified for AMHT, it will be necessary to meet with appropriate technical experts to consider how the alternative technologies currently available can accomplish what AMHT is to do. The technical requirements which will affect hardware and software specifications will include:

(1) *Patient Identification.* Accurate patient identification will usually require patient's name, sex, birthdate, and encoding by a unique number assigned by the AMHT which will serve as the basic linkage for patient data.

(2) *AMHT Data Base.* A data base must be provided of sufficient size to accept all AMHT data on all patients. The use of standard terms by all clinician users if they provide physical examination data, diagnoses and procedures will require the establishment of a data base dictionary defining acceptable data elements and a thesaurus of acceptable synonyms. Agreements will need to be developed as to the handling of natural language-free text entrees by technicians and other users.

(3) *AMHT Computer Medical Record.* To satisfy the functional requirements for handling a variety of forms of medical information for many AMHT examinations through time, it will be necessary to plan for a variable length, variable format, computer-stored medical record. The AMHT medical data base should be designed to be data independent; and ideally permit retrieval of all or any specified portion of any individual patient's record in a time-oriented sequence, or in chronological order by patient visits, or in a patient problem-oriented sequence as the physician-user may desire.

   The computer should have sufficient speed and memory to perform all required tasks with an acceptable response time to its users for online retrieval of data for any patient under active care by the department. Access to old data may be a problem when inactive patients again become active if the clinicians wish new data to be compared with old data in a computer integrated patient record.

(4) *System Configuration.* Improvements in new computers are being introduced continually, but at some time a decision will need to be made as to the system hardware configuration which hopefully will satisfy the functional requirements of the AMHT for perhaps 5 to 7 years. Generally, a small mini- or one or more microcomputers will be adopted for an AMHT. Current microcomputers have sufficient power to adequately handle some management applications (such as patient appointment or tracking systems) in addition to the AMHT primary patient data base.

   All programs should be written in a language compatible with other medical information systems with which the AMHT interfaces. If multiple applications exist communicating with multiple terminals, a sophisticated supervisor program will be required, which will impact the size of the AMHT computer. Documentation of all programs will be necessary to facilitate maintenance programming and program modifications. Programs must assure data integrity by decision rules for entry correction or removal of data. Data quality control programs are essential, especially for editing of incoming data or validity at the point of data entry, such as the automatically checking patient identifying data while the patient is still present. Error-checking programs should include checks for missing data, check digits for patient identification numbers, "validity" limits to check reasonableness of data when appropriate for numerical data, etc. To improve data quality, all data should be entered directly from source and avoid intermediary personnel. Online verification of data entered by the one who generated the data decreases errors in data entry.

(5) *Terminals.* Ergonomics, or human factors considerations, will determine the terminals best suited for the various specific users of AMHT. This man–machine interface may be the perception of the system to some users. Patients are not comfortable with a keyboard, but function well with a visual display of multiple choice "menu" formats. Technicians can readily adapt to special function keys for data entry; and clerks readily use the typewriter-type keyboard.

(6) *Reliability.* An AMHT requires a high degree of reliability (dependability) if it is processing in an online, realtime mode. In such cases, it will be necessary to have a proper mix of backup equipment. Fortunately for a AMHT using several microprocessors, it is often cost-effective to have an additional microprocessor available for backup. A backup procedure as a minimum should permit a degraded mode of operation until the system is restored. This will require collecting and storing data in a manner which will permit ready entry into the computer data base when the system again becomes operational. It is advisable to have more than one terminal at each user station to assure that one terminal will be operational if the other fails. Terminals should preferably be plug-in movable modules so they can be replaced quickly by a stand-by terminal.

All patient data within the AMHT must be subject to the same regulations of security, privacy and confidentiality as is the data in a hospital record. Programmers with access to the patient data base should be required to acknowledge the confidential nature of patient data. Research data files should require identification of the principal investigator prior to access. Research data should maintain confidentiality for individual patients.

(7) *Orientation and Training.* Representatives of all AMHT users (physicians, administrators, nurses, technicians) should participate in the planning for their orientation and training. When there is considerable user turnover, a continual orientation program may be necessary.

Personnel Requirements

For a Vendor Provided Turnkey System

If an "off-the-shelf", "turnkey" AMHT is available in the market which satisfies the functional requirements of the department, this is a fortunate and desirable solution. A computer scientist/engineer, experienced in computer hardware and software, and with a background in the life sciences will be able to serve as data processing manager and interface between the vendor and the users and between the users and the system. If maintenance on the system and program modifications or upgrades are provided by the vendor, this simplifies the manager's responsibilities; if not, a considerable technical expertise with the system will be required. An information scientist/data base manager will be necessary to monitor and control the uniformity of data terms being entered into the data base. For example, any user desiring to enter a new diagnostic or procedural term should be required by the data base manager to check the data base dictionary to see if the term is significantly different from, or is a synonym for, an already existing term.

For the Developmental AMHT

If it is decided to develop one's own "in-house" AMHT, it will always be a major project. The planning, programming and implementing of any AMHT will require a team with considerable competence and expertise, usually including a data processing supervisor or project chief, and information scientist/data base manager, a medical systems analyst, a systems programmer responsible for the operating system and the data base management system, and one or more applications programmers depending upon the number of computer applications and AMHT size. If hardware maintenance is not provided by the vendors, then a computer hardware engineer will be needed. Experience has shown that regardless of how thorough was the initial systems analysis of AMHT, deficiencies will become apparent after its implementation, and substantial modifications will be requested. Accordingly, a continuing staff for "maintenance" programming will be necessary. Experience also indicates that it will be prudent to budget for 25–50% more time and money than originally requested.

After completion of site preparation, installation, testing, acceptance and achieving an operational capability, it will be advisable for qualified personnel to periodically evaluate, revise and improve the AMHT.

References

[1] Collen, M. F., Ed., *Multiphasic Health Testing Services.* New York: John Wiley & Sons, 1978.

[2] Collen, M. F., Garfield, S. R., Richart, R. H., *et al.*, "Cost Analyses of Alternative Health Examination Modes," *Arch. Intern. Med.* **137,** 1977, pp. 73–79.

[3] Blum, B. I. and Lenhard, R. E.; "Experience in Implementing a Clinical Information System," in Lindberg, D. A. B. and Kaihara, S., Eds., *MEDINFO 80 Proceedings,* Amsterdam: North-Holland, 1980.

[4] Collen, M. F., Ed., *Hospital Computer Systems.* New York: John Wiley & Sons, 1974.

[5] Davis, L. S., "Data Processing for MHTS," chapter 8 in [4].

29

Clinical Computer Applications in Mental Health*†

John H. Greist, M.D., Marjorie H. Klein, Ph.D., Harold P. Erdman, Ph.D., and James W. Jefferson, M.D.

Historical Overview

In 1967, Howard Rome [1] predicted:

> Psychiatry is now on the threshold of a fourth quantum advance. Automation in information processing will achieve what never has been available heretofore—a valid data-base for psychiatry's assumptions, treatments, logistics, and at the same time if offers a potential solution for psychiatry's administrative complexities.

Because early programs commonly were time-consuming to prepare and often operated in batch mode, clinical computing mostly emphasized computing—often at the expense of clinical realities requiring more flexibility than computers and those programming them could provide. Data processing departments sprang up and their territory seemed foreign and forbidding to most clinicians. Programming staffs naturally cast their allegiance with whichever group used their products: administrators, managers, researchers, and only rarely clinicians.

The development and spread of on-line computing and better programming languages had made innovative application of computers in clinical settings possible, so that the boundary lines between clinical and management, administrative, educational and research computer applications are constantly being blurred, bent and at risk of being broken. Data collected directly from patients by computer interview can serve administrative, research and proper clinical functions ("pertaining to a clinic or bed side . . . or founded on actual observation and treatment of patients" [2]). Service utilization, educational expe-

*This work was supported in part by Dr. Greist's NIMH Research Scientist Development Award (MH70903-08) and by NIMH Grants MH16477 and MH16464-02 and MH32624-02 and by a grant from The National Library of Medicine LM03713-02.
†Reprinted with permission from *Journal of Medical Systems* 7, Plenum Publishing Corp., © 1983; modified from a version previously published in *Proceedings of the Sixth Annual Symposium on Computer Applications in Medical Care,* IEEE, 1982.

rience, billing, accounts receivable, planning, identification of subjects suitable for particular experiments and many other functions may all be provided by the same data base. Actual measurement of patient change may satisfy goals of treatment selection and modification, accountability and clinical research. The artificial barriers between administrative, research, educational and clinical functions are beginning to disappear in settings where workers are learning to use sophisticated patient interviewing and data base management systems. Programs for consultation, scheduling, word processing and bibliographic work are also being integrated, evaluated and implemented in clinical settings.

Clinical Computing

Computer Interviewing

Even before Rome's prediction, Slack and his colleagues had programmed a computer to conduct direct patient interviews [3]. Slack clearly recognized the revolutionary prospects of patient-computer interviewing and the potential effects of computer applications on physicians [4]. His group also anticipated questions about nonverbal communication [5] and struggled with ways to facilitate oral expression about emotional problems [6].

Many of the fundamental issues surrounding direct patient interviewing by computer have been clear for several years: patient acceptance of the technique [7] and patient characteristics that interfere with its use [8]; reliability and validity of data [3]; confidentiality [9]; cost; and problems of acceptance by clinicians [4].

Subsequent work on computer interviews has consistently confirmed and extended early findings. Patients do, with few exceptions, find computer interviews acceptable [10–20], often preferring to give information to the computer rather than a doctor. Preferences for a nonhuman interviewer seems particularly strong when the subject matter is sensitive [15,21,22]. Most patients, even those with major psychiatric disorders including some who are mute, can be interviewed by computer. With the exception of some patients with severe organic brain syndrome, patients with florid mania or antisocial personality disorder, they give data that are at least as reliable and valid and far more complete than those obtained by clinicians [3,6,23–26].

Confidentiality can be carefully protected and some interviews permit patients to decide after giving their information what uses will be made of it. Thus, a patient might decide to allow use of data for clinical purposes but not for research or might choose to delete the interview altogether.

Costs continue to decline because of less expensive and more powerful hardware and, more importantly, better programming languages and more efficient and effective interview driver programs [27] which permit nonprogrammers to construct and conduct computer interviews. A sophisticated single terminal interview system can now be purchased for $6,200 and additional stations added for $1,650.

Clinician acceptance of computer interviews has been slower than expected. This lag is less related to the quality of the programs available than to factors of marketing (researchers in academic institutions who predominantly develop, evaluate and describe clinical programs are reinforced most for those activities—not for marketing them) and problems involved in making changes within conservative professions.

Consultations

It has long been a goal of computer mavens to provide computer consultations to clinicians. Mathematical predictions regarding diagnosis, prognosis, treatment selection and patient outcomes have virtually always surpassed clinical predictions [28–30]. Bleich's elegant early work on the problems of fluid, electrolyte and acid-base abnormalities [31] provided both an indication of the potential of consultations and the problems of gaining clinician acceptance and of evaluating the impact of consultation programs [32]. The prospect of providing tertiary level consultation is secondary and primary setting remains attractive.

Predictions in psychiatry have dealt with diagnoses [33–39], medication prescription [40,41], elopement risk [42], interpretations of the Minnesota Multiphasic Personality Inventory [43] and dangerousness [44]. Predictions about dangerousness have been dropped from the Missouri System because of risks of false positives prolonging incarceration and false negatives leading to premature release.

Another generic model of computer consultations is the Lithium Information Center at the University of Wisconsin [45]. The Lithium Library [46] is the cornerstone of a growing number of clinical consultations and publications on the medical uses of lithium. More than 9,000 citations are contained in the Lithium Library and Paper Chase [47], the bibliographic program at the core of the Lithium Library permits searches by any combination of author, title word, subject, key or index word, journal, and/or year of publication. Because files are inverted, search times are short. The Lithium Information Center responds to written and telephone requests for information from clinicians, administrators, researchers, lawyers and patients. In 1981, there were 624 requests from 50 states, the District of Columbia and Puerto Rico, 9 Canadian Provinces and 24 foreign countries. In 1982, requests are running at a rate of 100 per month. More than 60 sites have direct on-line access to the Lithium Library program.

Availability of the Lithium Library and a grant from the National Library of Medicine has led to study of alternative ways of providing clinical information. A cryptic listing of citations can be time-consuming to pursue and frustrating when the sources are not readily available. The Lithium Index provides synopses of important clinical subjects and, to date, more than 90 have been written. On-line access by free text entry is rapid and cross referencing extensive. The Lithium Consultation organizes information about a single patient regarding diagnosis, pretreatment workup, and possible complications of lithium treatment and their management.

Access to the Lithium Library program has also facilitated publication of 4 books, 5 chapters, 15 articles and 6 letters to editors by Lithium Information Center personnel. Requests to the Lithium Information Center have targeted particular problem areas which have led to three focused research projects [48–50].

Data Base Management Systems

Rome's early appreciation of the organizing potential of computers [1] was initially impeded by the complexity and idiosyncratic nature of different psychiatric settings and the consequent cost of preparing different data bases to meet their disparate needs. Since different psychiatrists treat patients with similar disorders quite differently, the data they wish to keep is also different. A standardized data base for separate sites can not be assembled easily. Widespread application of computers in the management of psychiatric data has had to await either the development of data base management systems or consensus about the basic data set, still an elusive goal. In the meantime, many mental health facilities have gone ahead to develop stand-alone information systems [51–57]. With the development of the prototype NIMH Community Mental Health Center program, a well-designed and flexible set of programs will soon be available for many administrative, fiscal and clinical functions [58]. The outcome of this experiment will be of great interest.

Modern medical information systems have at their core a data base management system which permits flexible addition and deletion of data items. The most sophisticated of these data base management systems incorporate a dictionary driver concept in which the data base itself asks a series of questions about the data which, once answered, establish the "dictionary." The dictionary "drives" all other programs i.e. enter, edit, search, report, statistics, etc. Whenever a change is made in the dictionary, it drives required changes in all other programs.

Most data base management systems have been developed to deal with single encounters [59]. Only a few have been specifically designed to deal with data which changes over time, as patients change [60]. These time oriented data base management systems permit clinicians, administrators, researchers, and educators to track patients and their progress, follow a course of treatment, keep a service history, analyze staff or trainee activities, record fees, keep inventories, etc.

The ease of establishing discrete data bases with dictionary driven data base management systems leads to widespread use where they are available. In the Department of Psychiatry at the University of Wisconsin, we have set up seven EPIC [60] data bases (because of EPIC's generalizability, more than 100 other EPIC data bases have been created in other medical school departments and public institutions). Our outpatient data base contains on-line information about all patients seen in the clinic since 1975. Those in charge of residency education use this data base to assess the quantity and character of each trainee's clinical experience; administrators use the same data base to meet accountability requirements and for planning; clinicians have the option of using the data base to measure change in their patients; and researchers, with proper respect for confidentiality, ask and answer a variety of questions about clinical matters.

Treatment Programs

In psychiatry, work is under way to provide direct treatment services to patients using interactive computer programs. At the Institute of Psychiatry in London, England, Drs. Ghosh, Carr and Marks have already evaluated an interactive computer program for the assessment and treatment of agoraphobia [61,62]. Comparable outcomes were obtained when this program was compared with a human therapist and with self treatment guided by a book. At the University of Wisconsin, programs are being evaluated which provide cognitive behavior therapy to depressed patients, assist depressed patients in adhering more completely to their tricyclic antidepressant treatment regime, and generically define and manage emotional problems.

Person-Machine Interface

There is a substantial and growing literature about the problems of person-machine interface. Questions regarding keyboard design or the actual elimination of the keyboard through the substitution of a light pen or finger touch on a CRT screen are pointing the way toward an ultimate interface in which computers may be able to comprehend a number of vernaculars of human speech, "understand" the syntax and meaning expressed and respond through voice synthesizers. While improvements in terminal design are certainly possible, particularly for those who use them for extended periods, elimination of the keyboard also largely eliminates the possibility of incorporating free text responses. We and others have repeatedly found that patients manage quite well with standard terminal keyboards.

There is, however, a larger and more important "person-machine" interface problem: how computers are considered, evaluated, resisted and finally integrated into clinical practice. This problem exists for many professions and businesses [63]. It is difficult for those who work closely with computers and are most aware of their strengths (and limitations) to understand some of the reservations expressed and resistances raised to clinical computing applications. Some who have encountered resistance have written about the experience [64–70]. There appears to be a general covert fear on the part of many clinicians about losing rather than gaining something as computers are introduced into clinical practice. Openly expressed concerns range from losing money or status, to giving up parts of clinical practice that are most satisfying and enjoyable.

Part of the difficulty of introducing computers into clinical practice has been the limited range and utility of the programs now available. Clinicians, as most people who are laymen to computing, fantasize that computers are so powerful that they will do much more than they actually can. The beginnings we have made are modest and have been, we hope, modestly stated. Nevertheless, many clinicians find their expectations about computers frightening. Major changes are an ordeal for almost everyone and, as William Faulkner said, "What the heart loves becomes truth."

We are, perhaps, at a juncture analogous to the situation of American medical education at the end of the nineteenth century. Most medical schools were little more than diploma mills, teaching largely, if not exclusively, through lectures, and intent on continuing in the same manner. A few widely travelled and educated, innovative, open minded and determined colleagues at the Johns Hopkins Hospital and Medical School changed the program of instruction in that institution requiring a college degree for medical school entry and emphasizing clinical work during the junior and senior years. Through the mediation of the Flexner report [71], substantial alterations in medical education quickly followed throughout the United States. Welch, Osler, Halsted, Kelly and Hurd changed the profession in the span of a single generation and are properly honored for that contribution.

Osler is also revered for his humanitarian side and his great affection for older physicians. But he also stood on a chair at an annual meeting of the American Medical Association and shouted [72]:

> I stand here and say plainly and honestly before Dr. Atkinson what I and many other members have said behind his back, that he is not an efficient secretary of this Association, and that we have not found him so [hisses, followed by applause]. You may hiss if you will, but I unhesitatingly say that no more important step in advance will be taken by this Association than when it changes its secretary.

As psychiatrists, we know that some disorders are lifelong and wholely unresponsive to our ministrations. Darwin [73] wrote, at the conclusion of *The Origin of Species:*

> Although I am fully convinced of the truth of the views given in this volume . . . , I by no means expect to convince experienced naturalists whose minds are stocked with a multitude of facts all viewed, during a long course of years, from a point of view directly opposite to mine. . . . But I look with confidence to the future,—to young and rising naturalists, who will be able to view both sides of the question with impartiality.

Some of what we advocate is clearly radical and none of us can foresee all the consequences of the paths we have started down.

While the ethics of our profession require that we provide the best quality care to patients in sufficient quantity at a fair cost, this ideal is seldom attainable. There is often a substantial gap between what is known and what is practiced. Still, the ethics of individuals can anticipate and lead to changes in the ethics of the larger profession. There may be limits, however, to the improvement human clinicians can achieve, even when aided by computers. McDonald found that computer reminders enhanced physician compliance with their own self prescribed practice protocols from 20 to 50% and attributed the failure to approach complete compliance with their own protocols to the "nonperfectibility of man" [74].

One indication of ways in which computers can be integrated into clinical practice has come from experience with CAT and PET scanning which has, by and large, been introduced in departments of radiology. Radiologists are already somewhat remote from patients and accustomed to the use of technologic advances. These new scanning devices have increased radiologists' clinical effectiveness, monetary return, status in the profession and personal satisfaction with their work. To the extent that other computer applications are responsive to these important clinician reinforcers, they will also be likely to gain acceptance by clinicians.

The growing number of mental health computer applications developed in the past two decades will probably be surpassed in both quantity and quality in the five years ahead. Less expensive and more powerful computing hardware running better operating systems and higher level programming languages have led to the development of driver programs which permit naive computer users to bend the computer's power to mental health computing tasks. These technical advances promise radical changes *if* the tools are simply applied and evaluated in important problem areas and if the products developed are made acceptable to the mental health professions so that they are actually used.

The rapid evolution of computing hardware and software accompanied by a steadily decreasing cost and a lower learning threshold for beginners has actually led some to delay aquisition and use of computers to solve mental health problems. Waiting for the next generation of hardware and software to solve our problems is rapidly becoming a certifiable neurosis. Present hardware and software are already cost effective and we would argue that there is a greater cost to mental health institutions in delaying entry into computer usage in hopes of still cheaper hardware and software than in making use of presently available equipment and programs. Better software will be created as clinical users become more involved with computing.

With the high quality of present technology, the mental health field needs continued developmental efforts which are carefully evaluated. Many different approaches seem prudent at this time in a health care area with many different professions, belief systems, service needs and treatments. Although iterations are hard work and it would be ideal to design the ultimate system, total system designs are efficient only when one can clearly define the operations, goals, and output, a stage far away for mental health organizations and practitioners. It is time to proceed vigorously with the tools at hand. We are very much at the stage when aviation of the 1920's was developing and evaluating many different airframes, engines and instruments. In time, this process will yield the standardization in design, construction and use of clinical computing programs that has made modern air travel so swift, safe and economical.

Making computer applications acceptable to clinicians is a substantial problem. While the final outcome is not in doubt, the rate and extent of computer use may well be influenced by basic strategies adopted by computer developers. On the one hand, it makes sense to provide programs which fill a clearly recognized need and which make small demands for change in clinical practice. Administrative programs meet these criteria and are already in common and steadily spreading use. On the other hand, the greatest payoff from computer applications probably lies in their direct use for clinical functions such as patient interviewing, clinician consultation and actual patient treatment. All of these uses destabilize clinical practice and require substantial changes in the beliefs, thinking and behaviors of clinicians. Short of waiting for one or two generations of clinicians to die, attention will have to be paid to reinforcements for clinicians that will enable them to embrace clinical computing applications for mental health. The basic reinforcers appear to be status, money, clinical affectiveness and job satisfaction.

When there is a great deal of status associated with an activity, professionals are inclined to sanctify the activity and interpose educational, training and certification requirements between would be practitioners and patients. Psychoanalysis, neurosurgery and, to a lesser extent, all medical specialties are examples of this practice. When, however, a clinical service must be frequently repeated or provided to large numbers of patients, the status associated with these clinical activities is decreased and surrogates are acceptable. If insulin could be given yearly, it seems likely that the professors of medicine would retain that prerogative; if monthly, general practitioners would perform the function, with weekly administration nurses could serve; but because insulin must be administered daily and sometimes more often, patients are elevated in status and "trained" to self-administer insulin. When coronary care units were first opened in 1962 and for several years thereafter, physicians administered the antiarrhythmic drugs which yielded a reduction in hospital death rate from 30–40% to the present level of 10–20%. As the novelty of providing this treatment diminished and the status associated with staying awake all night to administer it paled, nurses were "trained" to provide this treatment [75]. Clean catch midstream urines are another low status procedure which patients have been "trained" to perform better when instructed by computer than by nurse or doctor [76]. When systematic desensitization was first developed, clinicians flocked to apply this effective treatment for phobias. When clinicians realized how boring the treatment process was, they rapidly substituted tape recorders, nurses [77] and computer programs [78]. When clinical procedures lose status, and/or cease to be enjoyable, clinicians become ingenious in finding and "training" surrogates to carry out the procedures.

There is legitimate concern about the effect of computers on clinician income. Present patient computer interactions can be conducted at a cost of less than $5 per hour, a far cry from the $100 per hour rate which psychiatrists commonly charge. For the present, however, some clinicians are using this financial leverage for their benefit, conducting psychological tests and other evaluations by computer interview and charging a moneymaking fee for this service. At a time when the health care budget is increasing at a rate greater than the base inflation rate, the economies associated with patient computer interactions should be shared between institutions, innovative practitioners and the public which ultimately supports all medical care.

Conclusion

To the extent that computer programs can substitute for clinical functions which are low in status and tedious to perform while increasing or at least not decreasing clinician income, they are likely to win acceptance. Programs that improve clinician functioning and through them, patient outcomes, should enhance clinician status and also gain acceptance. In time, even those programs which compete directly with clinicians and surpass them in effectiveness in direct patient treatment will be used because of both economic factors and ethics.

The path of progress is often steep, twisting and full of natural obstructions and man-made barriers. Clinical computing is a nascent force with at best a toehold on a sheer face of the clinical mountain. But the power of computing and the magnitude of mental health problems assures a steadily accelerating ascent.

References

[1] Rome, H. P., "Prospects for a Psi-net: The Fourth Quantum Advance in Psychiatry," *Compr. Psychiatry* **8,** 1967, pp. 450–454.

[2] Arey, L. B., Burrows, W., Greenhill, J. P., Hewitt, R. M., *Dorlan's Illustrated Medical Dictionary,* 23rd ed. Philadelphia and London: W. B. Saunders Company, 1957.

[3] Slack, W. V., Hicks, G. P., Reed, C. E., Van Cura, L. J., "A Computer Based Medical History System," *New England J. Med.* **274,** 1966, pp. 194–198.

[4] Slack, W. V., Van Cura L. J., Greist, J. H., "Computers and Doctors: Use and Consequences," *Comp. Biomed. Res.* **3,** 1970, pp. 521–527.

[5] Slack, W. V., "Computer Based Interviewing System Dealing with Nonverbal Behavior as well as Keyboard Responses," *Science* **171,** 1971, pp. 84–87.

[6] Slack, W. V., Slack, C. W., "Talking to a Computer about Emotional Problems; a Comparative Study," *Psychoth TRP.* **14,** 1977, pp. 156–164.

[7] Slack, W. V., Van Cura L. J., "Patient Reaction to Computer-Based Medical Interviewing," *Comp Biomed Res.* **1,** 1968, pp. 527–531.

[8] Coombs, G. J., Murray, W. R., Krahn, D. W., "Automated Medical Histories: Factors Determining Patient Performance," *Comp Biomed Res.* **3,** 1970, pp. 178–181.

[9] Laska, E. M., Bank, R., *Safeguarding Psychiatric Privacy: Computer Systems and Their Uses.* New York: John Wiley and Sons, 1975.

[10] Slack, W. V., Van Cura L. J., "Computer-based Patient Interviewing: Part 2," *Postgrad Med.* **43,** 1968, pp. 115–120.

[11] Grossman, J. H., Barnett, G. O., McGuire, M. T., Swedlow, D. B., "Evaluation of Computer-acquired Patient Histories," *JAMA* **215** 1971, 1286–1291.

[12] Hilf, F. D., Colby, K. M., Smith, D. C., Wittner, W. K., Hall, W. A., "Machine-mediated Interviewing," *J. Nerv. Ment. Dis.* **152,** 1971, pp. 278–288.
[13] Slack, W. V., Slack, C. W., "Patient-computer Dialogue," *New England J. Med.* **286,** 1972, pp. 1304–1309.
[14] Johnson, D. F., Mihal, W. L., "Performance of Blacks and Whites in Computerized versus Manual Testing Environments," *Am Psychol.* **28,** 1973, pp. 694–699.
[15] Greist, J. H., Gustafson, D. H., Stauss, F. F., Rowse, G. L., Laughren, T. P., Chiles, J. A., "A Computer Interview for Suicide Risk Prediction," *Am J. Psychiatry* **130,** 1973, pp. 1327–1332.
[16] Card, W. I., Nicholson, M., Crean, G. P., *et al.,* "A Comparison of Doctor and Computer Interrogation of Patients," *Int. J. Biomed. Comput.* **5,** 1974, pp. 175–187.
[17] Cole, E. B., Johnson, J. H., Williams, T. A., "When Psychiatric Patients Interact with On-line Computer Terminals. Problems and Solutions," *Behavior Research Methods and Instrumentation* **8,** 1976, pp. 92–94.
[18] Slack, W. V., Porter, D., Witschi, J., Sullivan, M., Buxbaum, R., Stare, F. J., "Dietary Interviewing by Computer," *J. Am. Diet Assoc.* **69,** 1976, pp. 514–517.
[19] Lucas, R. W., "A Study of Patients' Attitudes to Computer Interrogation," *Int. J. Man-Mach. Stud.* **9,** 1977, pp. 69–86.
[20] Angle, H. V., Ellinwood, E. H., Carroll, J., "Computer Interview Problem Assessment of Psychiatric Patients," in F. H. Orthner, Ed., Proceedings: The Second Annual Symposium on Computer Application in Medical Care (Washington, D.C.), *IEEE,* 1978.
[21] Greist, J. H., Klein, M. H., Van Cura L. J., Erdman, H. P., "The Computer Interview as a Medium for Collecting Questionnaire Data on Drug use in Predicting Adolescent Drug Use: Predicting Adolescent Drug Abuse," in D. J. Lettieri, Ed., Predicting Adolescent Drug Use: A Review of Issues, Methods and Correlates. Washington, D.C.: U.S. Government Printing Office, 1975.
[22] Greist, J. H., Klein, M. H., "Computer Programs for Patients, Clinicians, and Researchers in Psychiatry," in J. B. Sidowski, J. H., Johnson, T. A. Williams, Eds., *Technology in Mental Health Care Delivery Systems.* Norwood, N.J.: Ablex Publishing Corporation, 1980, pp. 161–182.
[23] Greist, J. H., Van Cura, L. J., Kneppreth, N. P., "A Computer Interview for Emergency Room Patients," *Comp Biomed Res.* **6,** 1973, pp. 257–265.
[24] Greist, J. H., Klein, M. H., Van Cura, L. J., "A Computer Interview for Psychiatric Patient Target Symptoms," *Arch. Gen. Psychiatry* **29,** 1973, pp. 247–253.
[25] Lucas, R. W., Mullins, P. J., Luna, C. B. X., McInroy, D. C., "Psychiatrists and a Computer as Interrogators of Patients with Alcohol-related Illnesses: A Comparison," *Brit. J. Psychiat.* **131,** 1977, pp. 160–167.
[26] Carr, A. C., Ancill, R. J., Ghosh, A., Margo, A., "Direct Assessment of Depression by Microcomputer: A Feasibility Study," *Acta Psychiat. Scand.* **64,** 1981, pp. 415–422.
[27] Bloom, S. M., White, R. J., Beckley, R. F., Slack, W. V., "Converse: A Means to Write, Edit, Administer, and Summarize Computer-based Dialogue," *Comput. Biomed. Res.* **11,** 1978, pp. 167–175.
[28] Sawyer, J., "Measurement and Prediction, Clinical and Actuarial," *Psychol. Bull.* **66,** 1966, pp. 178–200.

[29] Sines, J. O., "Acturial versus Clinical Prediction in Psychopathology," *Br. J. Psychiatry* **116,** 1970, pp. 129–144.
[30] Dawes, R. M., "The Robust Beauty of Improper Linear Models in Decision Making," *Am. Psychologist* **34,** 1979, pp. 571–582.
[31] Bleich, H. L., "Automated Instructional Programs for Advanced Medical Education," in *Computers in Biomedical Research, Vol. IV.* New York: Academic Press, Inc., 1974, pp. 289–300.
[32] Bleich, H. L., "The Computer as Consultant," *New England J. Medicine* **223,** 1973, pp. 308–312.
[33] Overall, J. E., Hollister, L. E., "Computer Procedures for Psychiatric Classification," *JAMA* **187,** 1964, pp. 583–585.
[34] Spitzer, R. L., Endicott, J., "DIAGNO: a Computer Program for Psychiatric Diagnosis Utilizing the Differential Diagnostic Procedure," *Arch. Gen. Psychiatry* **18,** 1968, pp. 746–756.
[35] Sletten, I. W., Altman, H. L., Ulett, G. A., "Routine Diagnosis by Computer," *Am. J. Psychiatry* **127,** 1971, pp. 1147–1152.
[36] Hirschfeld, R., Spitzer, R. L., Miller, R. G., "Computer Diagnosis in Psychiatry: A Bayes Approach," *J. Nerv. Ment. Dis.* **158,** 1974, pp. 399–407.
[37] Spitzer, R. L., Endicott, H., "Can the Computer Assist Clinicians in Psychiatric Diagnosis?," *Am. J. Psychiatry* **131,** 1974, pp. 523–530.
[38] Erdman, H. P., Greist, J. H., Klein, M. H., Jefferson, J. W., Salinger, R., Olson, W., "Clinical Usefulness of a Computer Program for Psychiatric Diagnosis," in B. Shriver, R. Walker, R. Grams, and R. Sprague, Eds., Proceedings of the Fourteenth Hawaii International Conference on System Sciences. North Hollywood: Western Periodicals, 1981, pp. 802–817.
[39] Laska, E., Siegel, C., Simpson, G., "Automated Review System for Orders of Psychotropic Drugs," *Arch. Gen. Psychiatry* **37,** 1980, pp. 824–827.
[40] Overall, J. E., Henry, B. W., Markett, J. R., "Decisions about Drug Therapy: I. Prescription for Adult Psychiatric Outpatients," *Arch. Gen. Psychiatry.* **26,** 1972, pp. 140–145.
[41] Sletten, I. W., Altman, H., Evenson, R. C., "Computer Assignment of Psychotropic Drugs," *Am. J. Psychiatry* **130,** 1973, pp. 595–498.
[42] Altman, H., Brown, M., Sletten, I. W., "And silently steal away: A Study of Elopers," *Dis. Nerv. Syst.* **33,** 1972, pp. 52–58.
[43] Fowler, R. D., "Automated Psychological Test Interpretation: The Status in 1972," *Psychiatry Ann.* **2,** 1972, pp. 10–28.
[44] Hedlund, J. L., Sletten, I. W., Altman, H., Evenson, R. C., "Prediction of Patients Who are Dangerous to Others," *J. Clinical Psychology* **29,** 1973, pp. 443–447.
[45] Jefferson, J. W., Greist, J. H., Marcetich, J. R., "The Lithium Information Center," in T. Cooper, S. Gershon, N. S. Kline, M. Schou, Eds., *Lithium: Controversies and Unresolved Issues.* Amsterdam: Excerpta Medica, 1979, pp. 958–963.
[46] Greist, J. H., Jefferson, J. W., Coombs, A. M., Schou, M., Thomas, M. A., "The Lithium Librarian—An International Index," *Arch. Gen. Psychiatry* **34,** 1977, pp. 456–459.
[47] Beckley, R. F., Bleich, H. L., "Paper Chase: A Computer-based Reprint Storage and Retrieval System," *Comput. Biomed. Res.* **10,** 1977, pp. 423–430.
[48] Jefferson, J. W., Kalin, N. H., "Effect of Long-Term Diuretic Use on Serum Lithium Levels in Man," *JAMA* **241,** 1979, pp. 1134–1136.

[49] Jefferson, J. W., Greist, J. H., Clagnaz, P. J., Eischens, R. R., Marten, W. C., Evenson, M. A., "The Effect of Strenuous Exercise on Serum Lithium Level in Man," *Am. J. Psychiatry* (in press).
[50] Jefferson, J. W., Kaufman, P. L., "Effects of Lithium on the Eye," unpublished data, 1982.
[51] Sletten, I., Ulett, G., Altman, H., "The Missouri Standard System of Psychiatry," *Arch. Gen. Psychiatry* **23,** 1970, pp. 73–79.
[52] Glueck, B. C., "Computers at the Institute of Living," in J. L. Crawford, D. W. Morgan, D. T. Gianturco, Eds., *Progress in Mental Health Information Systems: Computer Applications.* Cambridge, MA; Ballinger Publ. Co., 1974.
[53] Glueck, B. C., Erickson, R. P., Stroebel, C. F., "The Use of a Psychiatric Patient Record System," in W. Siler, D. A. B. Lindberg, Eds., *Computers in Life Science Research.* New York, Plenum Press, 1975.
[54] Williams, T. A., Johnson, J. H., Bliss, E. L., "A Computer-assisted Psychiatric Assessment Unit," *Am. J. Psychiatry* **132,** 1975, pp. 1074–1076.
[55] Greist, J. H., Klein, M. H., Van Cura, L. J., Faulkner J., "Computer Applications in Psychiatry at the University of Wisconsin: Current Status," *Curr. Con. Psychiatry* **3,** 1977, pp. 13–16.
[56] Laska, E. M., Siegel, C., Bank, R., "Management Information Systems in Mental Health," in J. B. Sidowski, J. H. Johnson and T. A. Williams, Eds., *Technology in Mental Health Care Delivery Systems.* N.J.: Ablex Publishing Corporation, 1980, pp. 27–42.
[57] Mezzich, J. E., Dow, J. T., Coffman, G. A., "Developing an Efficient Clinical Information System for a Comprehensive Psychiatric Institute: I. Principles, Design, and Organization," *Behavior Research Methods & Instrumentation* **13,** 1981, pp. 459–463.
[58] Wurster, C. R., Goodwin, J. D. "NIMH Prototype Management Information System for Community Mental Health Centers," in J. T. O'Neill, Ed., Proceedings of the 4th Annual Symposium Computer Applications in Medical Care. New York: *IEEE,* 1980.
[59] Entine, S. M., Friedman, R. B., "WISAR: A MUMPS Data Base System which Utilizes Non-prime Time," Proceedings of MUMPS Users Group Meeting, 1978.
[60] Faulkner, J. R., Klein, M. H., Greist, J. H., Taves, J. E., Bass, S., "EPIC: Information Management for Mental Health Clinicians, Administrators and Researchers," in J. L. Crawford, S. Vitale, J. Robinson, Eds., *Computer Applications in Mental Health.* Cambridge, MA: Ballinger Publ. Co., in press.
[61] Carr, A. C., Ghosh, A., "Response of Phobic Patients to Direct Computer Assessment," *Brit. J. Psychiat* **141,** 1982.
[62] Carr, A. C., Ghosh, A., "Accuracy of Behavioural Assessment by Computer," *Brit. J. Psychiat.* **141,** 1982.
[63] "Computers: Dealing with Terminal Phobia," *Time* **120,** July 19, 1982, p. 82.
[64] Friedman, R. B., Gustafson, D. H., "Computers in Clinical Medicine: A Critical Review," *Comp. Bio. Res.* **10,** 1977, pp. 199–204.
[65] Johnson, J. H., Williams, T. A., Klinger, D. E., Giannetti, R. A., "Interventional Relevance and Retrofit Programming: Concepts for the Improvement of Clinician Acceptance of Computer Generated Assessment Reports," *Behav. Res. Methods Instrum.* **9,** 1977, pp. 123–132.
[66] Johnson, J. H., Williams, T. A., Giannetti, R. A., Schmidt, L. J., "Strategies for the Successful Introduction of Computer Technology in a Mental Health Care

Setting: The Problem of Change," *Natl. Comput. Conf. Expos. Proc.* **46,** 1977, pp. 55–58.

[67] Johnson, J. H., Williams, T. A., Giannetti, R. A., Klingler, D. E., Nakashima, S. R., "Organizational Preparedness for Change: Staff Acceptance of an On-Line Computer-assisted Assessment System," *Behav. Res. Methods Instrum.* **10,** 1978, pp. 186–190.

[68] Hedlund, J. L., Vieweg, B. W., Cho, D. W., *et al.,* "Mental Health Information Systems: A State-of-the-art Report," Health Services Research Center/Health Care Technology Center, University of Missouri, Columbia, 1979, pp. 17–18.

[69] Teach, R. L., Suortliffe, E. H., "An Analysis of Physician Attitudes regarding Computer Based Clinical Consultation Systems," *Computers and Biomedical Research* **14,** 1981, pp. 542–558.

[70] Lindberg, D. A. B., "Diffusion of Medical Information Systems Technology in the United States," *J. Med. Sys.* **6,** 1982, pp. 219–228.

[71] Flexner, A., "Medical Education in the United States and Canada; A Report to the Carnegie Foundation for the Advancement of Teaching," with an introduction by Henry S. Pritchett. New York, 1910, pp. 1–346.

[72] Cushing, H., *The Life of Sir William Osler.* New York: Oxford Univ. Press, 1926, p. 415.

[73] Darwin, C., *On the Origin of Species.* London: J. Murray Publ., 1859.

[74] McDonald, C. J., "Protocol-based Computer Reminders, the Quality of Care and the Non-perfectibility of Man," *New England J. Medicine* **295,** 1976, pp. 1351–1355.

[75] Freitag, J. J., Miller, L. W., *Twenty-Third Edition of the Manual of Medical Therapeutic.* Boston: Litttle Brown and Company, 1980, p. 73.

[76] Fisher, L. A., Johnson, T. S., Porter, D., Bleich, H. L., Slack, W. V., "Collection of a Clean Voided Urine Specimen: A Comparison among Spoken, Written, and Computer-based Instructions," *Am. J. Pub. H.* **67,** 1977, pp. 640–644,

[77] Marks, I. M., Bird, J., Lindley, P., "Psychiatric Nurse-Therapy—Developments and Implications," *Behav. Psychother.* **6,** 1978, pp. 25–36.

[78] Lang, P. J., "The On-line Computer in Behavior Therapy Research," *Am. Psychol.* **24,** 1969, pp. 236–239.

30

Computers in Nursing Administration*

Rita D. Zielstorff

There are currently a variety of computer applications that assist nurse administrators in diverse settings with the everyday tasks of managing and planning nurse services. These applications vary widely in depth and complexity.

Administrative Systems

The most common application of computers in nursing administration is in collecting, aggregating and summarizing various data—the "administrative" or "data management" systems. Many institutions subscribe to commercial or nonprofit data processing services that process standardized data collection forms and return quantitative reports of census characteristics, financial measures, resource utilization and "productivity" measures.

Many nurse administrators take advantage of existing computer resources to develop customized applications "in-house." Although these applications are tailored to meet individual needs, they nonetheless serve as useful models for others, and they are occasionally even transferable to other institutions.

For instance, the Department of Nursing at Massachusetts General Hospital (MGH) has collaborated with the Laboratory of Computer Science to develop a system that supports nurses' registration for programs offered by Staff Education. After reviewing a previously distributed syllabus, nurses use computer terminals to register for specific courses. The system retains each nurse's identifying data (including courses previously taken) for at least two years, and makes sure that prerequisites are met before permitting registration. After completion of the course, Staff Education clerical personnel input grades and contact hours. The system uses this information to print various reports

 Adapted from Zielstorff, Rita D.: Computers for planning and evaluating nursing services. In Marriner, Ann, editor: *Contemporary Nursing Management*, St. Louis, 1982, The C.V. Mosby Co.

and certificates for nursing administration, the state nurses' association, and the participant.

The computer-based information system at El Camino Hospital in Mountain View, California, assists each head nurse to prepare a total budget for his or her care unit, including requirements for personnel, nonlabor expenses and capital expenditures. Detailed data from the previous operating year are used as the basis for forcasting the coming year's revenue and expenditures. Control is achieved through regular reviews of computer generated reports that compare actual figures with budgeted ones [1].

Public health and community health agencies are also turning increasingly to automated data processing [2].

The New Jersey State Department of Health has developed a data management system that at last report was used by 22 New Jersey Home Health Agencies, including voluntary, official and hospital-based agencies. In contrast with many other systems, there is only one basic input form that captures data for all patient transactions. Monthly and quarterly statistical reports include Patient Visits by Diagnosis, Patients Served by Guarantor, and Census/Caseload Summary [3].

The National League for Nursing, funded by the Division of Nursing, has conducted national conferences and regional workshops on the use of automated information systems in community agencies and prepared several very useful publications [4,5].

Computers have the potential to impact public/community health agencies not only by performing the usually labor-intensive tasks of statistical summarization and billing, but by stimulating and enabling systematic documentation, thus providing the data needed for planning and evaluation.

Management Information Systems

While administrative systems collect, summarize and format data needed for administrative decision-making, management information systems take a more active role in the management process.

For instance, an automated staffing system at Fairview General Hospital in Cleveland, Ohio periodically produces the attendance record of each individual employee. This is characteristic of an administrative system. However, the system has also been programmed to produce periodically a report of any employee who has taken an "unacceptable" number of sick or absent days within the previous three months, permitting "early identification action." This is a simple illustration of the difference between administrative systems and management information systems [6].

"HAS," or "Hospital Administrative Services," developed by the American Hospital Association has been expanded and named "Monitrend for Hospitals." This service provides monthly and quarterly reports, including a detailed section on the nursing department. The name change refers to its expanded capability for trending data based on previous internal measures. The system is also supported by educational material that assists administrators to make effective use of the data in management decisions [7].

Scheduling

Automated nurse scheduling is one of the most frequently asked about applications of computers in nursing. The systems that have been developed range in complexity from those that "fill in the blanks" according to predetermined staffing patterns, to systems that adjust staffing on a shift-to-shift basis according to both patient care workload and capacity of available personnel.

The centralized, computer-based staffing system at Fairview General Hospital produces an advance schedule for each care unit based on predetermined "core staffing requirements" as well as individual employee data such as education and requests for days off. Once the schedule is in effect, the system produces a number or reports that assist central staffing personnel to identify staffing deficits and make adjustments on a shift-by-shift basis [8].

One of the most comprehensive scheduling systems reported does long range staffing as well as shift-to-shift variable staffing based on "assignment difficulty"—a concept that encompasses several factors, including patient workload, capacity of staff, and organizational and environmental factors. Developed by Medicus Systems Corporation, the system requires several months of planning, including detailed analysis of work patterns at the subscribing institution. Incorporated into the system are several monthly and year-to-date management reports that assist administrative decision-making related to staffing and quality of care. System output can be combined with output from other hospital computer systems (such as personnel or payroll) to provide a single comprehensive staffing report [9].

Quality Assurance

Nowhere is the nurse administrator more confronted with the problems of record-keeping and data retrieval than in carrying out the many mandates for utilization review, quality assessment and quality assurance activities.

Again, there is a wide range of applications, from services that process abstracted medical record data to "total" systems that carry out automated concurrent review of care with automatic feedback.

The Commission on Administrative Services in Hospitals (CASH) is a nonprofit corporation that developed a system incorporating a 105-question audit instrument, each question individually weighted according to importance. The CASH system has established standard time and performance measures that are useful in planning staffing requirements and carrying out quality assessment activities [10].

The following examples of computer-assisted quality control and assessment depend on the existence of a patient database—that is, clinical data that are entered into the computer as part of the process of delivering daily care.

El Camino Hospital in Mountain View, CA has been using a hospital-wide Medical Information System (MIS) since 1972. A quality assurance system is now being designed to "integrate both concurrent and retrospective aspects of nursing audit, medical audit and utilization review into a coordinated patient care audit" [11]. A Nursing Care Planning Subsystem that incorporates standardized, diagnosis specific, outcome-oriented care plans and computer monitoring of time deadlines for patient progress is already in effect.

COSTAR, a Computer-Stored Ambulatory Record System developed jointly by MGH's Laboratory of Computer Science and the Harvard Community Health Plan in Boston, has been in use at HCHP since 1970 [12]. Data from pre-coded encounter forms (including diagnoses/problems, therapies and disposition) are transcribed into the computer, which stores the information for retrieval in a variety of formats. Because the data are machine readable, they can be manipulated according to provider specified rules for carrying out automated concurrent review with immediate feedback to the responsible provider. For instance, the computer is programmed to monitor whether patients with a throat culture positive for Group A Beta Strep receive treatment within four days. Failure to find notation of treatment causes computer printed reminders to the appropriate physician or nurse practitioner.

The COSTAR system has been rewritten and expanded to make it transferable to a variety of ambulatory care settings [13].

Difficulties and Challenges

There are many difficulties associated with developing, acquiring and using automated systems. These can be broadly classified as difficulties relating to:

(1) the nature of management science, particularly health care management;
(2) the nature of administrators, particularly nurse administrators; and
(3) the nature of computer technology and computer professionals.

In health care management, most of the decisions at top management levels currently fall in the "non-programmed" area. This is particularly true in administration of health care institutions, which Farlee calls "people processing" systems [14]. As she points out, productivity and efficiency are more difficult to define, measure and quantify in health care than in other industries. Before computers can be more widely applied to administrative decisions, there must be a concentrated effort to define standards of performance, identify the variables that determine each decision, and to quantify the ways in which these variables interact to affect a given situation.

Until this is done, we must relegate to the computer the task of sorting and printing the volumes of data that are thought to be needed in making decisions.

Here, too, there is urgent need for better conceptualization of information requirements so that the busy administrator does not become overwhelmed with myriad useless details. Too much information can be as detrimental to decision making as too little information, and it certainly costs more.

Administrators themselves often lack the skills to use technology to their advantage. Herzlinger indicts health care administrators for having "little acquaintance with the technical aspects of management, such as accounting, finance or data processing," and states that "this lack of training not only means that the managers are unable to take an active position in shaping the management control systems in their organizations, but also that they are insecure in doing so and, at times, . . . intimidated by the technical staff who work for them" [15].

Blair [16] writes of the the need for better prepared nurse administrators:

> During the past decade graduate students in nursing have had few formal opportunities to develop a conceptual base for administrative practice. . . . The results of these years of drought are now evident in the dearth of nurse administrators with the broad knowledge base and sophisticated skills that are necessary if nurses are to influence the scope and direction of health care delivery.

Nurse administrators, then, have the dual challenge of acquiring greater skills in management science and technology as well as encouraging research into identifying and quantifying the variables that affect nursing performance and the administration of nursing services.

Computer professionals have been guilty of seeing the computer as a solution to all problems, because of its astounding success in the business world. O'Desky and Ball wryly describe the average systems analyst as a person who is "energetic, anxious to be of help to the user, wants to be creative, is convinced that the computer can help solve the problem and doesn't have the slightest idea what the user's problem really is." [17].

There is an urgent need to develop and support programs that educate computer professionals in the unique problems of health care administration, and to educate health care administrators in the requirements and limitations of information science and computer technology.

Promises and Prospects

Future computer applications in nursing administration will likely have two characteristics: increased sophistication and increased scope.

Increasingly sophisticated systems will take the form of simulation and modeling programs that allow the administrator to "try out" alternative strategies and evaluate the results. They will also predict long range effects given existing or hypothetical variables. The nurse administrator could then ask such questions as: What would be the effect of an all-R.N. staff on budget and quality of care factors? What are the crucial variables that determine the nature and extent of staff turnover, and what combination of remedies would be most economical and effective? Given the present mix of staff characteristics and performance, what are the most pressing educational and development needs?

Computer systems will also become more sophisticated in terms of sharing databases and providing easy access to the data stored. In many installations, each set of application programs gathers and stores the data needed for that application. Since many applications use various elements of data in common (for example, patient name, unit number and location), the result is redundant data storage. Because each data file is structured for optimization within that application, trying to gather data from more than one file is cumbersome or impossible, resulting in poor communication between systems. With a Data Base Management System, data are stored independently of the programs that use them, thus there is little redundancy and improved flexibility for retrieving various combinations of data [18].

Another aspect of Data Base Management Systems is their ability to provide interfaces so that persons with no programming background can interact directly with the system in accessing the data stored [19]. Among the most frustrating problems for administrators who have automated systems are the expense, time delays and outright impossibility of analyzing data that they know they have collected and stored. The Laboratory of Computer Science at MGH is currently developing a Medical Query Language that allows administrators and providers to formulate structured questions about the data they have stored. The language is designed for systems programmed in the MUMPS language, especially for the COSTAR system [20].

The director of nurses with access to such a system could walk up to a terminal and type a few simple commands to answer such questions as: What are the ten most common primary diagnoses of patients seen at home by R.N.'s? What nursing interventions for a given problem are most associated with a positive outcome? What are the most frequently recorded nursing problems associated with a given medical diagnosis?

As health care organizations upgrade their abilities to use the technology, the computer industry will once again see medical systems as a viable commodity for marketing [21]. The availability of commercially developed systems for ancillary services and the technology for integrating the data from all of these subsystems will give the administrator a much richer database upon which to base decisions, without the time, cost and error rate associated with abstracting data.

A parallel and perhaps conflicting trend is the development of microsystems technology such as self-contained desktop computers. Microsystems will put circumscribed applications like staff scheduling and personnel data systems well within the individual administrator's reach; however, there is the danger that they will exacerbate the current problems of redundant data storage and inability to combine data from various systems for better synthesis of information and more meaningful management reports. One solution to this is to build in the capability of "networking" the individual systems so they are capable of "communication by command" [22].

Developments in management science, nursing, and computer hardware and software will obviously continue. The challenge is to see to it that they don't progress independently. Instead, nurse administrators and health care computer professionals must develop the skills to encompass each of these fields, so that progress in one can stimulate developments in the others.

With good communication among experts who have expanded their skills to be able to communicate effectively with each other and to have a good understanding of the rudiments of each other's professions, computers will prove to be an effective and perhaps indispensable tool in the management of nursing services.

References

[1] Fleming, J. "Planning and Budgets: Magic or Math," in Schmeid, E., Ed., *Maintaining Cost Effectiveness.* Wakefield, MA: Nursing Resources Inc., 1979.

[2] Saba, V., Levine E., "Management Information Systems for Public Health Nursing Services," *Pub. Health Rep.* **93,** 1978, pp. 70–83.

[3] National League fo Nursing: Selected Management Information Systems for Public Health Community Health Agencies. New York, 1978, pp. 113–127.

[4] National League for Nursing: Management Information Systems for Public Health Community Health Agencies: Workshop Papers. New York, 1975.

[5] National League for Nursing. State of the Art in Management Information Systems for Public Health Community Health Agencies: Report of the Conference. New York, 1976.

[6] Ballantyne, D., "A Computerized Scheduling System with Centralized Staffing," *J. Nurs. Adm.* **9,** 1979, p. 45.

[7] Koncel, J. "A New Tool for Meeting Management Information Needs," *Hospitals* **53,** 1979, pp. 62–65.

[8] Ballantyne, 38–45.

[9] Norby, R., Freund, L., Wagner, B., "A Nurse Staffing System Based Upon Assignment Difficulty," *J. Nurs. Adm.,* **7,** 1977, pp. 2–24.

[10] Griffith, J., *Quantitative Techniques for Hospital Planning and Control,* Lexington, MA: Lexington Books, 1972, pp. 304–321.

[11] Norwood, D. "A Patient Care Quality Assurance System Overlayed on a Hospital-wide Medical Information System," in Shires, D. B., and Wolf, H., *MEDINFO 1977,* New York: North-Holland Publishing Co., 1977, pp. 23–27.

[12] Gluck, J. "The Computerized Medical Record System: Meeting the Challenge for Nursing" *J. Nurs. Adm.* **9,** 1979, pp. 17–24.

[13] Barnett, G. O., Justice, N. S., Somand, M. E., *et al.*, "COSTAR–A Computer-stored Medical Information System for Ambulatory Care," *IEEE Proc.* **67,** 1979, pp. 1226–1237. [Reprinted as Chapter 19.]

[14] Farlee, C., "The Computer as a Focus of Organizational Change in the Hospital," *J. Nurs. Adm.* **8,** 1978, pp. 20–26.

[15] Herzlinger, R., "Management Noninformation Systems in Health Care Organizations," *Health Care Man. Rev.* **1,** 1976, pp. 71–76.

[16] Blair, E., "Needed: Nursing Administration Leaders," *Nsg. Outlook.* **24,** 1976, pp. 550–554.

[17] O'Desky, R., Ball, M., "Ten Rules to Bridge the Communication Gap between the Health Care Professional and his Computer Systems Department," *Hosp. Topics.* **54,** 1976, pp. 53–55.

[18] Veazie, S., Dankmyer, T., "HIS's, MIS's, DBMS's: Sorting out the Letters," *Hospitals* **51,** 1977, pp. 80–84.

[19] Bowie, J., "Computer Software," *J. Nurs. Adm.* **10,** 1980, pp. 32–42.

[20] Beaman, P. D., Justice, N. S., Barnett, G. O., "A Medical Information System and Data Language for Ambulatory Practices," *Computer* **12,** 1979, pp. 9–17.

[21] Grams, R., "The Current Status and Future Prospects for Computers in Hospitals," *Hospitals* **51,** 1977, pp. 187–193.

[22] Grams, *ibid,* 190.

31

An Information System for Oncology*

Raymond E. Lenhard, M.D., Bruce I. Blum, and Elizabeth E. McColligan, M.S., M.P.H.

Introduction

The Johns Hopkins Oncology Center (JHOC) was organized in 1976 as part of the National Cancer Program sponsored by the National Cancer Institute. It grew out of a series of separate departmental activities at Johns Hopkins and was designed to bring together a multidisciplinary, focused approach to cancer. Housed in a separate building in the Johns Hopkins Hospital complex, the Center has 56 high technology single patient beds, an active outpatient clinic capable of seeing five hundred patients per week, educational and administrative space and 36,000 square feet of laboratory research space.

The goals of the Oncology Center are:

(1) Basic and clinical research.
(2) Education of post graduate fellows, house staff, medical students community physicians, nurses, physician's assistants, radiation therapy technicians, and others.
(3) Specialized clinical management of patients with cancer.

To support the goals of the Center, a data system has been implemented which focuses on the individual patient and the medical provider. All data items collected are oriented toward identifying patient events without regard to where the patient may be treated (inpatient, outpatient). The computing system organizes data collection and correlates data items within specific time frames. It processes multiple data items within or between groups of patients to support clinical decision making, administration and research.

The Oncology Center Information System (OCIS) is managed by the Oncology Center administration. The OCIS supports several data activities such as the Johns Hopkins Hospital Tumor Registry and several national data exchanges, for example the reporting requirements for ECOG (The Eastern

*This chapter was prepared especially for this collection.

Cooperative Oncology Groups) and CCPDS (Centralized Cancer Patient Data System). There is a separate budget for data system support and the OCIS owns and operates its own computers with full responsibility for input/output decisions and operating practices.

The primary emphasis of the system is the collection and use of clinical information for prospective application in both clinical trials and individual therapy. Special tools have been developed to assist in identifying patients on study, displaying on-study summary data, and supporting retrospective analysis. The operation of the system is supported primarily from patient care revenues.

Information on all patients admitted to the Johns Hopkins Oncology Center inpatient or outpatient units is entered into the OCIS. Among these data are the routine demographic information, history of prior therapy, laboratory results such as hematology and chemistry data, and special purpose data from research laboratories. In addition, quantitative measurements of the patient's status such as the performance level, pertinent symptoms and physical findings are recorded. This information is collected at each clinic visit for outpatients and daily for patients on the inpatient services.

The data base is kept current by a data entry staff and clinical data coordinators. The former process results as they are delivered to the OCIS. Whenever possible, direct computer to computer transfer is used to transmit data. A data coordinator is part of the clinical unit's health care team and is responsible for the completeness and accuracy of the data base. The coordinator also assists the physician in obtaining the necessary outputs from the system. Most data products are delivered as printed output to the clinical units in time for use during morning rounds. CRT terminals are available at each nursing station and in the outpatient areas for on-line query. Data are available from the CRT within two hours of receipt or collection.

The OCIS operates on dedicated minicomputers (PDP-11/70s) and was developed in MUMPS. The system supports over 60 on-line terminals. The system design facilitates use by physicians and others without formal training in the use of computers. Implementation has been divided into two phases. The first concentrated upon collection and display of an accurate and timely data base. The second phase (currently being introduced) emphasizes interpretation of the data to produce protocol directed care plans and the development of statistical analytic capabilities.

Developmental History of the OCIS

The development of the OCIS was begun in the mid 1970's when plans were made to consolidate various oncology activities in a single facility. While there was no funding for a computer or software, the facility architecture and staffing profiles were predicated upon the assumption that a computer system would be available. Over 250 wall jacks were set into place with cables terminating in a computer room.

During the period of planning, it was decided to begin the development of a prototype system. At this time, outpatients were being treated in a Johns Hopkins clinic, and inpatients were being treated at Baltimore City Hospitals. In each case, procedures had been established for patient management based upon the careful evaluation and monitoring of clinical data. It was recognized that the labor intensive procedures in use for that population could not be extrapolated to support a facility with 56 beds and up to 100 outpatient visits a day. Clearly, computer support was required.

Based upon the Oncology Center's clinical practices, it was determined that an information system must:

(1) Organize and display the clinical data as an integrated report which combined inpatient and outpatient data in ways which could indicate trends and facilitate decision-making.
(2) Manage the basic patient record functions including a hospitalwide tumor registry, summary patient abstracts, current census and appointments.
(3) Schedule patient care activities and therapies as determined by clinical algorithms. Since most cancer therapies involve sequences of actions over long periods of time, the ability of the computer to provide reminders was considered an important tool for dealing with the problem of information overload.
(4) Maintain clinical data for retrospective analysis.

Consequently, in July 1975, one year before the Center was to open, work on a prototype began.

The first task in the development of a prototype was the conversion of an automated tumor registry file into a format which would be expanded to support the needs of the Center. The data base was transferred and programs were developed to produce patient abstracts such as that shown in Figure 1. At the same time, work began in the development of formats which could display clinical data effectively. Flow sheets and printer plots were designed and, after many iterations, the formats presented in the following sections were established.

The prototype system operated on a computer at the Applied Physics Laboratory. The tumor registry was a functional system with a current data base. It ran in the batch mode and was used for searches, reports, abstract generation, and quality assurance. The remainder of the prototype system was used on a limited basis. Because of the difficulty in establishing a current data base, the display system was most effective for retrospective studies; its primary contribution was the establishment of data base format requirements.

In 1976, in part as the result of a gift from the Educational Foundation of America, there were sufficient funds available to purchase a computer and support the development of a system for the Oncology Center. Called the Phase I system, this would convert the prototype to operate as an interactive system on a dedicated computer. A PDP-11 was selected with MUMPS as the development environment. Work on the prototype system stopped, and the development of the new system began.

```
          JOHNS HOPKINS HOSPITAL          HISTORY NO: 002826933
            ONCOLOGY CENTER                     NAME: **********
                                                DATE: 04/12/82
                                                                CLASSONCOLOGY
THURMONT, MD  21788                                             SEER 021/21788
SOCIAL SECURITY NO
SEX MALE                RACE CAUCASIAN      MAR. STATUS PEDIATRIC (16 OR UNDER)

:--------------------------------------------------------------------------:
: ADMINISTRATIVE INFORMATION  PRIMARY - 1                                  :
:--------------------------------------------------------------------------:
:ADMISSION DATE:  12/18/81          DISCHARGE DATE:  01/06/82              :
:PRELIMINARY: NO                    REGISTRY REPORTABLE: YES               :
:CCPDS COMPATABILITY FLAG: YES      ACCS NO: 812529                        :
:FIRST REPORTED TO CCPDS:                                                  :
:ABSTRACTOR: GMS                    MICROFILM: N                           :
:FIRST DATA ENTRY ID: JMK           DATE OF INITIAL GENERATION: 01/08/82   :
:LAST DATA ENTRY ID: JMK            DATE OF LAST GENERATION: 04/12/82      :
:--------------------------------------------------------------------------:
: DIAGNOSIS PRIMARY - 1                                                    :
:--------------------------------------------------------------------------:
:                                                                          :
:    JHH DIAGNOSIS        12/22/81  WILM'S TUMOR -STAGE IV (LUNG           :
:                                   METASTASES)                            :
:                                                                          :
:SITE:                    189.0     KIDNEY, NOS                            :
:MORPHOLOGY:              8960/39   NEPHROBLASTOMA, NOS                    :
:GRADE:                   NOT DETERMINED, N/S, N/A                         :
:LATERALITY:              LEFT ORIGIN OF PRIMARY                           :
:EXTENT:                  DISTANT                                          :
:    METASTATIC SITES:         PULMONARY                                   :
:BASIS OF DIAGNOSIS:      HISTOLOGY, X-RAY (CHEST)                         :
:                                                                          :
:REPORT SECTION                                                            :
:                                                                          :
:     8119934 1. LEFT KIDNEY (NEPHRECTOMY): NEPHROBLASTOMA (WILM'S TUMOR   :
:WITH CAPSULAR PENETRATION, EXTENSION TO ADJACENT NORMAL KIDNEY AND CALYCEAL :
:SYSTEM, VASCULAR AND URETERAL MARGINS NEGATIVE FOR TUMOR. 2. URETER:      :
:NEGATIVE FOR TUMOR. 3. RETROPERITONEAL LYMPH NODE: NEGATIVE FOR TUMOR.    :
:NOTE: HISTOLOGY IS FAVORABLE.                                             :
:                                                                          :
:                                                                          :
:SPECIAL NOTES:                                                            :
:                                                                          :
:     12/21/81 CHEST X-RAY SHOWS BILATERAL LUNG METASTASES                 :
:                                                                          :
:                                                                          :
:--------------------------------------------------------------------------:
:--------------------------------------------------------------------------:
: TREATMENT PRIMARY - 1                                                    :
:--------------------------------------------------------------------------:
:  FROM  :   TO   : LOC  : TYPE : DESCRIPTION                              :
:12/20/81:01/28/82: JHH  :  R   : LEFT FLANK AND RENAL BED 2100R X 30 DAYS :
:12/22/81:        : JHH  :  S   : LEFT NEPHRECTOMY (81-19934)              :
:12/28/81:        : JHH  :  C   : TREATMENT DAY #1 POG 8000 REGIMEN J -    :
:        :        :      :      : VINCRISTINE, ACTINOMYCIN D, ADRIAMYCIN,  :
:        :        :      :      : CYTOXAN                                  :
:--------------------------------------------------------------------------:
```

Figure 1. Sample abstract.

Table 1. Events in the Implementation of the OCIS

July, 1975	Begin Analysis and Prototype System
December, 1975	Prototype Tumor Registry Operational
July, 1976	Prototype System Complete
October, 1976	Oncology Center Facility Opened
August, 1977	Computer Delivered, Start New System
December, 1977	Data Base Moved to New Computer
April, 1978	Patient Data On-Line
April, 1979	Limited Daily Care Plans Used, 2 Units
August, 1979	Second Computer Installed
June, 1980	Phase I System Complete
January, 1982	Phase II new System in Partial Use
August, 1982	Phase II Conversion Complete

Table 1 contains a brief chronology of the system development activity. As can be seen, the prototype system was converted to an on-line system using the prototype's data base in a period of nine months. The reasons for this rapid implementation were:

(1) the existence of a prototype so that the design requirments were well known;
(2) the use of MUMPS with its powerful data access tools and interactive debugging; and
(3) the fact that all activity was devoted to development with no effort spent on system maintenance.

Development of the system continued and new functions were added: bacteriology reports sorted by date, organism or specimen; appointment system including scheduling of tests; pheresis system to manage blood product donation and transfusion; expansion of the administrative functions to include charge capture and level of care reporting; and the implementation of the daily care plans.

The success of the OCIS began to limit its effectiveness. The system became saturated at 20 jobs, daily production processing required 20 hours a day, it was difficult to add new functions to the existing system, and during periods of computer downtime there was no automated backup. It was therefore decided to add a second computer which would expand functionality and provide a backup capability.

The second computer was installed in 1980. At this time Standard MUMPS was available and the previous version of MUMPS was no longer supported. Because the two versions were sufficiently different, it was decided to reprogram the OCIS rather than convert it. This became known as the Phase II system. It also was decided to use a new programming tool, TEDIUM®, which would provide data base management system functionality and improve system maintainability. (Note: TEDIUM is a trademark of Tedious Enterprises, Inc.) Unfortunately, TEDIUM was only in its initial stage of development and the development staff was faced with:

(1) conversion of what was, by now a complex system;
(2) maintenance of the old MUMPS system; and
(3) the use of a new and poorly documented tool.

The result was a major slip in the schedule and some user frustration. However, the conversion was successful. Parts of the new system were put into operational use in 1981, and the old MUMPS system was retired in 1982.

The two OCIS computers are now linked with distributed data base software; continuous clinical and administrative support is provided. There is a direct link to the Department of Laboratory Medicine's computer system and all test results are automatically transferred to the OCIS. The OCIS computer is also used to support a prototype ambulatory care system and an Oncology

Table 2. OCIS Software and Hardware

OCIS Software (TEDIUM/MUMPS)
2,177 programs
456 relations
1,251 data elements
OCIS Data Base
37,000 Patients
33,000 Tumor Registry Entries
15,000 Abstracts
8,800 Patients with clinical data
216,000 Patient days
4,520,000 Data points
3,149 Patients with micro-bacteriology data
16,000 Patient days
200,000 Reports

Center pharmacy system. Data communication between these systems and the OCIS has been implemented.

Table 2 presents some survey information about the OCIS (not including the pharmacy system or the administrative system). The two-computer configuration supports 60 jobs at any one time. We are now at work on the Phase III system which will have enhanced capabilities for the management and statistical analysis of clinical data for retrospective studies.

This brief historical record has touched upon the software development process. Frequently, this is the critical link in system implementation. Nevertheless, it must be recognized that an information system involves more than computer programs. A staff must be hired and trained, users must be oriented, and the commitment to system use must be made. The installation of a HIS typically takes from six months to two years without any programming activity. In the case of the OCIS, we have been working on this problem for seven years and have developed (and thrown away) a prototype and a Phase I system. What exists today is used and is considered an integral part of patient care and Center operation. But it was neither easy nor fast; and it required the dedicated effort of a great many people to develop.

Management of Patient Data

Optimal clinical management requires an organized approach to medical decisions and the availability of a complete set of facts about the patient and his disease. The computer based data system lends itself directly to this need.

The Abstract

The patient's diagnosis must be documented and the level of reliability by which the diagnosis is known should be recorded. In addition, the sequence of any prior treatment and the response to that treatment must be known if a valid judgement as to the potential of response to the next treatment is to be made. Factors such as estrogen receptor status in breast cancer patients, histologic grade of the tumor and the depth of penetration of a superficial cancer such as a melanoma are variables which influence decisions both at the outset and during the course of a patient's treatment.

To meet this data need, the Oncology Center has developed a format called the Abstract. The Abstract contains the minimum amount of data retained on all patients seen in the Center. Every patient under the care of a physician at the Center has this data summarized from the medical record and stored in the Abstract format. All Abstracts are always available on-line to provide data for physicians who may be seeing the patient for the first time, to give a continuing reference for the reevaluation of treatment plans and to support any change in treatment which may be considered.

Figure 1 shows a sample Abstract. The patient's name, address and other identifiers are suppressed from this illustration. The diagnosis is shown both as an ICD-0 code and in English text. Treatments which have been given are indicated in the treatment section and show surgical procedures, special diagnostic tests, radiation therapy indicating both site and total rads of treatment given and the sequence of chemotherapy which has been administered. A brief section describing symptoms of the onset of disease and family and personal history is also included. A followup section of the Abstract (not shown) indicates the patient's referring physician and the patient's Oncology Center primary doctor. This section is of particular importance in assuring that the referring physician is kept informed as to the patient's status. It is Oncology Center policy to have periodic copies of clinical notes and other information sent to the referring physician so that the community doctor can continue to play an active part in the patient's management throughout the course of treatment and can coordinate the care of the patient in his home environment.

The Abstract is a statement about many important factors in the patient's care, but it is not meant to be a substitute for the complete hospital chart. However, the availability of these data in an on-line mode is an important adjunct to patient care since it provides immediately useful, focused data at any time of the day or night when the availability of the hospital chart may be limited. It also is a demonstration of the ideal minimum amount of information necessary for patient care and serves as an educational tool since the chronological organization of treatments strengthens the student's concept of the therapeutic decisions which have been made. The Abstract also often serves as the initial data base for new clinical research projects under consideration by staff or students.

Data Flows and Plots

The use of prospective clinical treatment protocols with evaluation parameters clearly defined at the outset is common in oncology clinical trials. An extension of this approach can be used to describe the management of many medical problems in which the relationship between laboratory tests, clinical findings and treatment must be known so that the changes brought about by treatment can be measured. This approach requires that specific physical findings, symptoms, laboratory tests and other measures of disease and treatment must be selected at the outset, measured and recorded, and then used in a before/after model to measure the outcome of treatment and the progression or regression of the disease under treatment. The recording and display of these data is carried out by the use of tables and graphs.

Flow Sheets. The OCIS allows physicians to request patient, disease and research protocol oriented flow sheets containing any data items stored in the system in one of two general flow sheet formats. The horizontal flow displays data with the dates in columns. It is the most common format used and is illustrated in Figure 2. A vertical flow is most commonly used to show changes in a few data elements over longer periods of time. A sample is illustrated in Figure 3.

Figure 2 shows seven patient encounters in chronologic sequence. Under each date, the following information is shown: The patient related clinical findings are provided first. This section shows temperature, weight, blood pressure and other "vital signs". This is followed by a graded quantification of symptoms and a direct measurement of the cancer shown in centimeters taken in cross diameter measurements. Hematology values, blood chemistries and treatments are then tabulated. By reading across columns one can see trends in blood counts, the size of measured tumor masses and the exact doses of medications which were administered. All clinical data can be displayed in this format. To shorten the display, physicians can request flow sheets of portions of this table which may be presented in differing sequences useful for specific medical needs. Data in the flow format can be reviewed or ordered on-line by a physician, nurse or other medical personnel. It also can be provided periodically by the system as a "standing order".

Figure 3 shows a vertical flow which allows for more than seven days of data to be displayed for up to seven data items. This figure shows serum and urine protein data for a patient with multiple myeloma. A variety of protein types are listed at the top including serum immunoglobulins and 24 hour urine measurements of total protein and of globulin light chain indicated as BJ protein. Changes in these measurements in relationship to time can be seen by reading down columns thereby providing an overview of the patient's disease as reflected in changes in the abnormal serum immunoglobulins. Although these tables are extremely useful for evaluation of patient course, this format has limitations in that the magnitude of the change is not immediately obvious

JOHNS HOPKINS ONCOLOGY CENTER HISTORY NO:
NAME:
DATE: 05/01/82

STANDARD FLOW

J7715		27JUL78 DAY 50	28JUL78 DAY 51	29JUL78 DAY 52	30JUL78 DAY 53	31JUL78 DAY 54	01AUG78 DAY 55	02AUG78 DAY 56
PAT STATUS								
TEMPERATUR	DEG C	36.9	36.9	37	37.1	37.2	37.1	37.2
BODY WT AM	KG	76.5	77.3	77.0		76.8	77.3	78.3
CARE LEVEL		3	3	3	3	3	3	3
TOTL INTAK	ML	1270	1500	2255	1730	2160	2580	2010
TOTL OUT	ML	2600	4050	2900	4000	2300	2700	2575
HEMATOLOGY								
WBC	#/CU	297*	132*	374*	297*	231*	649*	462*
HCT	%	22.0	24.0	26.0	22.5	21.5	21.5	21.5
PLATELETS	/MM3	143000*	173000	152000	161000	180000	161000	160000
POLYMORPHS	%	100*	100*	100*		100*	100*	500*
CHEMISTRY								
SER NA	MEQ/L	141				145		138
SER K	MEQ/L	3.9*				4.2		4.2
SER CL	MEQ/L	108*				107*		106
SER CO2	MEQ/L	26				26		25
SER UR NIT	MG/DL	15				16		14
SER CREAT	MG/DL	1.3				1.4		1.3
SER GLUC	MG/DL	131*				113		120*
TOT SER PT	GM/DL					5.8*		5.8*
SER ALBUMN	GM/DL					4.4		4.3
TOT BILI	MG/DL					0.7		0.7
SGOT	IU/L					8		7
SER ALKPHO	IU/L					65		64
SER CA	MG/DL					8.4*		8.1*
SER PO4	MG/DL					3.3		3.0*
SER PO4	MG/DL							2.6
SER CHOLES	MG/DL					128		124
SER UR ACD	MG/DL					6.8		6.9
SER MAG	MEQ/L							1.4*
DIAG TESTS								
HBS AG						0		
IGG	MG/DL					356		
IGA	MG/DL					5		
IGM	MG/DL					13		
24H URINE	ML		1900		3700		2900	
UR CREAT	MG/DL		53		57		76	
URINE NA	MEQ/L				70			
URINE K	MEQ/L				20			
URINE CL	MEQ/L				61			
UR TTL PRT	MG/24		1994*		3632*		3682*	
URINE MG	MEQ/L				1.5			
CHEMO								
ANTIBIOTIC								
OTHER MEDS								

Figure 2. Sample flow (horizontal).

JOHNS HOPKINS ONCOLOGY CENTER - 2N-BAH HISTORY NO:
PROVIDER: NAME:
DATE: 07/31/82

1980	IGA MG/DL	IGM MG/DL	IGG MG/DL	24BJ MG/24	24TP MG/24	SEMC GM/DL	SEAB GM/DL
04/16	42	30	4265	-------	-------	4.4	3.0
04/22	-------	-------	-------	848	1484	-------	-------
04/24	36	20	3048	779	1230	3.9	2.9
05/01	42	36	2596	-------	-------	3.0	3.2
05/05	42	44	2221	0	320	-------	-------
05/08	35	36	1864	0	126	2.3	2.9
05/20	-------	-------	-------	-------	-------	1.8	3.6
05/21	-------	-------	-------	-------	102	-------	-------

Figure 3. Sample flow (vertical).

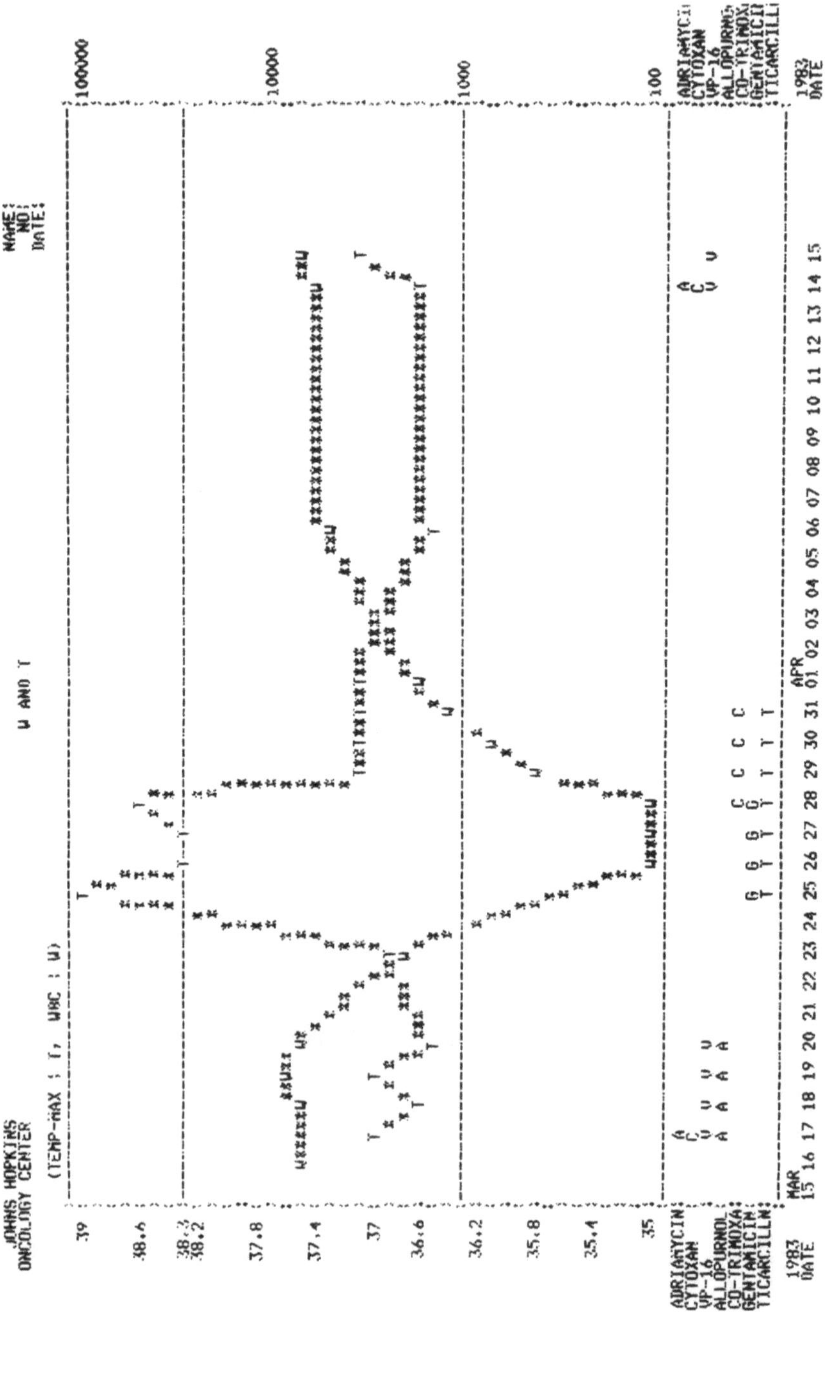

Figure 4. Sample plot.

on a rapid scan of the data and, more importantly, the rate of change is not apparent. The table compresses time intervals between samples and the user must take careful note of the date shown along the left column to be able to recognize temporal changes in the data. The OCIS has overcome this limitation by graphing data.

Plots. Graphs are being used more and more commonly to interpret the relationship between data items, change in data and rate of change. When viewed graphically, many medical problems are seen as a dynamic process and current events are more easily explained in relationship to those which have preceded it. Figure 4 shows data similar to that shown in Figure 2, in a far more understandable format. The relationship between blood counts, temperature, treatments and other events seen over a period of time indicate in a clear fashion the effect of medical treatment. This becomes a powerful tool for planning patient care and for teaching about disease and treatment.

Figure 4 shows the white blood cell count (W) and temperature (T) to be normal at the time that chemotherapy consisting of cyclophosphamide, adriamycin and VP 16-213 are given. Seven days later the white blood cell count has decreased to a level below one thousand per cubic millimeter, a point at which the risk of infection is greatly increased. The plot shows that the white blood cell count remained depressed below the one thousand level for seven days and then returned toward normal passing again through one thousand per cubic millimeter fourteen days after cytotoxic drug treatment. This graph also shows the temperature plotted in parallel. The Oncology Center infectious disease support protocol requires that patients be begun on antibiotics if they develop a fever while their white blood cell count is less than one thousand. Fever is defined as a temperature of 38.3 °C or higher. This graph shows that the temperature did cross that threshold on day 8 and the patient was started on antibiotic support of gentamycin and ticarcillin on that day. A rapid response to the antibiotics is indicated by the temperature fall toward normal. The temperature again rose through 38.3 °C, three days later. Antibiotic support was changed and trimethoprin-sulfamethoxazole was substituted for gentamycin and the temperature promptly responded. The patient was continued on antibiotics until the white blood cell count rose above one thousand at which time antibiotics were discontinued and when no recurrence of the fever was noted the patient was discharged home. A second cycle of treatment is also shown beginning on April 14.

This figure illustrates how clinical support plans and decision points can be defined and specific threshold parameters can be indicated for graphic aids for medical decisions. Using defined clinical care modules, a series of optimal care plans are being devised to help support blood product administration, the administration of intravenous fluids and the management of electrolytes, nutritional support and other medical problems.

Daily Care Plans

In order to understand how an automated system may be used as a therapy planning aid it is necessary to describe the general process of patient care in the Oncology Center. Patients are treated for their cancer as well as for disease or therapy-related medical complications. Much of this therapy follows predefined protocols (treatment sequences) which may be grouped into four categories.

(1) Research protocols. These define a set of therapeutic actions to be followed for a specific population in order to produce a consistent set of data to evaluate a hypothesis.
(2) Individual therapy. This involves the use of anti-tumor drugs generally in multi-drug combinations administered using complex time-sequenced relationships. Therapy may extend for months or years. An example is the use of a drug sequence demonstrated to be effective by a previous research protocol.
(3) General support and response to life threatening crisis. These involve the use of established procedures to deal with anticipated reactions and problems associated with the anti-tumor therapy. In many cases, these actions are documented as part of each research protocol. Examples are (a) the use of antibiotic drugs to treat infections when it is known that the anti-tumor drug will severely lower the white blood cell count and (b) management of fluid balance.
(4) Disease specific continuing care. This entails the recommended long and short-term care and patient monitoring associated with a specific disease independent of the therapy selected. These actions may also be included in the research protocol documentation. Examples are (a) routine six-month chest x-rays for all patients with breast cancer and (b) routine three-month monitoring of serum protein electropheresis for certain multiple myeloma patients.

At any given time an individual patient may be treated using one or more research protocols (e.g., an anti-tumor protocol and an antibiotic protocol). These research protocols may be supplemented by other support protocols (e.g., hypercalemia protocol to help manage high levels of serum calcium) and long-term follow-up (e.g., annual battery of tests). The situation is further complicated by the fact that:

(1) protocols may have different branches in which patients are given different doses of the same drug or combinations of different drugs;
(2) the Center is a teaching institution and thus subject to housestaff rotation;
(3) at any one time there are over 2,000 patients being actively treated by the Center; and
(4) there are presently more than 125 formal research protocols active in the Center.

The OCIS was designed to assist physicians in dealing with these problems by:

(1) maintaining abstracts which summarize previous and current therapies;
(2) presenting plots and flowsheets of current data to aid in decision-making; and
(3) producing daily care plans which integrate the requirements of multiple protocols to produce a patient specific plan.

To understand how this is accomplished, we first describe a research protocol and show how it is decomposed into computer processable segments. We next introduce the daily care plan. Finally, we discuss the operation of the system.

Protocol Description

Protocol 1476 is a multigroup protocol which is being carried out by the Eastern Cooperative Oncology Group (ECOG) for the treatment of advanced stages of Hodgkin's Disease. Thirteen patients at the Center are or have been treated according to the protocol; it is one of thirty-two ECOG protocols in use at the Center. The following discussion is based upon this protocol.

Once a protocol is approved, a document is issued which describes the objective of the study, an overview of the study, the patient selection criteria, identification of data to be collected, the evaluation criteria, modifications to be made in response to toxicity, etc. In the case of protocol 1476, this is incorporated in a 25 page document. To aid the physician in coordinating care, a schema is prepared which contains a pictorial overview of the treatment plan (Figure 5) [9].

Because of the large volume of information about each protocol, there is a need to prepare useful protocol synopses which can be easily accessed by the physician. A notebook containing all active protocol schema is kept in each physician area. More complete information is available from CRT's in the form of a synopsis which includes a general description, patient selection criteria, response evaluation criteria, drug dose modifications (Figure 6) and a schema. The objectives of the online synopsis are to (1) establish a mechanism for describing the protocol prior to its decomposition into a machine processable form and (2) provide a convenient, readily accessible source of information about the protocol. The synopsis is stored as text; some sections, such as dose modification, are organized into tables to permit machine processing.

Translation to a machine processable form begins with the schema which is written in the form of a primitive programming language. The objective of the schema is to show the relationships among blocks of therapy called treatment sequences. To illustrate, consider the portion of the schema shown in Figure 7. Lines which begin with an asterisk are comments; line labels start on the left. Two commands are used: START a treatment sequence and STOP a treatment sequence.

EST 1476
Revised

SCHEMA

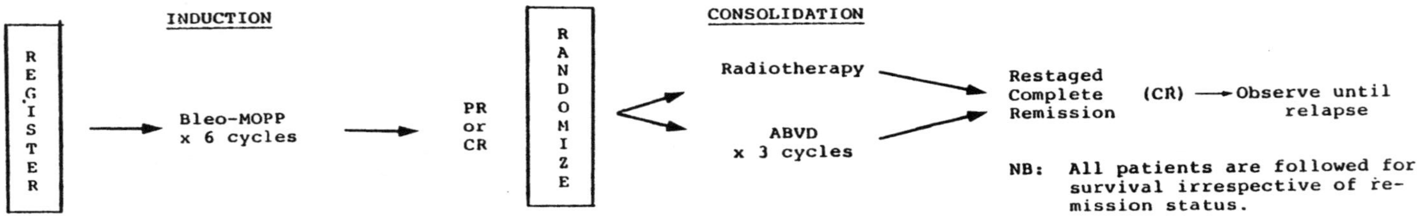

Registration Data:
- Age
- Performance Status
- Stage

Stratification:
- CR/PR
- Age
- Eligibility for Radiotherapy

Bleo-MOPP Regimen*

Bleomycin - 2 mg/M² I.V. days 1 & 8
Nitrogen Mustard - 6 mg/M² I.V. days 1 & 8
Vincristine - 1.4 mg/M² I.V. days 1 & 8
(Maximum dose not to exceed 2 mg)
Procarbazine - 100 mg/M²/day P.O. x 14 days
Prednisone - 40 mg/M²/day P.O. x 14 days
(Every fourth cycle only)

REPEAT CYCLES EVERY 28 DAYS

*See Section 5.11 for details.

ABVD Regimen*

Adriamycin - 25 mg/M² I.V. days 1 & 15
Bleomycin - 2 mg/M² I.V. days 1 & 15
Velban - 6 mg/M² I.V. days 1 & 15
DTIC - 250 mg/M² I.V. days 1 & 15

REPEAT CYCLES EVERY 35 DAYS

*See Section 5.121 for details.

Radiotherapy Regimen*

Sites of Pretreatment Involvement except Bone Marrow
Given over 2-3 weeks except divided doses if involved above & below diaphragm
Nodal Sites - 2000 rads
Organ Sites - 1500 rads at 150 rads/day

*See Section 5.122 for details.

Figure 5. Schema for Eastern Cooperative Oncology Group Protocol 1476.

```
          DRUG DOSE MODIFICATIONS                                           07/02/80

1. HEMATOLOGIC TOXICITY
 BLEO-MOPP/BLEO-MOP DOSE MODIFICATIONS
   WBC              PLTS                  BLEO      HN2      VINC     PROC     PRED
   >4000    AND   >100,000                100%      100%     100%     100%     100%
3000-3999   AND  80-100,000               100%       50%     100%      50%     100%
2000-2999   AND  60-79,000                100%       25%     100%      25%     100%
  <2000     AND    <60,000                OMIT      OMIT     OMIT     OMIT     OMIT

ABVD DOSE MODIFICATIONS
   WBC              PLTS                  ADRIA   BLEO    VELB    DTIC
   >4000    AND   >100,000                100%    100%    100%    100%
3000-3999   AND  80-100,000                50%    100%     50%    100%
2000-2999   AND  60-79,000                 25%    100%     25%    100%
  <2000     AND    <60,000                OMIT    OMIT    OMIT    OMIT

 Patients 60 years or older should be given 50% of the FULL DOSE of
  MYELOSUPPRESSIVE DRUGS for the FIRST CYCLE. IF WELL TOLERATED, FULL DOSE
  can be given in the SUBSEQUENT CYCLES.
 In patients randomized to ABVD CONSOLIDATION, the INVESTIGATOR may ELECT to
  ADMINISTER TWO-THIRDS DOSAGE of ABVD for the FIRST CYCLE IF PREVIOUS
  CYTOPENIAS have been a PROBLEM.
```

Figure 6. Portion of portocol dose modification synopsis.

Protocol 1476 begins by starting a set of baseline studies (treatment sequence "1476 INT") and stopping the sequence on day 2. The induction therapy involves two treatment sequences: "76 BMOPP" which defines the drugs to be given and "1476 FOL" which defines the follow-up studies to be conducted throughout the protocol.

Figure 8 displays the treatment sequence "76 BMOPP." Drugs and their baseline doses are identified and the administration schedule is given. Tests to be ordered in conjunction with the therapy are also specified. The system allows administration by day (or week) from the start of the treatment sequence, by day of week (e.g., M W F), or by days (weeks) since the last administration (e.g., Q2D or Q1W). The "1476 FOL" treatment sequence defines general protocol activities which might be conducted independent of the treatment

```
E1476      SCHEMA                                                          07/02/80

 *          BASELINE STARTS BASELINE WORK-UP (1476 INT)
 BASELINE   START       1476 INT
            END
 *          DAY 2 STOPS 1476 INT
 DAY 2      STOP        1476 INT
            END
 *          IN CYC1 BEGINS CYCLE 1 OF INDUCTION THERAPY (28 DAY CYCLE OF BMOPP)
 *          AND INDUCTION/CONSOLIDATION FOLLOW-UP (1476 FOL)
 *          BEGINS TREATMENT DAY 1
 IN CYC1    START       76 BMOPP,1476 FOL
            END
 *          IF PT IS > 60 Y.O. 50% OF BMOPP SHOULD BE GIVEN FOR 1ST CYCLE
 *          IN THIS CASE  >60 Y.O. SHOULD BE USED RATHER THAN IN CYC1
 >60 Y.O.   START       1476 >60,1476 FOL
            END
 *          IN CYC2 BEGINS CYCLE 2 OF INDUCTION THERAPY (28 DAY CYCLE OF BMOP)
 *          BEGINS AFTER COMPLETION OF INDUCTION CYCLE 1 (BMOPP)
 IN CYC2    STOP        76 BMOPP
            START       1476BMOP
            END
```

Figure 7. Initial portion of synopsis schema.

```
                              Oncology Information System                  06/07/83
                                 Treatment Sequences

(76 BMOPP)    1476 BMOPP CYCLE

  TREATMENT TYPE                                      SCHEDULE
-------------------------------------------------------------------------------------

    BLOOD PROCEDURES
      HEMOGLOBIN   (HGB)                              M,W,F
      HEMATOCRIT   (HCT)                              M,W,F
      WBC   (WBC)                                     Q2D
      PLT CNT   (PLTS)                                Q2D
      WBC DIFF   (DIFF)                               M,W,F
      S UR ACID   (SURC)                              1
      S ALK PHOS   (SALK)                             1,8
      SGPT   (SGPT)                                   1,8
      SGOT   (SGOT)                                   1,8
      SER CREAT   (SCR)                               1
      S UREA NIT   (SUN)                              1

    THERAPY
      BLEOMYCIN       2 MG/M2                         1,8
        I.V.
      NITR MUST       6 MG/M2                         1,8
        I.V.
      PROCARB         100 MG.M2                       1,2,3,4,5,6,7,8,9,10,11,12,13,14
        P.O.
      VINCRIST        1.4 MG/M2                       1,8
        I.V. MAX SINGLE DOSE 2.0  MG
      PREDNISONE      40 MG/M2                        1,2,3,4,5,6,7,8,9,10,11,12,13,14
        P.O.

    COMMENTS
      LEVEL  1     (GENERAL COMMENTS)
       (1)   TUM MEAS                                 1
         ENTER TUMOR MEASUREMENTS :
       (2)   TUM MEAS1                                1
         (1) --------      -----------------          --------- CM  X  -------- CM
       (3)   76 BMOPP13                               1
         MEASURE LONGEST DIAMETER & ITS PERPENDICULAR DIAMETER IN CM
```

Figure 8. Portion of treatment sequence 76 BMOPP.

cycle. It should be noted that each treatment sequence is independent and fully defined; this allows the use of an arbitrary number of active treatment sequences for any patient. It is also possible to define treatment sequences without a protocol.

The Care Plans

The daily care plan is produced by processing the active treatment sequences for a patient against his current therapy data. The daily care plan is printed in several different formats. A physician plan summarizes information about status and tests to be ordered, a general plan contains more complete information, and an order guide is used by the ward clerk to process requisitions and by the phlebotomist to obtain specimens.

The general daily care plan, as shown in Figure 9, contains:

Patient status data. Data elements include height, weight, body surface area, admission weight, and blood type.

Therapy summary. This contains protocols with associated start dates, physicians, and general comments.

Treatment sequence generated comments. These are grouped into several categories with the most important being printed in a box to command attention.

Tumor measurements. Where there are solid tumors, a section records initial and current values and computes total cross-section area and percent change.

Clinical findings. These are functions of the clinical data such as percent change in the admission weight, cumulative drug dose, or a maximum recorded laboratory value.

```
               JOHNS HOPKINS                    HISTORY NO:
              ONCOLOGY CENTER                         NAME:
                                                 PLAN DATE: 08/06/80
          GENERAL DAILY CARE PLAN                           WEDNESDAY

20 W F     HODGKIN'S DISEASE
HT  155          ADM WT  44       BAS  1.4    IDEAL WT  45    WT(08/05/80)  43.2
BLOOD TYPE  O+         HLA
PROTOCOL   E1475   BLEO-MOPP-XET     DAY 29       STARTED  07/09/80
                   SECOND CYCLE - BLEO MOPP  8/6/80

PROVIDER  LENHARD, DR R


PROTOCOL GENERATED INSTRUCTIONS

    ***********************************************************************
    *  MAX CUMLATIVE DOSE OF BLEOMYCIN IS 36 MG/M2                        *
    ***********************************************************************

       ENTER PERFORMANCE STATUS
       ENTER TUMOR MEASUREMENTS
       MEASURE LONGEST DIAMETER & ITS PERPENDICULAR DIAMETER IN CM
       SEE PROTOCOL (SECTION 5.21) FOR CHEMO DOSE MODIFICATION IF WBC <
         4000 OR PLTS < 100000
       SEE PROTOCOL (SECTION 5.12) FOR DOSE MODIFICATION IF TOTAL BLIL >
         1.5

MEASURABLE TUMOR SITES
                                                INITIAL               LAST

    ID   GRID  O/T    M/METHOD     SIZE (CM)      DATE     SIZE (CM)     DATE
    A                              9.7  7.4   07/09/80    9.7  7.4   07/09/80
                                   --------               --------
                   TOTAL AREA        71.78                  71.78

RECENT CLINICAL FINDINGS
   VALUE        DESCRIPTION                           LAST FINDING  STARTING   THROUGH
  5.6         CUMULATIVE TOTAL BLEOMYCIN                07/16/80    07/09/80   08/07/80
  44          INITIAL BODY WT AM                        07/09/80    07/09/80   07/09/80
  43.2        LAST BODY WT AM                           08/05/80    07/10/80   08/07/80
  97          PERCENT BODY WT AM BASED ON 44            08/05/80    07/10/80   08/07/80
  0.7         MAXIMUM SER CREAT                         08/04/80    07/09/80   08/07/80
  9100        MINIMUM WBC                               07/16/80    07/09/80   08/07/80
  704000      MINIMUM PLATELETS                         07/16/80    07/09/80   08/07/80

CHEMOTHERAPY DRUG HISTORY

    NAME                DOSE           GIVEN              % BASELINE
    NITR MUST           8.3 MG         07/16/80           100 % OF 6 MG/M2
    PROCARB             120 MG         07/22/80           85% OF 140 MG
    VINCRIST            1.9 MG         07/16/80           95% OF 2.0 MG
    PREDNISONE          56 MG          07/22/80           100% OF 40 MG/M2
    BLEOMYCIN           2.8 U          07/16/80           100% OF 2.8 U

DRUGS TO BE ORDERED ON 08/06/80

    2.80 MG BLEOMYCIN I.V., BASED ON 1.4 M2 X 2 MG/M2.
    8.40 MG NITROGEN MUSTARD I.V., BASED ON 1.4 M2 X 6 MG/M2
    1.96 MG VINCRISTINE I.V. MAX SINGLE DOSE 2.0 MG.,
       BASED ON 1.4 M2 X 1.4 MG/M2.
    140 MG PROCARBAZINE P.O., BASED ON 1.4 M2 X 100 MG/M2.
```

Figure 9. Sample general daily care plan.

```
          JOHNS HOPKINS                       HISTORY NO:
          ONCOLOGY CENTER                           NAME:
                                             PLAN DATE: 08/06/80
      GENERAL DAILY CARE PLAN                            WEDNESDAY

TESTS AND PROCEDURES WITH DATE LAST PERFORMED (> INDICATES ORDERED 08/06/80,
                                               * ORDERED STAT 08/06/80)

BLOOD AND SERUM TESTS
    >HCT   HEMATOCRIT      08/05/80          >DIFF  WBC DIFF        08/05/80
    >PLTS  PLT CNT         08/05/80          >WBC   WBC             08/05/80
    >HGB   HEMOGLOBIN      08/05/80           M12   SMA-12          08/01/80
     IGSV  IG SURVEY       07/04/80           SCU   SER COPPER      07/04/80
     M6    SMA-6           08/04/80          >SALK  S ALK PHOS      08/04/80
    >SGOT  SGOT            08/01/80          >SGPT  SGPT            08/01/80
    >TBLI  TOT BILI        08/01/80          >SUN   S UREA NIT      08/04/81
    >SCR   SER CREAT       08/04/80          >SURC  S UR ACID       08/01/80
RADIOLOGY PROCEDURES
     CXR   CHEST XRAY      07/08/80
PHYSICIAN PROCEDURES

    REPEAT ANY X-RAY TO DOCUMENT RESPONSE OR PROGRESSION
P.A. PROCEDURES
     MUMP  MUMPS SKIN      07/04/80           TRIC  TRICH SKIN      07/04/80
     PPD   PPD SKIN        07/04/80
DATA COORDINATOR INSTRUCTIONS

    HAS DISEASE PROGRESSED? CHECK WITH M.D. RE: STOPPING TX SEQUENCES
```

Figure 9 (continued). Sample general daily care plan. *Note:* (1) This plan is for the first day of the 2nd cycle of therapy. The comment in the box warns the physician that one of the drugs to be administered, Bleomycin, has a maximum cumulative dose. (2) The sections on measureable tumor sites, recent clinical findings and chemotherapy history are automatically updated from the clinical data base each day. The actual drug dose is computed on the basis of a body surface area (BSA, see top of plan) of 1.4. The general plan includes a list of all tests and procedures required for the protocol. Twelve tests are to be ordered from this plan. (3) General physician comments are listed at the beginning of the plan. Messages related to a procedure are listed in the procedure ordering section. Most messages tend to be reminders. The next version of the system will be able to generate warning messages based upon data trends.

Chemotherapy orders. Where a treatment sequence indicates that a drug is required, the fact is listed. The notice may also include the dose computed as a function of weight, body surface area or ideal body weight.

Dose modifications. When a drug is to be administered, the rules for dose modification may be printed in the format shown in Figure 6. (Not shown in Figure 9.)

Chemotherapy history. This is a flowsheet containing the date and dose of the last administration for each anti-tumor drug. If a baseline dose is available, the percent of the baseline is computed.

Tests and procedures. Depending on the format, this will contain tests and procedures to be ordered as determined by the active treatment sequence. STAT (immediate) tests are flagged and the date the test was last ordered is given.

Conclusion

The OCIS has been in operational use since 1977. All inpatient blood and urine tests have been ordered by care plans since early 1980. Support comes entirely from patient revenues. While there are no funds available to support a formal evaluation of the OCIS, it is clear that the system is used and is perceived to be useful. With the dramatic decrease in equipment costs, many of the OCIS functions now can be implemented with small, less expensive units. Hopefully, the concepts developed and tested at the Oncology Center will be transferred to this hardware for use by other tertiary care and chronic care facilities.

References

[1] Blum, B., and R. Lenhard, Jr., "Privacy and Security in an Oncology Information System," *Second Symposium on Computer Application in Medical Care,* November, 1978, pp. 5–9. Reprinted in V. Sondak, *Computers in Medical Administration.* Dedham, MA: Artech House, Inc., 1979, pp. 500–508.

[2] Blum, B. I., and R. E. Lenhard, Jr. "An Oncology Clinical Information System," *Computer Magazine* **12,** No. 11, November, 1979, pp. 42–50.

[3] Blum, B. I., and R. E. Lenhard, Jr., "Experience in Implementing a Clinical Information System," *MEDINFO 80 Conference,* Tokyo, Japan, September 29–October 4, 1980.

[4] Blum, Bruce I., "A Tool for Developing Information Systems," in H. J. Schneider and A. I. Wasserman, Eds., *Automated Tools for Information Systems Design.* New York: North-Holland Publishing Co., 1982, pp. 215–235.

[5] Blum, B. I., and R. E. Lenhard, "Displaying Clinical Data for Decision Making," *J. Clinical Eng.* **8,** 1983, pp. 61–68.

[6] Lenhard, R. E., Jr., B. I. Blum, J. M. Sunderland, H. G. Braine, and Rein Saral, "The Johns Hopkins Oncology Clinical Information System," *SCAMC-6,* 1982, pp. 28–43. Revised for *Journal of Medical Systems,* (7,2) April, 1983, pp. 147–174.

[7] Lenhard, R. E., Jr., and B. I. Blum, "Practical Applications of OCIS, A Clinical Information System for Oncology," *Computers in Biology and Medicine,* (in press).

[8] McColligan, E. E., B. I. Blum, R. E. Lenhard, and M. B. Johnson, "The Human Element in Computer Generated Patient Management Plans," *Tenth Annual Conf. of the Society for Computer Medicine,* San Diego, CA. September 24–27, 1980. Reprinted in *J. Clinical Computing* **IX,** No. 6, 1981, pp. 175–191, and also *J. Medical Systems* **6,** No. 3, 1982, pp. 265–276.

[9] Carbone, Paul P. Personal communication.

32

Data Management Systems in Clinical Research*

Chester King, Larry Manire, Robert M. Strong, Ph.D., and Larry Goldstein

Introduction

Data management consists of everything that happens to data from the process of defining what is to be collected up to its analysis. Computerized data management implies facilities for entering, storing, and retrieving data in a form suitable for analysis. The capabilities of data management systems fall into a continuum between those of file management systems and those of database management systems (DBMS). There is no sharp distinction between the two.

Traditional file management systems store data in unrelated files whose structure forces the user to be concerned with the physical layout of the underlying data storage system. Such systems may require a modification in the way the data is stored before it can be processed for new applications. This creates redundant files which lead to costly and error-prone data entry, update, or deletion.

The database management system has been recognized as an efficient and cost-effective tool in scientific, commercial, and military applications for over a decade. Its use in medical and public health research has lagged considerably because systems designed for other applications have not been able to capture the richness of medical data and have not featured sufficiently helpful or friendly user interfaces for the nonprogrammer. Researchers have not therefore been able to appreciate what a DBMS is and how it reduces the programming needed with older techniques.

A DBMS generally includes the following functions:

(1) defining and restricting the data items to be stored and organizing them into a database structure (database definition);
(2) entering values and checking their consistency with the item definition or the contents of the database (data entry);

*This chapter was prepared expressly for this book. A more complete discussion of the topics is contained in "Health Systems at Harvard University." Boston, MA: School of Public Health, Harvard University. A preliminary version was previously published in *Proceedings of the Fourth Annual Symposium on Computer Applications in Medical Care*, IEEE, 1980.

(3) editing previously entered values;
(4) controlling access to data;
(5) creating subsets of data objects, such as patients, characterized by combinations of conditions on data elements such as height, weight, diagnosis, etc., (querying); and
(6) generating reports, including reports in a tabular format suitable for statistical processing (retrieval).

The DBMSs we review perform most of these functions. All the systems permit a programmer to write further programs beyond the prepared package, though with a great variation in ease and cost.

A DBMS stores data into organized structures called databases. The data is stored with a minimum of duplication and can be used for a variety of purposes by means of the general functions—queries, report generation, or statistical analysis—and of special purpose programs. The data can have varying degrees of independence from the functions or programs that process it. At a minimum, users do not need to restructure their data for different kinds of applications. Symbolic referencing instead of positional referencing of data is one indicator of freedom from the complexities of physical storage structure. More independence is achieved when changes in the database structure, such as the addition of a new data item, do not require changes in the application programs. Data independence permits the kind of ad-hoc, unplanned interrogation of data which researchers need. Freedom from concerns with the underlying physical structure allows a more conceptual approach where users seek to represent their view of the world and not the computer's. Thus users can get much closer to their data.

We review four systems based on the file management concept: SAS, RS/1, PROPHET, and dBaseII. We review the DBMSs which run on minicomputers : CLINFO, MISAR, MEDUS/A, INGRES, and ORACLE.

We review two DBMSs running on mainframes, TOD and FOCUS. We review SIR, which runs on mainframes and large minicomputers such as DEC's VAX.

We review the most popular microcomputer data management system to date: dBaseII (with a 39% market share for the small micros). SIR, INGRES, ORACLE, MEDUS/A, TOD, and FOCUS also have or will soon have microcomputer versions.

Our review is based on published articles and manuals. Note that the Health Systems Project is responsible for the development and dissemination of one of the systems, MEDUS/A.

This article is a condensation of a booklet which gives readers an idea of the capabilities of each system, enabling them to compare the systems and seek more information from the designers.

For each system in the booklet we describe the file or database structure, the database definition process, the data entry, retrieval, and analysis process, and the installation considerations.

We have not reviewed a number of mainframe systems which require a large programming staff, such as TOTAL, DEC's DBMS, System 2000, and IDMS. We have also left out the myriads of systems which were designed for specific applications and are usually not portable to other sites.

We have had all of the systems descriptions in the booklet reviewed by users. However, even this is no guarantee of accuracy; we have learnt of both favorable and unfavorable aspects of systems long after the original 1980 survey was in print.

Criteria in Using File and Database Management Systems

Perhaps the most obvious difference between systems is how users interact with them:

(1) through a command language which supposes a familiarity with the commands and how they are used or
(2) through a menu driven dialog, which gives users a choice of options from a menu.

The system then goes through a script with a question and answer dialog which finds out what tasks the user wants done, explains to him how to do it, prompts him for the needed information, and finally performs the tasks. The latter method is far more friendly to the unsophisticates as it is less taxing, usually incorporates useful on-line documentation, or help messages, and relieves the user from numerous trips to the innards of a thick manual.

The command language found in most of the systems we review is termed nonprocedural. By contrast, procedural languages, such as Fortran, Cobol, Basic, and MUMPS, are frankly programmer-oriented languages which specify to the computer what has to be done to perform a task rather than what task the user wants done.

Database Definition

Once users decide how to organize data, they set up the tables in a file manager or enter the database definition describing their view of the data into the DBMS. Such a definition includes the overall structure and the characteristics of each data item, such as maximum length, range of values, list of values, data types, and so on. The data type determines the kind of data which can be stored for a particular type of item—e.g., integer, date, text, and so on. Systems differ in how natural the data types are. Some will have one type of integer, others will have several types of integers based on the underlying operating system's characteristics, which are irrelevant to researchers' needs. Systems also differ in their ability to represent data—some lack dates or times for instance. Figure 1 is a table comparing what data types are found in a selection of the systems reviewed.

Other parts of a DBMS use the database definition to customize their pro-

TYPE	SAS	RS/1	TOD	CLIN	MISAR	MEDUSA	SIR	INGRES	ORACL	FOCUS	DBAS2
NUMBER		X		X	X						
INTEGER	X		X			X	X	X	X	X	X
REAL NUMBER	X		X			X	X	X	X	X	X
TEXT	X	X	X	X	X	X	X	X	X	X	X
LIST			X	X	X	X		X	X		
LIST + OTHER						X			X		
DATE	X	X	X	X	X	X	X	X	X		
TIME	X	X*		X	X	X	X	X	X		
COMPUTED	X	X	X		X	X	X	X	X		

Figure 1. Data types supported by representative systems. *Note:* (1) Some systems differentiate between integers and real numbers, others represent both within one type. "List" refers to multiple choice list. "List + other" means that values which do not conform to one of the choices can also be represented, though systems differ as to the ease with which they manage this. For the time data type, * means that this is only supported on the VAX. The presence of a data type in a system does not necessarily mean the system can process data represented by the data type. This is particularly true of the text data type, for which Prophet, TOD, and CLINFO, among others have, have no retrieval syntax. (2) In this and other charts, the absence of a feature does not preclude programming it in, for a cost which depends on the efficiency of the underlying language and the state of the internals of the system (documentation, modularity, etc.).

grams for the user's application. DBMSs differ in how easy it is to manage this database definition: many create a number of the definitional characteristics through the data entry routines as part of validity checks. Users find that they have to keep track of far more elements with such systems. Each time a new questionnaire is set up involving some of the same data items as in a previous one, great care must be taken to duplicate the same item characteristics or risk losing data validity and integrity.

Other issues arise during querying, as the lack of an ability to check proposed queries against a full list of item characteristics reduces the system's ability to prevent nonsense queries, thereby taking up researchers' time.

The best database definition is one which is used by all parts of the system and thus provides the most self-documentation in the system as users can refer to it to find out an item's meaning no matter what task they are undertaking. This greatly reduces the memory burden on users. Such a database definition can be called "fully integrated."

Data Entry

Users enter data through a data entry facility or through a user-designed questionnaire. Some systems only allow for programmers to create data entry questionnaires. The DBMSs perform varying degrees of data checks. Figure 2

TYPE	SAS	RS/1	TOD	CLIN	MISAR	MEDUSA	SIR	INGRES	ORACL	FOCUS	DBAS2
DATA TYPE			X	X	X	X	X	X	X	X	X
RANGE			X	X	X	X	X	X	X	X	
LENGTH				X	X	X	X	X	X	X	
PATTERN						X		X	X	X	X
DUPLICATE			X			X		X	X	X	
RELATIONAL					X	X	X	X	X	X	
SECURITY						X			X		

Figure 2. Data checking options. *Note:* (1) PATTERN refers to pattern match, a check against a predefined pattern. DUPLICATE entry refers not to the entry of duplicate top-level cases, such as patients, which all systems implementing data checks do, but to entry of duplicate lower level items, which in TOD is limited to visits. (2) SECURITY refers not to access to a particular database, which all systems with any access controls implement, but rather to access to a particular case, which in a relational system means a row in a table and in a hierarchical one, a branch item. (3) SAS has a forms package which sets up data entry checking now available. Since there is no data dictionary in SAS, it does not reference the item definition for checking but rather has the user specify checks for each form. This forms package has not yet been reviewed.

TYPE	SAS	RS/1	TOD	CLIN	MISAR	MEDUSA	SIR	INGRES	ORACL	FOCUS	DBAS2
FACILITY		X	X	X	X	X					X
QUESTIONS			X		X	X	X	X	X	X	X
SEQUENCE			X	X	X	X	X	X	X	X	
HELP			X		X	X	X	X	X		
BRANCHING					X	X	X			X	
WORK SPACE			DB	DB	BI	LV	DB	BI	BI	LV	DB

Figure 3. Data entry options. *Note:* (1) FACILITY means a standard data entry facility available after database definition. QUESTIONS refers to language-based questionnaire. SEQUENCE refers to control of prompting sequence, whether it can be modified. (2) HELP refers to an interactive help function to give a description of the item during data entry. BRANCHING refers to branching logic, the built-in skipping based on previously entered data values. (3) WORK SPACE refers to the size of the work space (locking granularity) which any one user can access at the same time. (DB = data base, BI = branch item or table, LV = leaf value, value of an individual data item.)

shows how the systems reviewed differ in this regard. Some systems offer screen-oriented data entry; these should be evaluated in terms of how integrated the database definition is since clinical research has demanding quality control needs which are hard to meet if the database definition elements are in fact only found in the forms package, where multiple forms can thus set up inconsistent definitions for the same item.

Systems differ in how much on-line assistance is offered the data enterer, in the user's ability to define how the system prompts for data entry, in whether questionnaires can branch to different questions based on the data enterer's responses, and in their data editing styles. Such branching is useful in the world of medicine, where data is frequently sparse since no one patient has all diseases or is given all lab tests. Figure 3 shows some of these data entry features for a selected group of the systems reviewed.

Retrieval

Retrieval is accomplished through programs that present varying amounts of user options for formatting and selecting contents. These abilities are important if the activities of the DBMS are to be integrated with those of patient care. Such integration is desirable in order to maximize the incentives users have to keep good quality data.

Data entry and sometimes retrieval can be characterized by the number of users who can have access to the database at one time. A key consideration is the locking granularity, which is the granularity, or size, of the record unit from which others are isolated while a user works on it. The granularity ranges from the entire database to the item value. Thus there are systems which permit only one user to enter data for a database at a time to others which permit one user for a certain section to others which permit one user per data item. The locking granularity, or size of the work space, is reviewed in Figure 3.

Analysis

Analysis is usually accomplished in two stages: first, the selection of subsets of the data based on logical criteria and second, the analysis of such subsets via statistical programs. Systems vary widely in the ease and speed with which such queries are done. They also differ considerably in the richness of their query system, how detailed and varied the questions they support are. Figure 4 is a table comparing some query system features across systems. A key consideration for researchers and physicians is how easy such query systems are to use, whether they work through an interactive dialog or assume knowledge of a query language, be it procedural or nonprocedural.

Some systems have statistical programs or allow users to transfer selected data to full-fledged statistical packages such as SAS, BMDP, or SPSS.

Systems differ greatly in the ease with which they implement access control provisions and how extensive these are. The provisions range from control over users' access to capabilities and access into particular data items to ownership of individual cases within the same database. These provisions are naturally of great importance in many medical environments.

No matter how powerful the nonprocedural language or how extensive the user interfaces, there will always be a need in more complex projects for procedural programming. The main issue here is whether the system possesses a builtin language, known as a data manipulation language, or whether a conventional high level language such as MUMPS, Basic, or Fortran, must be resorted to. In the case of the built in language, this usually implies symbolic reference, or reference by name, to data items, as opposed to reference to a file location which implies knowledge of the system's internal structure. In the case of a high level language, the costs involved are dependent on the programming efficiency of the language and how disciplined and well documented the system's internal structure is.

Installation considerations include the hardware and operating system requirements and the degree of support and documentation provided by the designers. Such considerations are affected by the systems' ability to manage sparse data physically on the storage media; some have to reserve space on the media for data which is not entered—for those diseases the patient does not

TYPE	SAS	RS/1	TOD	CLIN	MISAR	MEDUSA	SIR	INGRES	ORACL	FOCUS	DBAS2
EQUALS					X	X	X	X	X	X	X
LESS,GREATER		X	X	X	X	X	X	X	X	X	X
IN RANGE			X	X		X			X	X	
IN STRING		X			X	X		X	X	X	
PATTERN					X	X	X	X	X		
ALPHA ORDER					X	X					
FIRST,LAST				X		X				X	X
MISSING			X		X	X	X				
DIALOG			X	X	X	X					

Figure 4. Query system features. *Note:* (1) SAS has no query system, though it does have a procedural retrieval language. LESS,GREATER is "less than" and "greater than", IN RANGE is within a range. (2) IN STRING, or string contains, asks whether a text item contains a specified string of characters. ALPHA ORDER means whether a string follows a given string alphabetically. MISSING refers to a test for a missing value code common to all data types. All systems have a null, but one of the nightmares of biostatistics is distinguishing the null value from the missing value. DIALOG refers to whether an interactive dialog is available to support users as they formulate questions on the database.

have. Systems also differ in their ability to efficiently manage the storage space in case data is deleted.

Database Structure

The file or database structure is the user's view of how the system organizes and represents information. Most file management systems organize their data into tables whose rows or columns represent data items. A "data item" is an abstraction, usually named, which takes on a value or values when associated with a particular object, such as the weight of a patient or the date of a visit.

The DBMSs we survey all have either a hierarchical or a relational database structure, or data model, which we describe below. The richness of the data model determines how close a representation of reality the stored data can be and therefore how spontaneous, easy, and doable future investigations can be.

Relational Model

Relational systems organize data into tables as do the file managers. However, they provide for ways of relating one table to another by means of "keys," or identifiers. Thus one table might have patient's demographic data while another might have the repeating visit data. The demographic table would have its repeating element patient identified by the patient's ID. The visit table would have its repeating element visit identified by both the ID and the visit number. For multilevel or multitable material, the repetition of numerous keys can be a burden.

Patient Table

ID	Name	Sex	DOB	City	Phone
001	Paul	M	1956	Boston	732-0977
002	Mary	F	1950	Devon	980-7856
249	Scott	M	1934	Gary	345-9085

Visit Table

ID	Visit	Date	Weight	Temp	Rate
001	1	1/3/80	145	98.0	65
001	2	1/4/81	156	98.6	71
249	1	2/3/82	189	114.7	86

Tables are easy data structures to understand and fine for certain applications. In particular, tables are useful for depicting flow sheets, laboratory data, and so on. One should distinguish between data representation and the data model; different data models may be able to represent data in the same way, so that both relational and hierarchical systems can represent flowsheets or other tabular data.

Whenever any significant amount of relationships have to be established between tables, the cognitive complexity increases considerably. A mathematically complete way of describing and manipulating tables is known as relational calculus.

Relational systems are ill suited to hierarchical data; update and deletion problems appear. For instance, if you delete a patient from the above table, his visits are not automatically deleted. Some relational systems allow users to set up links between tables on a permanent basis. However, not all of the system's functionality can work with these hierarchical-like structures. In particular, such joins are usually virtual tables that are copied over at the time of creation, so that any subsequent modifications to the database will not be automatically reflected in the join table. If the join tables are dynamically updated as their base tables are, there may be update problems: e.g., patients who are deleted will not see their visits deleted.

Even with one join, the typically multilevel medical data causes quite awkward representation problems because joins create new tables out of old ones. For this to happen, the demographic data is repeated for each row containing visit data and the tables become large and unwieldly and suffer from query inclusion/exclusion problems. The complexities associated with these issues increase considerably in multilevel situations, such as patients who have hospitalizations, each with several procedures.

The sparseness of medical data also creates problems: if there are several measurements of a data item, one column in the table must be reserved for each measurement. Typically not all of these are filled in and there will be problems in queries and reports.

Hierarchical Model

A hierarchy consists of data items organized into an inverted tree: the top item, such as "patient," forms the trunk, branching items, such as "visit," branch to further branching items or to leaf items, which contain data pertinent to the branching item above them. (See Figure 5.) The meaning of "data item" is extended in a hierarchical database to include entities such as patients or visits which are described by subordinate items.

Branching items are convenient ways of grouping logically related data. (See Figure 6.) Repeated branching items and their subitems represent repeat groupings of data characterizing the top item. Repeat groups of data are common in medical record data where one group might be laboratory data, another, clinic visit data, and a third, hospitalization data, all collected at dif-

```
                          TRUNK=PATIENT
                                |
                                |
                                |
 ------------------------------------------------------------
 |                 |                  |                  |
VISIT#1          VISIT#2           BRANCH#3           BRANCH#4
 |                 |                  |                  |
 |                 |                  |                  |
WEIGHT           LEAF#1            LEAF#1             LEAF#1
TEMP             LEAF#2            LEAF#2             LEAF#2
DIAG             LEAF#3            LEAF#3             LEAF#3
LEAF#4           LEAF#4            LEAF#4             LEAF#4
  .                .                 .                  .
  .                .                 .                  .
  .                .                 .                  .
```

Figure 5. A three level hierarchy. The trunk is often PATIENT, the branch, VISIT. Each branch acts as a heading for subordinate leaf items, which actually carry the data values. These repeat as a group when the branch does.

TYPE	SAS	RS/1	TOD	CLIN	MISAR	MEDUSA	SIR	INGRES	ORACL	FOCUS	DBAS2
MED. RES.			X	X	X	X					
DATA MODEL	F	F	H	H	H	H	H	R	R	H	F
DBDEF DRIVE			X	X	X	X					
DBDEF HELP						X					
DIALOG			DERQ	DERQ	DERQ	DERQ		EQ	EQ		DE
COMMAND LANG		X				X	X	X	X		
DATA LANG	X	X				X	X			X	X
HOST INTER	X					X	X	X	X	X	

Figure 6. Types of system interaction. *Note:* (1) MED. RES. means the system was designed specifically for medical research. DATA MODEL refers to the way data is structured in the system. F stands for file, as in file manager, H stands for hierarchical, and R stands for relational. These terms are explained in a section on database structure. (2) DBDEF DRIVE means the database definition drives, or customizes, the standard functions in the system—functions present without programming once the database has been defined. DBDEF HELP refers to whether the database definition is used to generate database specific help messages in data entry, report making, or querying. (3) The next four elements are arranged so that the ones requiring less training come first. DIALOG refers to whether an interactive dialog is available to aid users through a question and answer session, where D stands for database definition, E for data entry, R for reports, and Q for querying. (4) COMMAND LANG stands for nonprocedural command language, where a request is formulated using a precise syntax without having to specify how the request is to be performed other than in general, user-oriented terms. DATA LANG refers to data manipulation language, a built in programming language which refers to database items by name and protects the programmer from having to know the internals of the system. (5) HOST refers to a host language interface, the ability to use standard high level programming languages, such as Cobol, Fortran, or MUMPS, directly with a data base.

ferent times. A file manager could represent repeat groupings of data by using one table to index other tables. However, such structures have limitations.

The existence of the information below the top item in a hierarchy depends on that of the item above: if a patient is deleted, so are all his or her subitems' values—which is usually what you want to have happen.

The hierarchical systems we review differ in the number of branching levels they allow and in the way they are designed to appear to users. A number of them only go to three levels, which hampers efforts to represent such cases as patients who have multiple hospitalizations, and in each hospitalization multiple lab test and operations, each of which might consist of different procedures.

Medical data are replete with this sort of multilevel data wherever finer details are being explored and events repeat within the context of other events. If one can represent such repeating data at all in a three level system, one has to use a tabular data format to accomodate the maximum number of cases—one runs into all the problems of a relational or tabular structure (see below). For instance, when you have several possible instances of a measurement and associated comments, you have to say:

measure#1
comment#1
measure#2
comment#2
measure#3
comment#3

These item names tend to appear even if there is only one measurement and space is usually reserved for the maximum number. A multilevel hierarchy can better represent sparse medical data by only repeating as much as is needed.

An added problem with "shallow" systems is that the unit of analysis for statistical purposes is frequently restricted to the top level (patient) and possibly visit. This means that values for repeating data objects at lower levels, such as date-of-procedure for procedures within operations or size for kidney stones within interventions must be transferred to a statistical package with facilities for sorting and merging before they can be analyzed.

References

[1] Date, D. J., *An Introduction to Database Systems*. Reading, MA: Addison-Wesley, 1981.

[2] Health Systems Project, *Comparing Data Management Systems in Clinical Research*; 1983 Survey. School of Public Health, Harvard.

[3] Martin, J., *Computer Data Base Organization*. Englewood Cliffs, N.J.: Prentice Hall, 1977.

[4] Ullman, J. D., *Principles of Database Systems*. Rockville, MD: Computer Science Press, 1982.

[5] Wiederhold, G., *Databases for Health Care*. Berlin: Springer-Verlag, 1981.

33

Computerized Medical Information Systems for Hospital Departments*

Vincent F. Maturi, Ph.D., and Richard M. DuBois, Ph.D.

Introduction

Two years ago the authors conducted a limited study of the current state of commercially available, department specific medical information systems, including their relationship to hospital-wide communication systems [1]. This was a logical extension of the program in computer applications in hospitals supported by the National Center for Health Services Reesearch. The following procedure was used to carry out the study:

(1) Commercial vendors of department specific systems (laboratory, radiology, pharmacy, medical records) and vendors which offer both department specific and hospital-wide systems were identified.
(2) Requests were made for their marketing literature, for lists of hospitals that are customers, and for information of the performance of their systems in these hospitals.
(3) Several hospitals that are customers of the vendors were contacted for information on their system and their perception of performance.
(4) The information was normalized and synthesized to gain insight into the current role of computerized department specific systems.

At that time there appeared to be a trend toward the increased use of department specific systems with suitable interfaces so that they subsequently could be integrated into hospital-wide information systems. It was noted, however, that increasing sales of hospital-wide systems could have a counter impact on this trend in the future. Also, the problems facing potential users such as hospital administrators or department directors when purchasing such systems were described in terms of selection, flexibility, cost, services and interface.

*This chapter has been modified from a version previously published in *Proceedings of the Fourth Annual Symposium on Computer Applications in Medical Care,* IEEE, 1980.

Finally, the need for standards and the limited efforts in that direction were discussed.

Current Studies

In early 1980 the authors conducted another limited study to re-examine the current state of commercially available, department specific systems sold. The same procedure and essentially the same size of sample were used. Four specific questions were addressed:

(1) Are department specific systems currently being acquired at a greater rate than hospital-wide systems?
(2) Is there a trend for hospitals to acquire hospital-wide systems before acquiring department specific systems?
(3) Are currently active department specific system vendors providing greater interfacing capability than was reported in the previous study?
(4) Do the same problems facing potential users two years ago still exist?

Vendors were also asked to estimate what the role of department specific systems would be in five years from the present.

The results of the study are shown in Tables 1, 2, and 3. Due to the subjective nature of the performance data, it is not included in the Tables; but information obtained appears in a general way in the text to follow. Table 1 contains a brief description of what typical department specific systems are designed to accomplish and what major vendors are involved. Tables 2 and 3 describe the characteristics of typical systems available from vendors and their projection for use in the next five years.

Rate of System Acquisition

There are still about as many vendors of department specific systems as before—some new vendors replaced retiring ones; and the acquisition of such systems have been increasing to some extent. However, more vendors are marketing hospital-wide systems, and the rate of acquisition of these systems by hospitals, as judged by the number of sales in the last two years, is increasing more rapidly than the rate of increase of department specific systems (see Tables 2 and 3).

Trend in Acquiring Hospital-Wide Systems First

Vendors were asked to identify hospitals that have both department specific and hospital-wide systems. They indicated that in the majority of cases, the department specific systems were acquired by the hospital before a hospital-

Table 1. Department Specific Functional Applications

System Name	Operations Performed	Accomplishments	Major Vendor
Clinical Laboratory	Collecting, organizing, formatting final test results for medical staff, administration, and business office. Includes: recording of test requisitions, scheduling of specimen collections, calculating test results, periodically preparing summary reports of all test runs for a given patient, preparing statistical reports, keeping records for quality control, and administrative control of laboratory operations, and building a data base for patient care and research [1;4,5].	Reduces clerical errors; results more rapidly available; increases productivity	Becton Dickinson; Technicon; Medlab; CHC Co.
Radiology	Scheduling patients; organizing and storing the dictated radiologist's report of findings; making the report available to the referring physician; tracking x-ray films; producing statistical, management, and financial reports [1,4,5].	Better management; increased productivity	Siemens; General Electric
Pharmacy	Controls dangerous drugs; provides for drug ordering and inventory control; controls drug distribution to patients; stores and retrieves drug information; constructs patient profiles; maintains hospital formulary; generates charges for patient billing [1,4,5].	Reduces medication errors; provides inventory control.	Becton Dickinson; IBM; SMS
Medical Records	Maintains and keeps track of medical charts; tracks the completion status of records; generates notices for the attending physician when final treatment information and diagnosis is needed as a reminder to complete documentation; indexes the record by patient identification number, by disease classification and other demographic factors for medical audit and research [1,4,5].	Reduces lost medical records	McDonald Douglas; Medicus

Table 2. Data on Commercial Information Systems (Vendors Selling Department-specific Systems)

Name of System	Office	Typical Hospital Using System[a]	Does Vendor Provide Interface	Main Function	Estimated Units Sold[b] 1976–78	Estimated Units Sold[b] 1979–80	Vendors Projection (5 yrs) on Rate of Acquisition
CHC [2]	Houston, Texas	Barnes, MO Montefiori, N.Y.	yes	Laboratory	14	16	Moderate
DATA STAT [3] Nat. Data Corp.	Atlanta, Georgia	Kaiser, California	yes	Pharmacy	180	220	Moderate
LCI [6]	Madison, Wisconsin	Ohio State Univ.	yes	Laboratory	35	37	Moderate
MEDLAB [7]	Salt Lake City, Utah	Hillcrest, Tulsa Oklahoma	yes	Laboratory	24	28	Moderate
RUBICON [8]	Dallas, Texas	Univ. of Chicago	yes	Laboratory	5	9	Moderate
SIEMENS [9]	Washington, D.C.	John Hopkins, MD.	no	Radiology	6	10	Moderate
BECTON DICKINSON [10] CLAS	Sharon, Mass.	Univ. of Washington Seattle	yes	Laboratory Pharmacy	12 5	2 9	Discontinuing Moderate

[a]Also the typical hospital contacted.
[b]Estimates reported by manufacturer's representatives and from sales literature.

Table 3. Data on Commercial Information Systems (Vendors Selling both Department-specific and Hospital-wide Systems)

Name of System	Office	Typical Hospital Using System[a]	Does Vendor Provide Interface	Main Function	Estimated Units Sold[b] 1976–78	Estimated Units Sold[b] 1979–80	Vendors Projection (5 yrs.) On Rate of Acquisition
HBO [11] MEDPRO	Atlanta, Georgia	St Thomas, Nashville, Tenn.	yes	Pharmacy Hospital-wide	— 80	— 160	Slow Rapid
IBM [12]	Atlanta, Georgia	Ft. Myers Community, Florida Variety Children, Miami	yes	Laboratory- Pharmacy Hospital-wide	50 15	52 30	Slow Rapid
McDonald [3] Douglas	St. Louis, MO	D.C. General, Washington, D.C.	yes	Pharmacy- Lab Hospital-wide	232 18	260 36	Slow Rapid
MEDITECH [14] MR_{II}, HSCS	Boston, Mass.	Univ. Rochester, N.Y. St. Joseph's, Redding, PA	yes	Lab.- Pharmacy Hospital-wide	30 2	31 4	Slow Rapid
SMS [15]	King of Prussia, PA.	Wuesthoff, Florida New York Hospital, N.Y.	yes	Pharmacy Hospital-wide	— 50	— 100	— Rapid
TECHNICON [16] LDM, TMIS	Tarrytown, N.Y. Madison, Wisconsin	Youngstown, Ohio El Camino, CA	yes	Laboratory Hospital-wide	10 6	15 9	Slow Moderate
Electronic [17] Data Systems	Dallas, Texas	North Arundel, MD. Holy Cross, MD	yes	Laboratory Hospital-wide	1 2	1 4	Slow Rapid
Whittaker-MEDICUS SPECTRA	Evanston, Ill.	University of Texas	yes	Laboratory Pharmacy Hospital-wide	10 6	11 12	Moderate Rapid

[a] Also the typical hospital contacted
[b] Estimates reported by manufacturer's representatives & sales literature.

wide communication system was added—estimating the ratio to be 3:1. They attributed this current situation to the historical sales of department specific systems. However, vendor projections indicate that in the future hospitals without department specific systems already in place are more likely to acquire hospital-wide communication systems first. Inquiries to a sample of the hospitals which acquired department specific systems in the last year confirmed the projection. Although the tendency is for the acquisition of a hospital-wide system first, hospitals reported that other factors can influence the trend. As an example, it was brought out that the clinical laboratory is a significant profit-making part of the hospital, and an influential laboratory director often is in a position to persuade the administration to acquire a laboratory system first.

System Interfacing Activity

The interface problem between computer systems appears to have become less severe over the past two years. The survey two years ago indicated that about one-half of the vendors in the sample offered their customers a suitable interface. In the recent survey, nearly all of the vendors offered suitable interfaces. In a hospital where a hospital-wide communication system already exists, vendors of department specific systems are expected to provide the interface; but when the department specific system already exists, the vendors of the communication system generally are responsible for providing the interface. Many of the department specific systems have now become part of networks, and their vendors claim that microcomputers and specially developed software have alleviated the problem. Thus, it appears that industry will provide expanded interfacing capability, a trend which may encourage building distributed systems involving department specific and hospital-wide systems.

System Acquisition Problems of Hospitals

It was observed during the study that the problems facing hospital administrators or department directors when purchasing or using department specific systems are still present today as they were two years ago. These problems include *selection, cost, flexibility,* and *standards.* Some progress has been made on some of these problems as indicated below:

(1) *Selection.* The hospital administrators contacted indicated that their selection of computerized systems is more cautious now than in the past. The increasing literature on the subject and the drive for cost containment may have been a dominant factor for such caution.

(2) *Cost.* In the last two years, the cost of hardware continued to decrease dramatically while the cost of software which involves professional labor increased substantially. The high cost of software and the low cost of hardware has reduced the number of "homegrown" systems.

(3) *Flexibility.* In the past, the department specific systems acquired by hospitals usually were "turnkey" systems; as such they tend to be rather inflexible. However, the vendors of turnkey systems are beginning to compensate for this shortcoming by offering unique software that allows some limited change in use without reprogramming.
(4) *Standards.* No significant progress appears to have been made on developing standards. As a result the interfacing between systems can still be costly and troublesome especially when the involved vendors do not cooperate.

Conclusion

Although the sales of department specific systems still are significant (slightly increasing) more vendors are currently marketing hospital-wide systems and the rate of acquisition of the latter is increasing more rapidly than that of department specific systems. The opposite appeared to be the situation when the authors made the survey two years ago. The same problems that existed two years ago are still facing the hospital community when purchasing or using department specific systems, although these problems tend to be less severe. The need for standards does not appear to be as urgent as previously thought in view of the trend for improved interfacing between systems.

Abbreviations

The following abbreviations are used:

CHC	Community Health Computing, Inc.
HBO	Huff, Barrington, Owen Corporation
IBM	International Business Machines
LCI	Laboratory Computing, Inc.
LDM	Laboratory Data Manager (Technicon)
SMS	Shared Medical Systems
TMIS	Technicon Medical Information Systems

Acronyms

The following acronyms are used:

CLAS	Clinical Laboratory System (Becton Dickinson, Inc.)
HDCS	Hospital Data Collection System
MR_{II}	Medical Record System (McDonald-Douglas)
MEDPRO	Medical Processing (HBO Corporation)
SIREP	Siemens Reporting Radiology (Siemens Inc.)

References

[1] Maturi, V. and R. DuBois, "Trends in Computerized Medical Information Systems for Hospital Departments," *Proceedings of the Society For Computer Medicine,* Eighth Annual Conference, October 14, 1978.
[2] Communications from CHC Corporation, July 1980
[3] Communications from National Data Corporation, August 1980
[4] Austin, J. and Green, B., "Hospital Information Systems: A Current Perspective," *Inquiry* **XV,** June 1978.
[5] Staff Report, "Computer Applications in Health Care," National Center for Health Services Research, April 26, 1978.
[6] Communications from LCI Corporation, July 1980.
[7] Communications from MEDLAB Corporation, August 1980.
[8] Communications from Rubicon Corporation, July 1980.
[9] Communications from Siemens Corporation, August 1980.
[10] Communications from Becton Dickinson Co., July 1980.
[11] Communications from HBO Corporation, August 1980.
[12] Communications from IBM Corp., July 1980.
[13] Communications from McDonald Douglas Corp., August 1980.
[14] Communications from MEDITECH Corporation, July 1980.
[15] Communications from Shared Medical Systems, August 1980.
[16] Communications from Technicon Corporation, July 1980.
[17] Communications from Electronic Data Systems, August 1980.
[18] Communications from Spectra Whittaker Corp., September 1980.

Editor's Note: The original paper, published in SCAMC-4, contains an appendix which has not been reproduced here.

V. EVALUATION

The evaluation of an information system is an essential but difficult task. In an era where complete and correct computer systems are the exception and not the rule, is it surprising that we find it awkward to evaluate their benefits? The three chapters in this final part address issues in evaluation. The first, by Drazen, considers the methodologies used and the second, by Bradley and Campbell, addresses the problem of comparing service benefits. The final chapter, by Kerr, demonstrates how one might evaluate the cost-benefits of an office practice system; the topic of his study was a computer dental management system, but the techniques are transferable.

Methods for Evaluating Costs of Automated Hospital Information Systems*†

Erica L. Drazen

Background

Computerization offers the modern hospital a means of controlling the increasing volume of clinical and administrative information. Since computerized hospital information systems were first used in the 1970s, numerous products offering varying scope, functions, and system configurations have been marketed; some hospitals have even developed their own systems. Consequently decisionmakers considering automated information handling are faced with numerous options, each offering different benefits.

Computerized information handling can be very costly. A comprehensive hospital-wide system—one that integrates patient records and reports for all hospital departments—for a moderate size hospital may cost as much as ten-million dollars. A smaller system processing information for one hospital department, though less expensive, will have more limited impacts to hospital operations. Thus, decisionmakers need to consider the costs and the cost impacts of the various forms of computerized information handling and weigh these against the benefits of each option.

The cost impacts of Automated Hospital Information Systems (AHIS) have been examined by several studies. Some have been sponsored by the federal government and others by vendors of commercial systems and individual hospitals. Since no compendium of the methodologies used in these studies was available, the National Center for Health Services Research sponsored a study to identify and review these methodologies and to prepare a report summarizing and critiquing them [1]. This chapter summarizes some of the results of

*This research was supported by the Health Care Information Systems and Technology Group of the National Center for Health Services Research under contract 233-79-3000, Modification 2.

†This chapter has beeen modified from a version previously published in *Proceedings of the Fourth Annual Symposium on Computer Application in Medical Care,* IEEE, 1980.

that project. The original study focused only on systems designed to manage data used in patient care. Therefore, studies of purely clinical systems, such as CT scanners or systems designed only for fiscal data processing, were excluded.

The AHIS cost evaluation studies were identified by a literature search, which included use of computerized abstracting services; by a survey of vendors of systems performing clinical information handling functions; by a review of available information concerning applications for Certificate-of-Need; and by contacts with researchers in the field. Studies dealing explicitly with cost and cost impacts were reviewed. Those selected for detailed analysis [2–12] include most of the published work in the field.

Cost Evaluation Methods Identified

The primary determinant of the scope and content of an AHIS cost evaluation is the purpose for undertaking it—why is it being performed and what use is intended for the results? This is largely a function of the perspective of the decisionmaker. A hospital administrator is interested in the cost of hospital operations; whereas a government regulator is concerned with the impact of an AHIS on the cost of health care delivery.

The timing of the study *vis-à-vis* the decision about an AHIS also determines the method used. If the evaluation is in anticipation of adopting a system, the basic approach must be predictive. Conversely, if the interest is in using the experience from one hospital as the basis for developing policy or guidelines for other installations, the logical approach is to compare actual costs before and after the introduction of an AHIS.

The institutional setting, the nature of the AHIS, and the perspective of the evaluator regarding cost impacts also affect the methodologies used. In some cases, cost savings alone are likely to be the major impetus for adopting a computer system. In others, cost considerations may be secondary to other benefits, but the evaluator may want to know what portion of the system costs will be offset by labor savings.

Consistent with the purpose of the evaluation and the objectives for the AHIS, specific impact areas must be identified and questions and/or hypotheses relating to the impacts posed.

The most common reasons for implementing an AHIS are to increase staff productivity (by removing time-consuming paperwork) and to make information flow more timely and reliable. Because of the difficulty of translating patient care benefits into dollar savings, most cost studies have been focused on changes in manpower productivity and the impact of these changes on personnel costs. Some studies have also assessed the nature and magnitude of other impacts, such as increased accuracy of communications, but not the cost implications of these benefits. Therefore, most cost studies conducted to date have been focused on a limited number of the potential cost impacts of an AHIS.

The final step in the study design is identifying a measure of change that will provide answers to the study questions. So many variables affect the cost of hospital operations that it is difficult to find a single evaluation measure that both captures and isolates the cost impacts of an AHIS.

The discrete labor impacts of automation that are most predictable and measurable are at the level of individual task or job performance—e.g., posting physician orders in patient chart, preparing charge input for billing. Demonstrating that timesavings in task performance will be translated into dollar savings (labor force reduction) is a classic problem of studying this impact.

An alternative approach examines trends in costs before and after AHIS use (e.g., cost per patient-day or procedure) or productivity (labor hours per unit of service). With this type of statistical measure, however, predicting future trends or attributing the changes observed over time solely to the introduction of an AHIS are problematic.

The level of detail and the scope of a study are limited by the resources available. Collecting new data solely for use in the evaluation can be very costly. Most studies are therefore focused on the major impacts for which sufficient data can be obtained.

In the past, most studies of AHIS costs have been performed by hospitals to determine the net costs of implementing an AHIS and to delineate benefits in order to decide whether to automate, and which type of system to choose.

A few studies have attempted to characterize and quantify multiple areas of impact of a comprehensive, hospital-wide AHIS by measuring and comparing functional aspects, costs, and patient care effects of information handling activity throughout the hospital before and after the system was introduced. This type of study is a massive undertaking and has been conducted with sponsorship from the federal government.

In past work on AHIS costs, all types of systems are represented from departmental to modular distributed, to comprehensive hospital-wide systems. Commercial systems are included, as well as developmental ones. The studies also cover the various hospital settings: individual department, community hospital, and university teaching hospital. The specific methodologies employed are determined primarily by the purpose for the study. Therefore, the review of the studies indentified was oriented toward examination of specific methodological issues and the generic approaches available for analyzing each issue.

Methodological Issues

Although the study we are reporting on included examination of many issues, two methodological points have been most problematic in AHIS cost studies to date and have often contributed to erroneous or misleading results: techniques for analyzing labor savings and valuing savings from changes in billing practices. These two issues will be discussed fully here. The full report on the project provides an analysis of other issues [1].

Predicting Labor Savings

Labor savings, achieved through the reduction of clerical workload, are a major economic rationale for automating information handling. Savings of nursing labor were the largest cost reductions identified (or predicted) in most of the studies conducted to date. A great many of the cost evaluations of AHIS have included labor savings, and several different approaches have been tried to quantify these impacts.

Four different techniques have been used:

(1) Task analysis was performed by Gall *et al.* at El Camino Hospital [3], by Technicon [11], by Mishelevich *et al.*, for a radiology department system at Dallas County District Hospital [9], and by Brian *et al.*, at Santa Monica Hospital [5];
(2) Job content analysis was used by the Fairfax Hospital Association [6];
(3) Work sampling was performed at El Camino to document total staff time devoted to information handling [12] and in the retrospective study by Huff and Bond [7] at the hospitals of the Sisters of the Third Order of St. Francis; and
(4) Trend analysis of labor utilization was applied in the nursing cost avoidance method used by Technicon [11].

Task Analysis. In task analysis, individual work activities that will be affected by automation are identified, e.g., filling out an x-ray requisition, telephoning for a laboratory result. Then the time spent in the activity in the manual system is measured. The time spent in the activity with automation is measured (in a retrospective study) or predicted (in a prospective evaluation). This information is combined with information on the volume of activity and the salary of personnel performing the task in order to calculate labor savings. Table 1 illustrates this technique.

Table 1. Example of Task Analysis in an Admitting Department

Function/Job Title	Time Savings (Min. Per Unit)	Volume (Units Per Day)	Potential FTE Workload Reduction
Process Pre-Admission/ Admitting Clerk	8.0	26	0.6
Bed Assignment for Daily Admissions/Dept. Head (60%)	4.0	31	0.4
Asst. Dept Head	4.0	21	0.2
Admitting Clerk	1.0	51	0.2
Subtotal			**0.8**
Process Emergency Admissions/Admitting Clerk	5.0	17	0.3

Task analysis provides a complete inventory of potential impacts but translating labor savings in specific tasks to reductions in payroll is difficult. A simple summation of all labor savings for all departments and labor categories and across shifts is obviously unrealistic. Specific assumptions need to be adopted. Some realistic assumptions are:

(1) Assume savings can only be summed within each department.
(2) Assume transfer of tasks from lower to higher level staff to realize time savings, RN's taking on functions previously performed by aides or clerks, technicians replacing clerks in ancillary departments.
(3) For nursing departments assume small amounts of time savings (less than 0.2 FTEs) can be transferred between shifts.
(4) Assume that 50% of the time savings of 0.5 FTE or more can be realized by use of parttime staff. (In some cases savings won't be possible because of core staffing requirements, e.g., at least one ward clerk must be on duty to answer phones).
(5) Round remaining partial FTEs down to next lowest whole FTE.

When these assumptions were applied instead of a simple addition, the predicted labor force reduction in one study [3] would decrease to 38 from 47 full time equivalent staff. It is easy to see that unrealistic assumptions can greatly influence the predicted labor savings.

Job Content Analysis. Job content analysis overcomes some of the problems of task analysis since it focuses on the employee, rather than the task. The steps in this process include identifying jobs that will be performed differently with AHIS, estimating (or measuring) the time that will be saved, and redefining (combining) jobs to realize the savings. Table 2 shows the results of a job content analysis.

Job content analysis is usually performed in collaboration with the supervisor of a department, since thorough knowledge of the area is required, and

Table 2. Example of Job Content Analysis for a Radiology Department

	Affected Tasks	Saved Hours Per Day
Technician 1	perform exams	0
Technician 2	perform exams	0
Technician 3	scheduling, perform exams	1
Technician 4	perform exams, retrieve files	3
Clerk 1	transcription of reports	7
Clerk 2	scheduling/phone	4
Clerk 3	phone/x-ray filing	6
Clerk 4	phone/scheduling/report distribution	2
Clerk 5	phone/requisition processing/receptionist	3

since supervisors must agree to implement any changes. Job content analysis is less time consuming to perform than task analysis and results are easier to translate into realistic staff reductions. It may also be less accurate since judgements of time saving are used and there may be incentives to inflate or underestimate savings.

Work Sampling. Work sampling (also referred to as activity sampling) can also be used to estimate labor savings. In this case the work performed by individuals is documented by measuring periodically the activity of personnel in a unit over an extended period. The data are used to describe how individuals or categories of staff spend their time. Then the activities affected by AHIS can be identified and the potential labor savings can be estimated (in predictive studies) or measured (in retrospective studies). Since work sampling provides information on how time is allocated to work activities, it can be used to determine how activities have changed (use of freedup time) and how much "downtime" exists in manual and automated systems, etc. Table 3 provides results of work sampling of information handling activities in a surgical intensive care unit.

Work sampling is especially useful in analyzing labor impacts in nursing units. Nursing staff perform many information handling tasks in conjunction with other patient care activities but may not devote large blocks of time solely to information handling. Sometimes information handling and patient care tasks are performed simultaneously. Therefore, changes in nursing staff activities can be best captured and quantified by direct observation (as in work sampling).

Trend Analysis. Trend analysis involves examining trends in cost variables (i.e., nursing cost/patient day) prior to AHIS with those after AHIS to identify and quantify changes. The technique is used most often in retrospective studies but can also be used prospectively if there is information available to project

Table 3. Results of Nursing Staff Work Sampling in a Surgical Intensive Care Unit

	Hours per Shift		
Activity	Day	Evening	Night
Patient measurements	1.4	0.8	0.5
Nondrug ordering	0.8	0.4	1.1
Drug recording	0.1	0.0	0.1
Nondrug recording	4.9	3.0	4.4
Review patient records	3.9	3.4	3.3
Communication with nurses	9.6	4.1	3.6
Communication with doctors	3.2	1.6	3.1
Research	—	—	—
Teaching or Education	4.3	0.3	0.4
Total	**28.2**	**13.6**	**16.5**

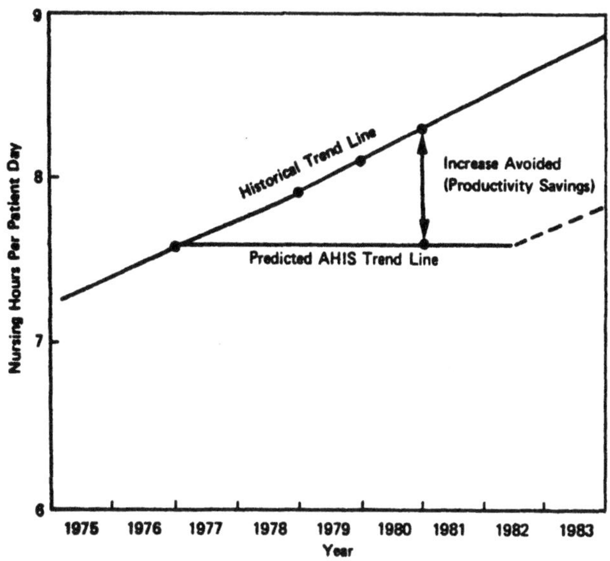

Figure 1. Sample trend analysis of nursing productivity.

trends. Performing a trend analysis to estimate labor savings involves first collecting historical data on trends in labor productivity and extrapolating these trends into the future. In retrospective studies productivity after AHIS would be plotted, in prospective studies future productivity must be estimated. Once the trends with and without the AHIS have been plotted, then the two productivity lines are compared in order to estimate labor productivity differences at different points in time. Figure 1 shows an example of trend analysis of productivity of nursing labor on an inpatient unit.

Results from trend analysis are difficult to interpret because changes in labor cost or productivity can be the result of many causes other than changes in information handling tasks. Major influences on staff productivity may result from changes in the patient mix, the occupancy rate, regulatory and payment policies, or the addition of new services (especially intensive care services). Identifying and adjusting for these other variables to isolate the impact of an AHIS is a difficult (sometimes probably impossible) undertaking. Furthermore, the large number of institutional variables make it impossible to generalize the results of trend analysis performed in one institution to other hospitals.

Importance of Method Used. In many published cost evaluations of AHIS, the cost savings have been overstated because of use of an unrealistic assumption. As an illustration of the magnitude of this problem, consider the data presented in Table 4, which contains study results of three evaluations. Study 1 was a major evaluation study in which several methods were used to estimate cost savings. This table shows the results of task analysis. Study 2 also used task

Table 4. Labor Savings From Three Predictive AHIS Cost Studies (Original Assumptions)

	Study 1	Study 2	Study 3
Nursing personnel	917,304	2,107,975	excluded
Nonnursing personnel	209,616	350,765	533,107
Total Labor Savings	**1,126,920**	**2,458,740**	**533,107**
Total labor savings per patient day	9.16	19.67	2.80

analysis (with different assumptions than Study 1). However, Study 2 included an assumption that productivity changes at a previous AHIS installation would be realized in this new site. Study 3, which was conducted by a hospital considering AHIS implementation, used job content analysis to analyze the labor savings of nonnursing personnel. Nursing personnel were excluded from the initial analysis because of the previously discussed difficulties in predicting work force reductions in nursing staff. Table 4 presents the study results as reported.

In Table 5 these results have been adjusted to reflect more realistic assumptions. Trends in productivity were assumed not to be transferable from one site to another, and labor savings were recalculated with the assumptions discussed previously in this paper for predicting labor savings based on task analysis.

It is obvious from this example that the assumptions used in analyzing AHIS labor impacts can greatly affect the results of the analysis. In many cases these assumptions have not been described in the published studies so it is not possible to estimate their influence on the study results. Without a knowledge of the assumptions used, the results of cost evaluations are of no use to future decisionmakers.

Even when the most realistic assumptions are used, predicted cost savings will not be realized unless the institution is committed to a definite program of change specifically designed to adjust procedures and staffing to utilize the potential time savings resulting from AHIS implementation.

Table 5. Predicted Labor Savings From Three Predictive AHIS Cost Studies (Revised Assumptions)

	Study 1	Study 2	Study 3
Nursing personnel	661,166	604,999	excluded
Nonnursing personnel	209,616	350,765	533,107
Total labor savings	**870,782**	**955,764**	**533,107**
Total labor savings per patient day	7.08	7.65	2.80

Valuing Impacts of Automated Billing

If an Automated Hospital Information System replaces a manual or semimanual billing system, in almost all cases this increases the accuracy and efficiency of billing. Several types of cost impacts have been attributed to this change.

Often several categories of personnel—e.g., keypunch operators—can be eliminated, and although these savings are partially offset by additional computer personnel, a net payroll savings is possible.

An AHIS may also reduce the outstanding receivables by reducing the lag time between service delivery and billing. This can be characterized as a cost saving because it reduces a working capital loan (and hence the interest owed) or as a one-time revenue increase. Either way, it is viewed as a reduction in receivables and is a tangible financial benefit to the hospital and the health care system and a valid cost saving.

Some previous studies [4,5] have also predicted that increased billing accuracy will mean that charges previously not recorded will be captured and that this will increase revenue. These recovered charges have been added to the cost savings calculated from labor savings, reduction in receivables, supplies, etc. Since a change in the accuracy of billing does not affect the *cost* of the service being delivered (merely how it is paid for), recovered charges do not represent a cost savings and should not be included in a cost analysis.

Charge recovery has frequently been included as a cost saving in some early cost evaluations of AHIS. Since this can greatly inflate the total predictions of cost savings, care should be taken to identify and delete these items before using the evaluation results.

Conclusions Regarding AHIS Cost Evaluations

This chapter has discussed two specific methodological issues in cost evaluations: valuing labor savings, and AHIS impacts on hospital revenue. However, it may be useful to discuss briefly the conclusions of the entire study.

We found that few rigorous studies have been conducted of cost impacts of implementation of automated hospital information systems. Most of the work done in this area in the past has involved predicting cost impacts. The assumptions used in some predictions have tended to overstate the cost saving impacts of an AHIS. Insufficient attention has been paid to enumerating and quantifying the costs of installing and operating an AHIS. Overstated cost savings and understated system costs may have produced unrealistic estimates of the net cost impacts of AHIS.

Little work has been done on validating the results of predictive methodologies. A few retrospective studies have been performed, but the results have been equivocal, and attributing the results observed to automation alone is difficult in the complex environment of the hospital. Therefore, there is very little documentation of actual impacts of an AHIS on the productivity of hospital staff or on overall cost changes to guide the conduct of cost assessments.

Through improvements in information flow, an AHIS offers benefits to the quality of service delivery, and these are usually major motivations for a hospital to implement a system. Cost studies in the past have been concentrated on direct cost impacts, such as labor savings, that are easier to identify and quantify in economic terms. More work needs to be done exploring the cost implications of improvements in the quality of service delivery, such as improved turnaround time for test results reporting and decreased loss of information in the system.

One type of cost impact that has often been included in the past is changes in hospital revenue resulting from increased accuracy of billing for services. Revenue recovery is not a measure of changes in the costs of operating a hospital and, therefore, should not be counted among the impacts.

Cost studies have been conducted for different hospital settings (departments, community hospitals, teaching hospitals) and system configurations (dedicated departmental, modular distributed, and comprehensive). The specific methodologies employed are generally applicable, however, to any setting or system type. The purpose in undertaking the cost study determines the approach that is used. Individual hospitals typically are interested in *predicting* cost impacts before system implementation; federally sponsored studies have focused on measuring actual impact by comparing costs before and after system implementation.

References

[1] Drazen, E. L., R. Feeley, J. Metzger and H. Wolfe, *Methods for Evaluating Costs of Automated Hospital Information Systems,* Report by Arthur D. Little, Inc., to National Center for Health Services Research NTIS PB-80-178-593, March 1980.

[2] Barrett, J. P., P. L. Hersch and R. J. Caswell, *Evaluation of the Implementation of the Technicon Medical Information System at El Camino Hospital, Part II, Economic Trend Analysis.* Columbus, Ohio: Battelle, May 14, 1979.

[3] Gall, J. E., *et al., Demonstration and Evaluation of a Total Hospital Information System,* National Center for Health Services Research, NTIS PB 262 106, 1975.

[4] Arenson, R. L. and J. W. London, "Radiology Operations Management Computer System, Hospital of the University of Pennsylvania," in *Proceedings of the American College of Radiology Sixth Conference on Computer Applications in Radiology,* March 1979.

[5] Brian, W., M.D., *et al., Cost Benefit Analysis of Installation of a Spectra System at Santa Monica Hospital,* University of Southern California, Center for Health Service Research, January 1976.

[6] Fairfax Hospital Association, *Cost-Benefit Analysis of Data Care System* (unpublished), Falls Church, Virginia, 1979.

[7] Huff, W. S. and E. J. Bond, *Demonstration of a Shared Hospital Information System,* The Sisters of the Third Order of St. Francis, Peoria, Illinois, 1970.

[8] Kim, S. H. and H. H. Schmitz, "A Cost-Benefit Analysis of a Modular Hospital Information System," in M. W. Hopfe, Ed., *American Institute for Decision Sciences Proceedings,* Boston, Massachusetts, November 14–16, 1973.

[9] Mishelevich, D. J., W. G. Gipe, J. R. Roberts, C. Denny, A. D. Stern and M. W. Day, "Cost-Benefit Analysis of a Computer-Based Hospital Information System," in *Proceedings of the Third Annual Symposium in Computer Applications in Medical Care,* 1979.

[10] Schmitz, H. H., "An Evaluation of the Immediate Financial Impact of the Hospital Information System at Deaconess Hospital," in R. C. Koza, ed., *Health Information Systems,* Colorado Associated University Press, Boulder, Colorado, 1974, pp. 265–282.

[11] Technicon, Medical Information Systems Corp., *Cost-Benefit Analysis of the Technicon Medical Information System (MIS) for (unnamed hospital),* December 1975.

[12] Battelle Memorial Institute, *Evaluation of the Implementation of a Medical Information System in a General Community Hospital, Vol. II, Evaluation Studies and Analyses,* NTIS PB-232 785, September 30, 1973.

35

Information System Selection: Methods for Comparing Service Benefits*

Evelyn Bradley, Dr.P.H., and James G. Campbell

Introduction

Installing an automated hospital information system is a major undertaking, involving substantial time, effort, manpower and economic investment. Nonetheless, it is estimated that half the larger short term general hospitals will install information systems by the end of the decade [1].

Justification for investing in an automated information system includes both anticipated *economic* savings, and improvement in hospital *operations* unrelated to cost.

A rigorous information system selection process requires that expected economic costs and benefits be quantified for each system which is considered during selection. A method for identifying and quantifying specific information system contributions to hospital operations would support a more comprehensive system selection process. Those in the position of selecting a system would be aided by a summary of each proposed system's relative contribution to departmental and overall hospital performance. If each system's potential to provide service benefits were made explicit, this would support more objective decision-making, reducing the number of "intangible" variables to be weighed subjectively by the decision-maker.

This chapter discusses the concept of service benefits, and why they should be considered in system selection. Then problems are presented concerning development of a summary figure for each proposed system's potential impact on services. Presentation of these problems is followed by discussion of solutions. In particular three methods for translating service benefits into a single unit of measure are described and discussed. One method, labor equivalents, is original to the authors.

Service Benefits

Service Benefits are information system contributions to hospital performance, to the efficiency and effectiveness of services provided in the hospital. Many of these effects are of direct benefit to patient care.

Improvements in efficiency and effectiveness may be improvements in the *internal* procedures of a department, the way work is performed. Some service benefits are improved departmental *outputs*—improvements in the services provided by the department.

Examples of Service Benefits in specific departments which may result from implementation of an information system include:

(1) *Laboratory*:

Reduce the number of misplaced specimens;
Reduce results reporting turnaround time.

(2) *Nursing Service:*

Provide services to patients in a timely manner;
Maintain compliance with established nursing procedures.

(3) *Medical Records:*

Reduce number of incomplete records;
Reduce time required to locate charts;
Increase compliance with JCAH standards.

Service benefits focus on improvements in departmental procedures and outputs rather than focusing on improvements in information. Timely and accurate information is a tool which can be used by departments to improve their operations—service benefits show that operations have been improved.

Considering Service Benefits In System Selection

In system selection, estimates of costs and cost savings from various systems are expressed quantitatively. In a rigorous selection process, cost-benefit or return-on-investment comparisons are carefully considered.

Potential service benefits are also considered in system selection, even when they are not quantified. Hospital decision makers select a system which provides desired capabilities for improving hospital operations; they do not simply select the lowest-cost system available.

The A. D. Little study concerning evaluation of information system costs states, "If a system is being implemented primarily to improve quality of services, the evaluation must include assessment of the system's impact on services in addition to analysis of costs" [2]. If service benefits should be considered in retrospective system evaluation, they deserve even more vigorous treatment during system selection. Hospital managers purchasing an information system may be as concerned about information system support to staff in performing their jobs as they are about cost savings.

Potential service benefits are frequently mentioned in documents justifying purchase of information systems. These benefits are not, however, quantified. Quantification of potential service benefits and development of a composite measure of each proposed system's service benefits would provide data for a more objective decision making process.

There are several barriers to developing such a composite measure of service benefits. The first has been the absence of rigorous research to identify key potential information system impacts on service in each department. Recent research has identified key service benefits, using a research protocol which combined review of impacts cited in literature and telephone interviews, with review of specific application programs [3,5].

A second barrier to developing a composite measure is that quantification of service benefits requires hospital staff effort to quantify anticipated benefits. For example, an objective for an information system such as "improve patient care through responsive and error-free clinical data communication between patient care units and ancillary areas," to be quantified, would have to be specified—resulting in objectives such as "reduce errors in orders received by 20%," and "reduce time delay between order transmittal and results reporting by 15 minutes."

A third problem is that service benefits are expressed in a variety of units such as time saved and number of errors. To develop a composite measure for rapid comparison of several systems, a common denominator, or common unit of measure must be selected. Three methods for translating service benefits into a common unit of measure are presented below.

Developing A Composite Measure

There are several circumstances in which a composite measure of service benefits would be useful. One is for comparison of potential service benefits during system selection. A second is for comparison of the experience of various hospitals as determined in post-installation evaluation.

Three alternative methods are presented here for translating service benefits into a single unit so they can be summed to form a composite measure. Each method has strengths and weaknesses. Evaluators have different opinions on which method they find most meaningful. Therefore, the three methods are compared according to key criteria. These criteria are:

whether the method incorporates the value or worth of service benefits to the hospital;

whether the method relies on objective or subjective measurement, and whether the measure is reliable (whether different evaluators would achieve the same results);

cost to apply the method.

Point System

The point scoring approach has been applied in the American Hospital Association's manual for preparing requests-for-proposals (RFP) [4]. In this approach a total number of points, for example 1,000 points, are allocated among the following system features: quantitative and qualitative costs and benefits, functional requirements, technical requirements, and vendor characteristics. The hospital manager decides how important each of these features is, relative to the others, in selecting an information system; if service benefits are considered most important they might receive 400 of the 1,000 points. Then the service benefit points are suballocated first among departments and then, within departments, among specific anticipated benefits according to their importance to the hospital.

Once the decision-maker has identified specific anticipated service benefits, and has allocated points to each in accordance with its importance, vendors are asked in the request for proposal, to indicate the extent to which their system provides service benefits (along with technical and other system features). Vendor systems are compared based on "how well the hospital is satisfied with the way each qualitative benefit can be provided by the system."

Market Approach

A second approach to evaluating service benefits which has been proposed in a "market" approach. The market this refers to is the hospital budget. Each department head whose area is affected by information systems operations is presented with a list of potential service benefits. The department head then decides how much existing budget he or she would be willing to commit in order to obtain each benefit. These "market" values of service benefits promised by vendors of each proposed system are summed and used for comparison.

Labor Equivalent Approach

A third method, developed by the authors, is the labor equivalent approach. Once service benefits of interest have been identified, industrial engineers or management consultants estimate the labor which would be required to produce each benefit in the current environment—that is without any change in the level of automation available.

For example, one service benefit objective of the hospital might be to reduce order transmittal delays by fifteen minutes on average. If, in order to achieve this objective without installing a new information system, hiring an additional messenger would be necessary, this benefit yields one labor equivalent. Note that labor equivalents are not labor savings but a measure of the information system's contribution to improving hospital performance.

Vendors presenting proposals are asked to identify service benefits provided by their systems. As with the point system, hospital decision-makers must judge the reliability of vendor claims, by looking for specific applications and technical features, and by conferring with other purchasers of the system.

The hospital would then have a profile of expected service benefits for each proposed system. The labor equivalents associated with each proposals can be summed, providing a composite measure of each proposed system's contribution to hospital efficiency and effectiveness. A system which costs 1 million dollars and provides .9 million dollars in economic return and 200 labor equivalents can be compared with a system which costs 1 million dollars, provides 1.1 million dollars in economic returns and 100 labor equivalents.

Labor equivalents can be translated into dollars to make the comparison even clearer. It would not be appropriate to assign each equivalent a value equal to a salary; if service benefits were worth that much to hospital management, employees would have been hired to produce them. However the decision-maker can set a dollar figure for how much each labor equivalent is worth. A decision might be that each equivalent is worth $2,000/year. Once this decision is made the value of anticipated labor equivalents for each proposed system can be combined with anticipated economic impacts for cost-benefit analysis.

Comparison

Each of the three translation methods provides a means for expressing service benefits in a single unit of measure. Hospital decision-makers wishing to consider various sytems' potential effects on efficiency and effectiveness in a structured way have several alternative methods for doing so. The appropriate method is that method which provides the most meaningful information to the decision maker. For some, points may be meaningless; for others, points may speak clearly. While the market method expresses service benefits in dollars (a readily understood unit of measure), there is some danger that these dollars, which represent a *hypothetical* sacrifice, will be mistakenly considered as hard dollars. Labor equivalents are expressed either as number of worker-equivalents or as a dollar value for those workers, as assigned by the top management.

Meaningfulness, one criteria for selecting a method, is subjective and will vary from one decision-maker to another. Three other criteria worth considering when selecting a method, were mentioned earlier: the method's ability to reflect hospital values, its ability to yield reliable and objective results, and the cost to apply the method. The relative strengths (+) and weaknesses (−) of each method with regard to the criteria are summarized in Figure 1.

The point system *may* reflect the value of service benefits to the hospital. Since the process of allocating points is subjective, the ability of a point allocation system to reflect hospital values depends on who allocates points and whether that individual is considered to represent the hospital's values. The Board and CEO, being responsible for prioritizing hospital values, should decide on overall point allocation between economic impacts, service impacts and other selection features. The CEO and Board may also wish to review suballocation of Service Benefit points among and within departments. If

METHOD / CRITERIA	POINT	MARKET	LABOR EQUIVALENT
REFLECTS VALUE OF BENEFIT TO HOSPITAL	+/-	+/-	+
RELIABILITY/ OBJECTIVITY	-	-	+
COST TO APPLY	+	+	+/-

Figure 1. Comparison of three translation methods.

department directors assign points, overall hospital perspective may not be reflected. However, it is difficult to command the top management time required for a detailed point allocation review.

A major limitation of this approach, if point allocation decisions are made by managers of different departments, is that the allocation of points may not be reliable. Managers' evaluations may vary over time, leading to unreliable estimates. Managers in different areas may value the same benefit very differently resulting in uneven estimates of service benefits' worth. However, because this method is based on opinion or judgement as to relative value rather than to objective measurement, it is not costly to apply.

The market approach shares with the point system the advantage that it is easy to apply. Asking department heads for subjective opinion should not use substantial resource. The market method's primary problem is, as with the point system, the subjective nature of the estimates. Department heads' evaluations may vary according to changes in their budgetary constraints, leading to unreliable estimates. A manager may give low estimates of the worth of service benefits in order to express disapproval of the prospect of introducing an information system. Additionally, a manager might not, in fact, be willing to sacrifice budget if given hard rather than hypothetical choices. The market approach shares with the point system the problem that top management, responsible for representing the overall view of hospital values, does not set the proxy measure in this method. Departmental managers set market values; their viewpoint is narrower, focused on the concerns of their particular departments.

One major strength of the labor equivalent approach is that equivalents are established objectively. A department head, management consultant, or industrial engineer is called upon to estimate methods and labor requirements for achieving service benefits in the current environment. These estimates are likely to be more consistent between observers than are subjective estimates of the "worth" of a benefit used in the other two methods.

Essentially, the labor equivalent method moves subjective decision-making to the level of the individual responsible for selecting the information system. The process of translating service benefits to a common term is objective; the labor equivalent method provides an objective means of arriving at a single figure representing each proposed system's service benefit contribution. A figure is placed, by the decision-maker, on the value of this single number. By contrast, with both point and market methods each service benefit must be weighed individually, a complex valuing process. Labor Equivalents provide a means by which top management's valuation of the worth of service benefits can be acquired without requiring a great deal of top management time.

The labor equivalent method may be more costly to apply than methods which ask for subjective opinions. It takes time to analyze departmental operations and then develop methods and estimate labor required to achieve service benefits. However if the hospital maintains a base of standard industrial engineering data, this reduces the cost of estimating labor equivalents.

If a hospital elects to develop a composite measure of service benefits, the method of choice depends on the importance of each of these criteria to the decision-maker. Individuals vary both on which method is easiest to understand, and which method provides the most meaningful picture of different systems' service benefit contribution.

Implications for the Future

The trends in evaluation of information systems have shifted over the years. In the 1960's and early 1970's efforts were concentrated on exploring the capabilities of information systems in the health care field. There was concern for showing that hospital information systems could actually be made to work.

Current efforts focus on demonstrating the economic viability of hospital information systems. However, the trends in hospital information systems, like those elsewhere in the information system field, have shown increasingly powerful techniques to be available at a lower cost each year. Vendors now offer a greater variety of relatively inexpensive program packages in many health care areas.

As a greater variety of economical information systems compete with one another in the marketplace, the evaluation issue of the 1980's will pass beyond cost-benefit analysis to assessing cost-effectiveness. Hospitals will choose from a variety of systems which are acceptable in economic terms; they will select the system which best helps them achieve their mission. The final decision in selecting an information system will always weigh unquantified factors. However, quantification of service benefits provides a data for debating the merits of various systems.

The material presented here encourages hospitals to quantify service benefits as an aid to system selection. In this way system selection will reflect the hospital's multiple goals, particularly the service mission which has always been the hallmark of health care enterprises.

References

[1] United States General Accounting Office, "Computerized Hospital Medical Information Systems Need Further Evaluation to Ensure Benefits from Huge Investments," GAO Report AFMD-81-3, Gaithersburg, Md., 1980.
[2] Drazen, Erica, *Methods for Evaluating Costs of Automated Hospital Information Systems.* Cambridge, MA: A. D. Little, Inc., 1980.
[3] Bradley, Evelyn, "Evaluating Information System Impacts on Hospital Performance," *Proceedings, Fifth IEEE Symposium on Computer Applications in Health Care,* 1981.
[4] American Hospital Association, *Hospital Computer Systems Planning: Preparation of Request for Proposal,* Chicago, Ill., 1980.
[5] Lutheran Hospital Society of Southern California, *Evaluating Automated Hospital Information Systems,* Center Publications, Los Angeles, 1982.

Quantifying the Cost-Benefits of Computer Dental Management Systems*

David R. Kerr, D.D.S., M.B.A.

Introduction

There is little literature that investigates the cost benefits of computers in dental practice. Most of this literature is anecdotal and does not attempt to quantify a computer's benefits in practice [1–7]. However, a study prepared by Arthur Young and Company [8], which investigated the potential uses of the computer in dental practice attempted to determine the rate of return of various types of computer systems such as service bureaus, time sharing systems and in-house computers. The methodology used in their study was to examine the labor savings of using a computer for a range of tasks that would be performed in the office (Table 1). However, when calculating the labor savings of using a computer in the dental office, the Arthur Young study assumed full automation of all clerical tasks.

There were two purposes of this study: to survey dental management programs and, using the methodology of the Arthur Young study, to calculate the internal rate of return based on the labor savings that would result from using the most commonly available dental programs.

Method

Survey Method

Vendors of dental computer management programs were determined through advertisements found in major dental journals, popular computer magazines, and at two major dental meetings, including the annual session of the American Dental Association, in an attempt to survey those vendors that were most

Table 1. Tasks Included in Arthur Young and Company Study

Miscellaneous Lookups
Patient records and plans
Ledgers
Other data
Preparation
W'2s, 941's
Utilization review reports
Ad Hoc reports
Patient ledgers
Treatment plans
Aging and deliquency reports
Practice analysis and financial analysis
Schedules
Patient statement
Patient recall
Insurance claim forms
Payroll checks and ledgers
Accounts payable
Inventory reports
Wordprocessing

likely to have contact with the dentist interested in using a computer in their office. Eighteen vendors were surveyed.

Quantitative Method

Replicating the methodology of the Young study, the internal rate of return method was used to calculate the return on investing in an in-house computer based on an estimated adjusted initial investment of $19,336. Although, systems surveyed varied in a price range of about $15,000 to $120,000, a $19,336 investment fairly represents the average base price of most dental management systems.

The internal rate of return method determines the discount rate (rate of return) that equates the expected net present value of all cash values to zero. The expected cash flow was estimated to be the salary savings that would result from using a computer to perform various office tasks less annual computer operating costs.

Assuming equal future cash flows, this can be mathematically stated as:

$$SS = \left(\sum_{x=1}^{n} Xn/TW \times S\right) - \left(\sum_{y=1}^{n} Yn/TW \times S\right) \tag{1}$$

$$CF = SS - O \tag{2}$$

$$-I + \sum_{t=1}^{n} \frac{CFt}{(1+r)^t} = 0 \tag{3}$$

where Xn = minutes used for completing manual task n;
Yn = minutes used for completing computer task n;
S = clerical salary;
O = operations cost;
TW = total work minutes;
SS = salary savings using computer;
CF = cash flow at time "t";
I = adjusted investment cost;
which is: investment cost − investment credit − 1st year depreciation.

The cost analysis was based on the same assumptions as used in the Young study which were:

patient record base of 1,000;
three hundred patients treated monthly;
patients are sent monthly bills;
practice analysis and financial analysis done monthly;
approximately 50 checks were written per month;
operating cost of $3,004 which includes system maintenance, supply costs and postage costs;
a ten percent investment credit allowed on the initial purchase price;
depreciation credit deducted from the investment price using straight line depreciation;
a five year investment return period;
clerical salary—$10,000;
total work minutes = 115,200 (240 days × 480 minutes per day).

Table 2. Base Case Tasks

Function	No Computer Use		Computer Use
	(in minutes)		
Misc. ledger lookups	1,800		720
Utilization review reports	2,880		180
Ad hoc reports	5,760		360
Patient ledger	18,000		5,400
Aging analysis	1,440		180
Practice analysis	5,760		180
Patients statements	18,000		5,400
Recall	15,000		3,000
Insurance claims	18,000		3,600
Word processing	5,750		2,800
Total work minutes	**92,400**		**21,900**
Calculated discount rate (rate of return)		−6.3%	

Task Determination

Rather than assuming a fully automated office, as in the Young study, only those labor tasks which are commonly used by the first time user were included in the analysis. These tasks were determined by the promotional literature of the surveyed vendors and in discussions with the vendors who indicated typical applications of their systems.

Once this base case was established, other available, though less commonly used accounting and clerical tasks were cumulatively added to the base case. The internal rate of return was calculated for each case.

Two types of sensitivity analyses were performed. Using the base case, sensitivity analysis to changes in clerical salary was determined, and sensitivity to adjusted initial investment was performed for the base case and the maximum expanded tasks case.

Discussion

All vendors offered programs that performed accounts receivable, insurance processing, patient billing and daily, weekly, and monthly productivity reports. Eighty-nine percent offered the capacity to have a patient recall system, while thirty-three percent of the vendors offered patient scheduling capability. Thirty-nine percent of the vendors offered wordprocessing, though most focused on its use in formletter generation. Accounts payable, general ledger and payroll were either offered optionally or not at all.

Though level of program sophistication varied, all vendors satisfy the demand to automate the three areas most frequently desired by dentists, namely accounts receivable, insurance processing and billing. The lack of emphasis on using the computer for other accounting tasks may reflect the practitioners lack of knowledge in the financial management area. As will be discussed, this lack of use in other areas of financial management remarkably affects the return on investment.

Based on the survey results and using the minutes that the Young study estimated were needed to perform related tasks such as ledger lookups, a set of tasks likely to be included in a typical dental computer management system was used as a base case to estimate the labor savings that would result in using a computer (Table 2).

The Young study found that a fully automated office based on an initial adjusted investment of $19,336, would give a discount rate of 15.2%. However, only automating those tasks included in the base case, results in a discount rate of −6.3%.

The internal rate of return, calculated by cumulatively adding expanded tasks to the base case (Table 3) varied from a discount rate of −0.2% to a maximum of 6.3%.

Sensitivity to changes in the adjusted initial investment was calculated for

Table 3. Expanded Tasks Cases

Function Added to Base Case (Base Case + Added Function)	Total Return
Accounts Payable	−.2%
Payroll Checks and Ledgers	2.0
W/2's Preparation	2.1
Patient Scheduling	6.3

Table 4. Sensitivity to Adjusted Initial Investment

Investment	Base Case	Maximum Expanded Case
$17,402	−3.0%	10.0%
15,469	−0.2	15.1
11,602	11.0	28.5
9,668	18.0	38.4
7,734	29.0	52.6

Table 5. Sensitivity to Salary Using the Base Case

Salary	Return
$11,500	1.4%
12,000	3.0
12,500	6.6
13,000	8.9
13,500	11.2
14,500	15.7

both the base case and maximum expanded case (Table 4) and using the base case, sensitivity to salary was calculated (Table 5). The discount rate varied from −3% for an adjusted initial investment of $17,402 to 18% for an adjusted initial investment of $9,968. In the expanded case, a $17,402 adjusted initial investment would give a discount rate of 10%.

The discount rate is quite sensitive to changes in salary. Using the base case tasks, a salary of $13,500, a $3,500 increase over the base case, gives a discount rate of 11.2% as compared to the base case of −6.3%. While this salary may be high for the average clerical salary, it is feasible that the combined salary of those persons performing these tasks is at least this amount.

How does one determine what discount rate is acceptable? One method is to invest only in those projects that give a rate of return greater than your cost of capital. This cost of capital has historically varied from 10 to 15% for most firms [9]. Taking ten percent as the minimum required rate of return, an adjusted initial investment would need to be about $11,000 for the base case and in the maximum expanded case, $17,000. With the present high interest rates, a higher discount rate would be required. Another method to determine what discount rate to use is to evaluate the return on equally risky investments and choose whichever gives the higher return. Assuming that a computer project is as risky as common stock investment, the average rate of return for common stocks has been approximately twelve percent, assuming a five percent inflation rate. No matter which discount rate is used, this study has demonstrated very low and even negative rates of return result when one invests in the typical dental management computer system that performs limited tasks.

This study is an initial investigation to develop a method to quantify the benefits that most computer systems sold today would give the practitioner. Thus, it attempted to apply a theoretical model to a real world problem. As a model, there are simplifying assumptions. This study only estimates the labor savings cost of a computer system. It is questionable whether the minutes used in the task analysis are truly additive nor does it capture the time needed to maintain a computer system for example, to backup data files.

The analytic method used in the Young study and replicated in this study does not include in the cash flow the depreciation benefits of a purchased computer system or the interest deduction if a loan were used to purchase the system. Consequently the cashflow stream is understated, resulting in a lower discount rate.

The future cash flow period is assumed to be five years. This is the shortest depreciation life of computer equipment as allowed by the Internal Revenue Service. In many cases, a computer system will be used for a longer period, which would increase the payback period. This would result in a higher rate of return on investment.

This model does not attempt to measure the qualitative value of having accurate and timely information. The value of having this information will depend on the user's skill and propensity to use the information gained from a management of information system. For some users, the information gained from the computer will become an effective management tool. This person will probably use the information that is generated from the computer effectively and may demand new forms of information from the computer. For others, the computer will be one more problem to be managed, resulting in more problems than it is worth. However, this model can give the potential computer user a yardstick to judge the soundness of investing in a computer system.

Conclusion

The quality of computer packages varied tremendously. Making an adequate comparison of the quality and capability of these systems is difficult. The practitioner unfamiliar with computers and finance would have a difficult time evaluating these systems. As computers have a great potential to aid as well as harm the practice of a health practitioner, an appropriate function for medical computing organizations may be to investigate establishing minimum standards that computer management systems should meet.

The methodology used in this study does not quantify the value of having timely and accurate information, nor does it quantify the value of having information that in most cases is not readily available without the use of a computer. However, the methodology presented here does give the decision-maker a minimum parameter with which to evaluate the soundness of investing in a computer management system for the health practice office.

Two investment strategies become apparent from this study:

(1) If one is to invest in the typical dental management computer system, one should consider automating the entire financial process if one is to receive an adequate return on one's investment. This strategy is particularly appealing to those who are knowledgeable in financial management techniques or are willing to learn how to use the information and control that one could gain from a comprehensive computer system.
(2) An alternative strategy would be to automate the most time consuming task using a less expensive microcomputer system. This strategy will require a lower initial investment, familiarize the first time user with computers and is likely to give an adequate return on one's investment. But most importantly, this study reinforces the need to objectively analyze the goals of using a computer in a health practice before one invests in any type of computer management system.

References

[1] Chaikin, Richard. "Computer Billing," *Practice Administration,* May 1977, p. 65.
[2] Dimitry, J. and Hyton, H., "Computerized Practice Management," *Ontario Dentist* **56,** No. 4, April 1979, p. 33.
[3] Editors, "Choosing a Dental Computer," *Computer News for Dentists* **1,** No. 6, August, 1980, p. 1.
[4] Editors. "Computerized Recall," *Computer News for Dentists* **1,** No. 8, October, 1980, p. 1.
[5] Editors. "A Microcomputer for the Office," *Dental Management,* September, 1980, p. 61.
[6] Neiburger, E. "A Guide to Computer Equipment for the Dental Office," *Illinois Dental Journal,* September–October, 1980, p. 466.
[7] Sacks, R. "A New View of Computer Billing," *Dental Management* March, 1979, p. 19.
[8] U.S. Department of Health, Education and Welfare, Public Health Service, Health Resources Administration, *Investigation of the Application of Computer Technology to Private Dental Practice,* Prepared by Arthur Young and Co., DHEW Publication No. (HRA) 79-21.
[9] Weston, F. and Brigham, E. *Managerial Finance,* The Dryden Press, 1978, p. 296.

Index

MIX
Papier aus verantwortungsvollen Quellen
Paper from responsible sources
FSC® C105338
FSC
www.fsc.org

If you have any concerns about our products,
you can contact us on
ProductSafety@springernature.com

In case Publisher is established outside the EU,
the EU authorized representative is:
Springer Nature Customer Service Center GmbH
Europaplatz 3, 69115 Heidelberg, Germany

Printed by Libri Plureos GmbH
in Hamburg, Germany